Continuous Random Variables

Chi-Square PDF   (p. 5-32)

$$f_X(x) = \frac{x^{(n-2)/2}e^{-x/2}}{2^{n/2}\Gamma(n/2)} \quad x > 0$$

$$\mathcal{E}\{X\} = n; \quad \text{Var}\,[X] = 2n$$

$$M_X(s) = \frac{1}{(1-2s)^{n/2}}$$

Exponential PDF   (p. 3-15)

$$f_X(x) = \lambda e^{-\lambda x} \quad x \geq 0$$

$$\mathcal{E}\{X\} = 1/\lambda; \quad \text{Var}\,[X] = 1/\lambda^2$$

$$M_X(s) = \frac{\lambda}{\lambda - s}$$

Gamma & Erlang PDF   (p. 5-31)

$$f_X(x) = \frac{\lambda^a x^{a-1}}{\Gamma(a)} e^{-\lambda x} \quad x \geq 0$$

$$\mathcal{E}\{X\} = a/\lambda; \quad \text{Var}\,[X] = a/\lambda^2$$

$$M_X(s) = \frac{\lambda^a}{(\lambda - s)^a}$$

The Erlang density occurs when $a = n$, a positive integer; then $\Gamma(n) = (n-1)!$.

Gaussian PDF   (p. 3-16)

$$f_X(x) = \frac{1}{\sqrt{2\pi\sigma^2}} e^{-(x-\mu)^2/2\sigma^2}$$

$$\mathcal{E}\{X\} = \mu; \quad \text{Var}\,[X] = \sigma^2$$

$$M_X(s) = e^{\mu s + \sigma^2 s^2/2}$$

Laplace PDF   (p. 4-19)

$$f_X(x) = \frac{\alpha}{2} e^{-\alpha|x-\mu|}$$

$$\mathcal{E}\{X\} = \mu; \quad \text{Var}\,[X] = 2/\alpha^2$$

$$M_X(s) = \frac{\alpha^2}{\alpha^2 - s^2} e^{\mu s}$$

Rayleigh PDF   (p. 5-40)

$$f_X(x) = \alpha^2 x e^{-\alpha^2 x^2/2}; \quad x \geq 0$$

$$\mathcal{E}\{X\} = \sqrt{\pi/2}\Big/\alpha; \quad \text{Var}\,[X] = (2 - \pi/2)/\alpha^2$$

Uniform PDF   (p. 3-14)

$$f_X(x) = \frac{1}{b-a} \quad a \leq x \leq b$$

$$\mathcal{E}\{X\} = \frac{a+b}{2}; \quad \text{Var}\,[X] = \frac{(b-a)^2}{12}$$

$$M_X(s) = \frac{e^{bs} - e^{as}}{(b-a)s}$$

PROBABILITY
for ELECTRICAL
and COMPUTER
ENGINEERS

PROBABILITY
for ELECTRICAL
and COMPUTER
ENGINEERS

CHARLES W. THERRIEN
MURALI TUMMALA

CRC PRESS

Boca Raton   London   New York   Washington, D.C.

Library of Congress Cataloging-in-Publication Data

Therrien, Charles W.
   Probability for electrical and computer engineers / Charles W. Therrien, Murali Tummala.
      p. cm.
   Includes bibliographical references and index.
   ISBN 0-8493-1884-X
      1. Electric engineering—Mathematics. 2. Probabilities. 3. Computer
   engineering—Mathematics. I. Tummala, Murali, II. Title.

   TK153.T44 2004
   621.3'01'5192—dc22   2003065421

Visit the CRC Press Web site at www.crcpress.com

© 2004 by CRC Press LLC

No claim to original U.S. Government works
International Standard Book Number 0-8493-1884-X
Library of Congress Card Number 2003065421
Printed in the United States of America 1 2 3 4 5 6 7 8 9 0
Printed on acid-free paper

To all the teachers of probability – especially Alvin Drake, who made it fun to learn and Athanasios Papoulis, who made it just rigorous enough for engineers.

Also to our families for their enduring support.

Preface

Beginnings

About ten years ago we had the idea to begin a course in probability for students of electrical engineering. Prior to that electrical engineering graduate students at the Naval Postgraduate School specializing in communication, control, and signal processing were given a basic course in probability in another department and then began a course in random processes within the Electrical and Computer Engineering (ECE) department. ECE instructors consistently found that they were spending far too much time "reviewing" topics related to probability and random variables and therefore could not devote the necessary time to teaching random processes.

The problem was not with the teachers; we had excellent instructors in all phases of the students' programs. We hypothesized (and it turned out to be true) that engineering students found it difficult to relate to the probability material because they could not see the immediate application to engineering problems that they cared about and would study in the future.

When we first offered the course Probabilistic Analysis of Signals and Systems in the ECE department, it became an immediate success. We found that students became interested and excited about probability and looked forward to (rather than dreading) the follow-on courses in stochastic signals and linear systems. We soon realized the need to include other topics relevant to computer engineering such as basics of queueing theory. Today nearly every student in Electrical and Computer Engineering at the Naval Postgraduate School takes this course as a predecessor to his or her graduate studies. Even students who have previously had some exposure to probability and random variables find that they leave with much better understanding and the feeling of time well spent.

Intent

While we had planned to write this book a number a years ago, events overtook us and we did not begin serious writing until about three years ago; even then we did most of the work in quarters when we were teaching the course. In the meantime we had developed an extensive set of PowerPoint slides which we use for teaching and which served as an outline for the text. Although we benefited during this time from the experience of more course offerings, our intent never changed. That intent is to make the topic interesting (even fun) and *relevant* for students of electrical and computer engineering. In line with this, we have tried to make the text very readable for anyone with a background in first-year calculus, and have included a number of application topics and numerous examples.

As you leaf through this book, you may notice that topics such as the binary communication channel, which are often studied in an introductory course in communication theory, are included in Chapter 2, on the Probability Model. Elements of coding (Shannon-Fano and Huffman) are also introduced in Chapters 2 and 4. A simple introduction to detection and classification is also presented early in the text, in Chapter 3. These topics are provided with both homework problems and computer projects to be carried out in a language such as MATLAB. (Some special MATLAB functions will be available on the CRC web site.)

While we have intended this book to focus primarily on the topic of probability, some other topics have been included that are relevant to engineering. We have included a short chapter on random processes as an introduction to the topic (*not* meant to be complete!). For some students, who will not take a more advanced course in random processes, this may be all they need. The definition and meaning of a random process is also important, however, for Chapter 8, which develops the ideas leading up to queueing theory. These topics are important for students who will go on to study computer networks. Elements of parameter estimation and their application to communication theory are also presented in Chapter 6.

We invite you to peruse the text and note especially the engineering applications treated and how they appear as early as possible in the course of study. We have found that this keeps students motivated.

Suggested Use

We believe that the topics in this book should be taught at the earliest possible level. In most universities, this would represent the junior or senior undergraduate level. Although it is possible for students to study this material at an even earlier (e.g., sophomore) level, in this case students may have less appreciation of the applications and the instructor may want to focus more on the theory and only Chapters 1 through 6.

Ideally we feel that a course based on this book should directly precede a more advanced course on communications or random processes for students studying those areas, as it does at the Naval Postgraduate School. While several engineering books devoted to the more advanced aspects of random processes and systems also include background material on probability, we feel that a book that focuses mainly on probability for engineers with engineering applications has been long overdue.

While we have written this book for students of electrical and computer engineering, we do not mean to exclude students studying other branches of engineering or physical sciences. Indeed, the theory is most relevant to these other disciplines, and the applications and examples, although drawn from our own discipline, are ubiquitous in other areas. We would welcome suggestions for possibly expanding the examples to these other areas in future editions.

Organization

After an introductory chapter, the text begins in Chapter 2 with the algebra of events and probability. Considerable emphasis is given to representation of the sample space for various types of problems. We then move on to random variables, discrete and continuous, and transformations of random variables in Chapter 3. Expectation is considered to be sufficiently important that a separate chapter (4) is devoted to the topic. This chapter also provides a treatment of moments and generating functions. Chapter 5 deals with two and more random variables and includes a short (more advanced) introduction to random vectors for instructors with students having sufficient background in linear algebra who may want to cover this topic. Chapter 6 groups together topics of convergence, limit theorems, and parameter estimation. Some of these topics are typically taught at a more advanced level, but we feel that introducing them at this more basic level and relating them to earlier topics in the text helps students appreciate the need for a strong theoretical basis to support practical engineering applications.

Chapter 7 provides a brief introduction to random processes and linear systems while

Chapter 8 deals with discrete and continuous Markov processes and an introduction to queueing theory. Either or both of these chapters may be skipped for a basic level course in probability; however we have found that both are useful even for students who later plan to take a more advanced course in random processes.

As mentioned above, applications and examples are distributed throughout the text. Moreover, we have tried to introduce the applications at the earliest opportunity, as soon as the supporting probabilistic topics have been covered.

Acknowledgments

First and foremost, we want to acknowledge our students over many years for their inspiration and desire to learn. These graduate students at the Naval Postgraduate School, who come from many countries as well as the United States, have always kept us "on our toes" academically and close to reality in our teaching. We should also acknowledge our own teachers of these particular topics, who helped us learn and thus set directions for our careers. These teachers have included Alvin Drake (to whom this book is dedicated in part), Wilbur Davenport, and many others.

We also should acknowledge colleagues at the Naval Postgraduate School, both former and present, whose many discussions have helped in our own teaching and presentation of these topics. These colleagues include John M. ("Jack") Wozencraft, Clark Robertson, Tri Ha, Roberto Cristi, (the late) Richard Hamming, Bruno Shubert, Donald Gaver, and Peter A. W. Lewis.

We would also like to thank many people at CRC for their help in publishing this book, especially Nora Konopka, our acquisitions editor. Nora is an extraordinary person, who was enthusiastic about the project from the get-go and was never too busy to help with our many questions during the writing and production. In addition, we could never have gotten to this stage without the help of Jim Allen, our technical editor of many years. Jim is an expert in LaTeX and other publication software and has been associated with our teaching of Probabilistic Analysis of Signals and Systems long before the first words of this book were entered in the computer.

Finally, we acknowledge our families, who have always provided moral support and encouragement and put up with our many hours of work on this project. We love you!

Charles W. Therrien
Murali Tummala
Monterey, California

Contents

1 Introduction

In many situations in the real world, the outcome is uncertain. For example, consider the event that it rains in Boston on a particular day in the spring. No one will argue with the premise that this event is uncertain, although some people may argue that the likelihood of the event is increased if one forgets to bring an umbrella.

In engineering problems, as in daily life, we are surrounded by events whose occurrence is either uncertain or whose outcome cannot be specified by a precise value or formula. The exact value of the power line voltage during high activity in the summer, the precise path that a space object may take upon reentering the atmosphere, and the turbulence that a ship may experience during an ocean storm are all examples of events that cannot be described in any deterministic way. In communications (which includes computer communications as well as personal communications over devices such as cellular phones), the events can frequently be reduced to a series of binary digits (0's and 1's). However, it is the *sequence* of these binary digits that is uncertain and carries the information. After all, if a particular message represented an event that occurs with complete certainty, why would we ever have to transmit the message?

The most successful method to date to deal with uncertainty is through the application of probability and its sister topic, statistics. The former is the subject of this book. In this chapter we visit a few of the engineering applications for which methods of probability play a fundamental role.

1.1 The Analysis of Random Experiments

A model for the analysis of random experiments is depicted in Fig. 1.1. This will

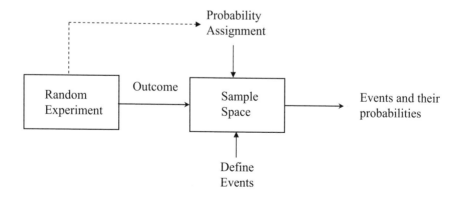

Figure 1.1 The probability model.

be referred to as the basic probability model. The analysis begins with a "random experiment," i.e., an experiment in which the outcome is uncertain. All of the possible outcomes of the experiment can be listed (either explicitly or conceptually); this complete collection of outcomes comprises what is called the *sample space*. Events of interest are "user-defined," i.e., they can be defined in any way that supports analysis of the experiment. Events, however, must always have a representation in the sample space. Thus formally, an *event* is always a single outcome or a collection of outcomes from the sample space.

Probabilities are real numbers on a scale of 0 to 1 that represent the likelihood of events. In everyday speech we often use percentages to represent likelihood. For example, the morning weather report may state that "there is a 60% chance of rain." In a probability model, we would say that the probability of rain is 0.6. Numbers closer to 1 indicate more likely events while numbers closer to 0 indicate more unlikely events. (*There are no negative probabilities!*) The probabilities assigned to outcomes in the sample space should satisfy our intuition about the experiment and make good sense. (Frequently the outcomes are equally likely.) The probability of other events germane to the problem is then computed by rules which are studied in Chapter 2.

Some further elements of probabilistic analysis are random variables and random processes. In many cases there is an important numerical value resulting from the performance of a random experiment. We may consider the numerical value to be the outcome itself of the experiment, in which case the sample space could be considered to be the set of integers (if appropriate) or the real line. Frequently, however, the experiment is more basic and the numerical value, although important, can be thought of as a derived result. If we choose the sample space as the listing of more basic events, then the number of interest is represented by a *mapping* from the sample space to the real line. Such a mapping is called a *random variable*. If the mapping is to n-dimensional Euclidean space the mapping is sometimes called a *random vector*. Generally we deal with just a mapping to the real line because a random vector can be considered to be an ordered set of random variables. A *random process* is like a random variable except each outcome in the sample space is mapped into a function of time (see Fig. 1.2). It will be seen later that a random process is an appropriate representation

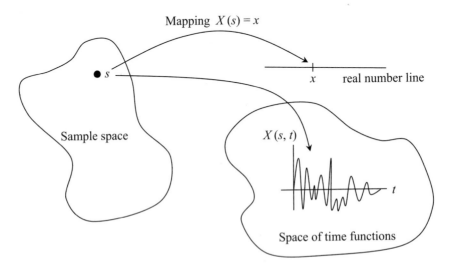

Figure 1.2 Random variables and random processes.

of many signals in control and communications as well as the interference that we refer to as noise.

The following example illustrates the concept of a random variable.

Example 1.1: Consider the rolling of a pair of dice at a game in Las Vegas. We could represent the sample space as the list of pairs of numbers

$$(1,1), (1,2), \ldots, (6,6)$$

indicating what is shown on the dice. Each such pair represents an *outcome* of the experiment. An important numerical value associated with this experiment is the "number rolled," i.e., the sum of total number of dots showing on the two dice. This can be represented by a random variable N. It is a mapping from the sample space to the real line since a value of N can be computed for each outcome in the sample space. For example, we have the following mappings:

outcome		N
$(1,2)$	$\longrightarrow$	3
$(2,3)$	$\longrightarrow$	5
$(4,1)$	$\longrightarrow$	5

The sample space chosen for this experiment contains more information than just the number N. For example, we can tell if "doubles" are rolled, which may be significant for some games.

□

Another reason for modeling an experiment to include a random variable is that the necessary probabilities can be chosen more easily. In the game of dice, it is obvious that the outcomes (1,1), (1,2), ..., (6,6) should each have the same probability (unless someone is cheating), while the possible values of N are not equally likely because more than one outcome can map to the same value of N (see example above). By modeling N as a random variable, we can take advantage of well-developed procedures to compute its probabilities.

1.2 Probability in Electrical and Computer Engineering

In this section, we present several examples of interest to electrical and computer engineering to illustrate how a probabilistic model is appropriate.

1.2.1 Signal detection and classification

A problem of considerable interest to electrical engineers is that of signal detection. Whether dealing with conversation on a cellular phone, data on a computer line, or scattering from a radar target, it is important to detect and properly interpret the signal.

In one of the simplest forms of this problem, we may be dealing with digital data represented by an analog voltage where we want to decode the message bit by bit. The receiver consists of a demodulator, which processes the radio frequency signal, and a decoder which observes the analog output of the demodulator and makes a decision: 0 or 1 (see Fig. 1.3).

Let us assume that in the absence of noise a logical 0 is represented by -1 volt and a logical 1 is represented by $+1$ volt. However due to interference and noise in the transmission the analog signal s at the output of the demodulator is perturbed so that it may take on a whole range of values. Let us suppose that the demodulator output

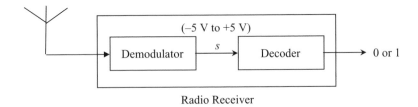

Figure 1.3 Reception of a digital signal.

is hard-limited so that s is known to remain between -5 and $+5$ volts. The job of the decoder is then to decide whether the transmitted signal was a logical 0 or a logical 1.

Several elements of probability are evident in this problem. First of all, the transmission of a 0 or 1 is a random event. (If the receiver knew for certain what the transmission was going to be, we would not need to transmit at all!) Secondly, the intermediate signal s observed in the receiver is a random variable. Its exact relation to the sample space will depend on details of the probabilistic model; however, the numerical value that this voltage takes on is uncertain. Finally, a decision needs to be made based on the observed value of s. In the best situation the decision rule will minimize the *probability* of an error.

Elements of this problem are dealt with in Chapters 3 and 5 of the text and have been known since the early part of the last century. The core of the results lies in the field that has come to be known as *statistical communication theory* [1].

1.2.2 Speech modeling and recognition

The engineering analysis of human speech has been studied for decades. Recently significant progress has been made thanks to the availability of small powerful computers and the development of advanced methods of signal processing. To grasp the complexity of the problem, consider the event of speaking a single English word, such as the word "hello," illustrated in Fig. 1.4. While the waveform has clear structure in some areas, you can see that there are significant variations in the details from one repetition of the word to another. Segments of the waveform representing distinct sounds or "phonemes" are typically modeled by a random process (the subject of Chapter 7), while the concatenation of sounds and words to produce meaningful speech also follows a probabilistic model. Techniques have advanced through research such that synthetic speech *production* is fairly good, while speech *recognition* is still in a relatively embryonic stage and its current success in the commercial world has been limited. In spite of difficulties of the problem, however, speech is believed by some to become the standard interface for computer/human communication that will ultimately replace the mouse and keyboard in present day computers [2]. In whatever form the future research takes, you can be certain that probabilistic methods will play an important if not central role.

1.2.3 Coding and data transmission

The representation of information as digital data for storage and transmission is rapidly replacing older analog methods. The technology is the same whether the data represents speech, music, video, telemetry, or something else. Simple fixed-length codes

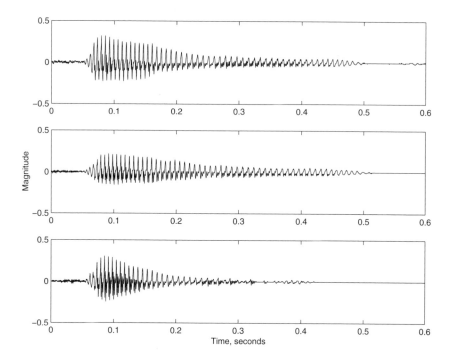

Figure 1.4 Waveform corresponding to the word "hello" (three repetitions, same speaker).

such as the ASCII standard are inefficient because of the considerable redundancy that exists in most information that is of practical interest.

In the late 1940's Claude Shannon [3, 4] studied fundamental methods related to the representation and transmission of information. This field, which later became known as *information theory*, is based on probabilistic methods. Shannon showed that given a set of symbols (such as letters of the alphabet) and their probabilities of occurrence in a message, one could compute a probabilistic measure of "information" for a source producing such messages. He showed that this measure of information, known as entropy H and measured in "bits," provides a lower bound on the average number of binary digits needed to code the message. Others, such as David Huffman, developed algorithms for coding the message into binary bits that approached the lower bound.[1] Some of these methods are introduced in Chapter 2 of the text.

Shannon also developed fundamental results pertaining to the communication channel over which a message is to be transmitted. A message to be transmitted is subject to errors made due to interference, noise, self interference between adjacent symbols, and many other effects. Putting all of these details aside, Shannon proposed that the channel could be characterized in a probabilistic sense by a number C known as the *channel capacity* and measured in bits. If the source produces a symbol every T_s seconds and that symbol is sent over the channel in T_c seconds, then the source rate and the channel rate are defined as H/T_s and C/T_c respectively. Shannon showed that if

---

[1] Huffman developed the method as a term paper in a class at M.I.T.

the source and channel rates satisfy the condition

$$\frac{H}{T_s} \le \frac{C}{T_c}$$

then the message can be transmitted with an arbitrarily low probability of error. On the other hand, if this condition does not hold, then there is no way to achieve a low error rate on the channel. These ground-breaking results developed by Shannon became fundamental to the development of modern digital communication theory and emphasized the importance of a probabilistic or statistical framework for the area.

1.2.4 Computer networks

It goes without saying that computer networks have become exceedingly complex. Today's modern computer networks are descendents of the ARPAnet research project by the U.S. Department of Defense in the late 1960's and are based on the store-and-forward technology developed there. A set of user data to be transmitted is broken into smaller units or packets together with addressing and layering information and sent from one machine to another through a number or intermediate nodes (computers). The nodes provide routing to the final destination. In some cases the packets of a single message may arrive by different routes and are assembled to produce the message at the final destination. There are a variety of problems in computer networks including packet loss, packet delay, and packet buffer size that require the knowledge of probabilistic and statistical concepts. In particular, the concepts of queueing systems, which preceded the development of computer networks, have been aptly applied to these systems by early researchers such a Kleinrock [5, 6]. Today, electrical and computer engineers study those principles routinely.

The analysis of computer networks deals with random processes that are *discrete* in nature. That is, the random processes take on finite discrete values as a function of time. Examples are the number of packets in the system (as a function of time), the length of a queue, or the number of packets arriving in some given period of time. Various random variables and statistical quantities also need to be dealt with such as the average length of a queue, the delay encountered by packets from arrival to departure, measures of throughput versus load on the system, and so on.

The topic of queuing brings together many probabilistic ideas that are presented in the book and so is dealt with in the last chapter. The topic seems to underscore the need for a knowledge of probability and statistics in almost all areas of the modern curriculum in electrical and computer engineering.

1.3 Outline of the Book

The overview of the probabilistic model in this chapter is meant to depict the general framework of ideas that are relevant for this area of study and to put some of these concepts in perspective. The set of examples is to convince you that this topic, which is basically an area of mathematics, is extremely important for engineers.

In the chapters that follow we attempt to present the study of probability in a context that continues to show its importance for engineering applications. All of the applications cited above and others are discussed explicitly in later chapters. Our goal has been to bring in the applications as early as possible even if this requires considerable simplification to a real world problem. We have also tried to keep the discussion light (although not *totally* lacking in rigor) and to occasionally inject some light humor.

The next chapter begins with a discussion of the probability model, events, and probability measure. This is followed in Chapter 3 by a discussion of random variables. Averages, known as statistical expectation, form an important part of the theory and are discussed in a fairly short Chapter 4. Chapter 5 then deals with multiple random variables, expectation, and random vectors. We have left the topic of theorems, bounds, and estimation to fairly late in the book, in Chapter 6. While these topics are important, we want to present the more application-oriented material first. Chapter 7 then provides an introduction to random processes and the final chapter (Chapter 8) continues to discuss the special classes of random processes that pertain to the topic of queueing.

Topics in mathematics are not always easy especially when old paradigms are broken and new concepts need to be developed. Perseverance leads to success however, and success leads to enjoyment. We hope that you will come to enjoy this topic as well as we have enjoyed teaching and writing about it.

References

[1] David Middleton. *An Introduction to Statistical Communication Theory*. McGraw-Hill, New York, 1960.

[2] Michael L. Dertouzos. *What WILL be*. HarperCollins, San Francisco, 1997.

[3] Claude E. Shannon. A mathematical theory of communication. *Bell System Technical Journal*, 27(3):379–422, July 1948.

[4] Claude E. Shannon. A mathematical theory of communication (concluded). *Bell System Technical Journal*, 27(4):623–656, October 1948.

[5] Leonard Kleinrock. *Queueing Systems - Volume I: Theory*. John Wiley & Sons, New York, 1975.

[6] Leonard Kleinrock. *Queueing Systems - Volume II: Computer Applications*. John Wiley & Sons, New York, 1976.

2

The Probability Model

This chapter describes a well-accepted model for the analysis of random experiments which we refer to as the *Probability Model*. We also define a set algebra suitable for defining sets of events, and describe how measures of likelihood or *probabilities* are assigned to these events. Probabilities provide quantitative numerical values to the likelihood of occurrence of events.

Events do not always occur independently. In fact, it is the very *lack* of independence that allows us to infer one fact from another. Here we give a mathematical meaning to the concept of independence and further develop relations to deal with probabilities when events are or are not independent.

Several illustrations and examples are given throughout this chapter on basic probability. In addition, a number of applications of the theory to some basic electrical engineering problems are given to provide motivation for further study of this topic and those to come.

2.1 The Algebra of Events

We have seen in Chapter 1 that the collection of all possible outcomes of a random experiment comprise the *sample space*. Outcomes are members of the sample space and events of interest are represented as *sets* (see Fig. 2.1). In order to describe these events

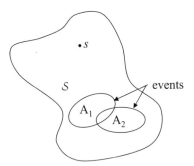

Figure 2.1 Abstract representation of the sample space S with element s and sets A_1 and A_2 representing events.

and compute their probabilities in a consistent manner it is necessary to have a formal representation for operations involving events. More will be said about representing the sample space in Section 2.1.2; for now, we shall focus on the methods for describing relations among events.

2.1.1 Basic operations

In analyzing the outcome of a random experiment, it is usually necessary to deal with events that are derived from other events. For example, if A is an event, then A^c, known as the *complement* of A, represents the event that "A did not occur." The complement of the sample space is known as the *null event*, $\emptyset = \mathcal{S}^c$. The operations of multiplication and addition will be used to represent certain combinations of events (known as intersections and unions in set theory). The statement "$A_1 \cdot A_2$," or simply "$A_1 A_2$" represents the event that *both* event A_1 and event A_2 have occurred (intersection), while the statement "$A_1 + A_2$" represents the event that *either* A_1 or A_2 *or both* have occurred (union).[1]

Since complements and combinations of events are themselves events, a formal structure for representing events and derived events is needed. This formal structure is in fact a set of sets known in mathematics as an *algebra* or a *field* and referred to here as the *algebra of events*. Table 2.1 lists the two postulates that define an algebra $\mathcal{A}$.

1. *If* $A \in \mathcal{A}$ *then* $A^c \in \mathcal{A}$.
2. *If* $A_1 \in \mathcal{A}$ *and* $A_2 \in \mathcal{A}$ *then* $A_1 + A_2 \in \mathcal{A}$.

Table 2.1 Postulates for an algebra of events.

Table 2.2 lists seven axioms that define the properties of the operations. Together these

$A_1 A_1{}^c = \emptyset$	Mutual exclusion
$A_1 \mathcal{S} = A_1$	Inclusion
$(A_1{}^c)^c = A_1$	Double complement
$A_1 + A_2 = A_2 + A_1$	Commutative law
$A_1 + (A_2 + A_3) = (A_1 + A_2) + A_3$	Associative law
$A_1(A_2 + A_3) = A_1 A_2 + A_1 A_3$	Distributive law
$(A_1 A_2)^c = A_1{}^c + A_2{}^c$	DeMorgan's law

Table 2.2 Axioms of operations on events.

tables can be used to show all of the properties of the algebra of events. For example, the postulates state that the event $A_1 + A_2$ is included in the algebra. The postulates in conjunction with the last axiom (DeMorgan's law) show that the event "$A_1 A_2$" is also included in the algebra. Table 2.3 lists some other handy identities that can be derived from the axioms and the postulates. You will find that you use many of the results in Tables 2.2 and 2.3 either implicitly or explicitly in solving problems involving events and their probability. Notice especially the two distributive laws; addition is distributive over multiplication (Table 2.3) as well as *vice versa* (Table 2.2).

Since the events "$A_1 + A_2$" and "$A_1 A_2$" are included in the algebra, it is easy to

---

[1] The operations represented as multiplication and addition are commonly represented with the intersection $\cap$ and union $\cup$ symbols. Except for the case of multiple such operations, we will adhere to the former notation introduced above.

$$\mathcal{S}^c = \emptyset$$

$A_1 + \emptyset = A_1$	Inclusion
$A_1 A_2 = A_2 A_1$	Commutative law
$A_1(A_2 A_3) = (A_1 A_2)A_3$	Associative law
$A_1 + (A_2 A_3) = (A_1 + A_2)(A_1 + A_3)$	Distributive law
$(A_1 + A_2)^c = A_1{}^c A_2{}^c$	DeMorgan's law

Table 2.3 Additional identities in the algebra of events.

show by induction for any finite number of events A_i, $i = 1, 2, \ldots, N$, that the events

$$\bigcup_{i=1}^{N} A_i = A_1 + A_2 + \cdots + A_N$$

and

$$\bigcap_{i=1}^{N} A_i = A_1 A_2 \cdots A_N$$

are also included in the algebra. In many cases it is important that the sum and product of a countably infinite number of events have a representation in the algebra. For example, suppose an experiment consists of measuring a random voltage, and the events A_i are defined as "$i - 1 \leq \text{voltage} < i; \; i = 1, 2, \ldots$." Then the (infinite) sum of these events, which is the event "voltage ≥ 0," should be in the algebra. An algebra that includes the sum and product of an infinite number of events, that is,

$$\bigcup_{i=1}^{\infty} A_i = A_1 + A_2 + A_3 + \cdots$$

and

$$\bigcap_{i=1}^{\infty} A_i = A_1 A_2 A_3 \cdots$$

is called a sigma-algebra or a sigma-field. The algebra of events is defined to be such an algebra.

Since the algebra of events can be thought of as an algebra of sets, events are often represented as Venn diagrams. Figure 2.2 shows some typical Venn diagrams for a sample space and its events. The notation '$\subset$' is used to mean one event is "contained" in another and is defined by

$$A_1 \subset A_2 \iff A_1 A_2{}^c = \emptyset \tag{2.1}$$

2.1.2 Representation of the sample space

Students of probability may at first have difficulty in defining the sample space for an experiment. It is thus worthwhile to spend a little more time on this concept.

We begin with two more ideas from the algebra of events. Let $A_1, A_2, A_3, \ldots$ be a finite or countably infinite set of events with the following properties:

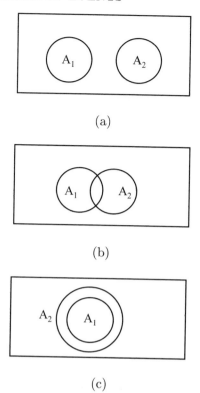

(a)

(b)

(c)

Figure 2.2 Venn diagram for events. (a) Events with no commonality $(A_1 A_2 = \emptyset)$. (b) Events with some commonality $(A_1 A_2 \neq \emptyset)$. (c) One event contained in another $(A_1 \subset A_2)$.

1. The events are *mutually exclusive*. This means that only one event can occur at a time, i.e., the occurrence of one event precludes the occurrence of other events. Equivalently,

$$A_i A_j = \emptyset \ \text{ for } i \neq j$$

2. The events are *collectively exhaustive*. In other words, one of the events A_i must always occur. That is,

$$A_1 + A_2 + A_3 + \cdots = \mathcal{S}$$

A set of events that has *both* properties is referred to as a *partition*.

Now, for an experiment with discrete outcomes, the following provides a working definition of the sample space [1]:

The Sample Space is represented by the finest-grain, mutually exclusive, collectively exhaustive set of outcomes for an experiment.

You can see that the elements of the sample space have the properties of a partition; however, the outcomes defining the sample space must also be *"finest-grain."* This is important, since without this property it may not be possible to represent all of the events of interest in the experiment. A Venn diagram is generally not sufficient to represent the sample space in solving problems, because the representation usually needs to be more explicit.

A discrete sample space may be just a listing of the possible outcomes (see Example

2.1) or could take the form of some type of diagram. For example, consider the rolling of a pair of dice. The sample space might be drawn as shown in Fig. 2.3.

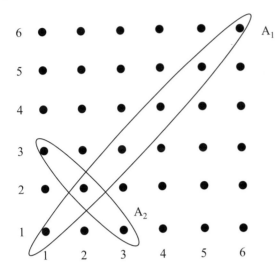

Figure 2.3 Sample space corresponding to roll of the dice. A_1 is the event "rolling doubles"; A_2 is the event "rolling a '4'."

   The black dots represent the outcomes of the experiment, which are mutually exclusive and collectively exhaustive. Some more complex events, such as "rolling doubles" are also shown in the figure. It will be seen later that the probabilities of such more complicated events can be computed by simply adding the probabilities of the outcomes of which they are comprised.

   For an experiment in which the outcome is a real number or a set of real numbers, the sample space is usually chosen as a subset of the real line or a subset of N-dimensional Euclidean space ($\mathcal{R}^N$), as appropriate. This is the case for some of the examples in Chapter 1. If the outcome of the experiment were complex numbers, then you would probably define the sample space as a subspace of the space of complex numbers ($\mathcal{C}^N$). These are examples of *continuous* sample spaces. We shall emphasize that in the solution of most problems involving probability, a first step is *to find an appropriate representation for the sample space.*

2.2 Probability of Events

2.2.1 Defining probability

We have seen that probability represents the likelihood of occurrence of events. The probability model, when properly formulated, can be used to show that the *relative frequency* for the occurrence of an event in a large number of repetitions of the experiment, defined as

$$\text{relative frequency} = \frac{\text{number of occurrences of the event}}{\text{number of repetitions of the experiment}}$$

converges to the probability of the event. Although probability could be *defined* in this way, it is more common to use the axiomatic development given below.

   Probability is conveniently represented in a Venn diagram if you think of the area covered by events as measures of probability. For example, if the area of the sample space $\mathcal{S}$ is normalized to one, then the area of overlap of events A_1 and A_2 in Fig. 2.2(b) can be thought of as representing the probability of the event $A_1 A_2$. If the probability of this joint event were larger, the events might be drawn to show greater overlap.

Probability can be defined formally by the following axioms:

(I) The probability of any event is nonnegative.

$$\Pr[A] \geq 0 \qquad (2.2)$$

(II) The probability of the universal event (i.e., the entire sample space) is 1.

$$\Pr[\mathcal{S}] = 1 \qquad (2.3)$$

(III) If A_1 and A_2 are mutually exclusive, then

$$\Pr[A_1 + A_2] = \Pr[A_1] + \Pr[A_2] \quad (\text{if} \quad A_1 A_2 = \emptyset) \qquad (2.4)$$

(IV) If $\{A_i\}$ represent a countably infinite set of mutually exclusive events, then

$$\Pr\left[\bigcup_{i=1}^{\infty} A_i\right] = \sum_{i=1}^{\infty} \Pr[A_i] \quad (\text{if} \quad A_i A_j = \emptyset \quad i \neq j) \qquad (2.5)$$

Although the additivity of probability for any finite set of disjoint events follows from (III), the property has to be stated explicitly for an infinite set in (IV). These axioms and the algebra of events can be used to show a number of other properties, some of which are discussed below.

From axioms (II) and (III), the probability of the complement of an event is

$$\Pr[A^c] = 1 - \Pr[A] \qquad (2.6)$$

Since by (I) the probability of any event is greater than or equal to zero, it follows from (2.6) that $\Pr[A] \leq 1$; thus

$$0 \leq \Pr[A] \leq 1 \qquad (2.7)$$

for any event A.

If $A_1 \subset A_2$ then A_2 can be written as $A_2 = A_1 + A_1{}^c A_2$ (see Fig. 2.2(c)). Since the events A_1 and $A_1{}^c A_2$ are mutually exclusive, it follows from (III) and (I) that

$$\Pr[A_2] \geq \Pr[A_1]$$

From (2.6) and axiom (II), it follows that the probability of the null event is zero:

$$\Pr[\emptyset] = 0 \qquad (2.8)$$

Thus it also follows that if A_1 and A_2 are mutually exclusive, then $A_1 A_2 = \emptyset$ and consequently

$$\Pr[A_1 A_2] = 0$$

If events A_1 and A_2 are not mutually exclusive, i.e., they may occur together, then one has the general relation

$$\Pr[A_1 + A_2] = \Pr[A_1] + \Pr[A_2] - \Pr[A_1 A_2] \qquad (2.9)$$

This is not an addition property; rather it can be derived using axioms (I) through (IV) and the algebra of events. It can be intuitively justified on the grounds that in summing the probabilities of the event A_1 and the event A_2, one has counted the common event "$A_1 A_2$" twice (see Fig. 2.2(b)). Thus the probability of the event "$A_1 A_2$" must be subtracted to obtain the probability of the event "$A_1 + A_2$".

These various derived properties are summarized in Table 2.4 below. It is a useful excercise to depict these properties (and the axioms as well) as Venn diagrams.

$$\Pr[A^c] = 1 - \Pr[A]$$

$$0 \leq \Pr[A] \leq 1$$

If $A_1 \subseteq A_2$ then $\Pr[A_1] \leq \Pr[A_2]$

$$\Pr[\emptyset] = 0$$

If $A_1 A_2 = \emptyset$ then $\Pr[A_1 A_2] = 0$

$$\Pr[A_1 + A_2] = \Pr[A_1] + \Pr[A_2] - \Pr[A_1 A_2]$$

Table 2.4 Some corollaries derived from the axioms of probability.

As a final consideration, let $A_1, A_2, A_3, \ldots$ be a finite or countably infinite set of mutually exclusive and collectively exhaustive events (see Section 2.1.2). Recall that such a set of events is referred to as a *partition*. The probabilities of the events in a partition satisfy the relation

$$\sum_i \Pr[A_i] = 1 \qquad (2.10)$$

and if B is any other event, then

$$\sum_i \Pr[A_i B] = \Pr[B] \qquad (2.11)$$

The latter result is referred to as the *principle of total probability* and is frequently used in solving problems. The relation (2.11) is illustrated by a Venn diagram in Fig. 2.4. The event B is comprised of all of the pieces that represent intersections or overlap of event B with the events A_i.

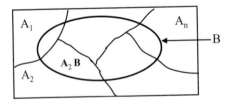

Figure 2.4 Venn diagram illustrating the principle of total probability.

Let us consider the following example to illustrate the formulas in this section.

Example 2.1: Simon's Surplus Warehouse has large barrels of mixed electronic components (parts) that you can buy by the handful or by the pound. You are not allowed to select parts individually. Based on your previous experience, you have determined that in one barrel, 29% of the parts are bad (faulted), 3% are bad resistors, 12% are good resistors, 5% are bad capacitors, and 32% are diodes. You decide to assign probabilities based on these percentages. Let us define the following events:

Event	Symbol
Bad (faulted) component	F
Good component	G
Resistor	R
Capacitor	C
Diode	D

A Venn diagram representing this situation is shown below along with probabilities of various events as given:

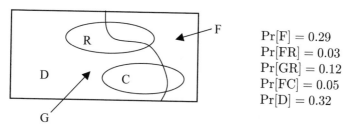

$$Pr[F] = 0.29$$
$$Pr[FR] = 0.03$$
$$Pr[GR] = 0.12$$
$$Pr[FC] = 0.05$$
$$Pr[D] = 0.32$$

We can answer a number of questions.

1. What is the probability that a component is a resistor (either good *or* bad)?
   Since the events F and G form a partition of the sample space, we can use the principle of total probability (2.11) to write
   $$Pr[R] = Pr[GR] + Pr[FR] = 0.12 + 0.03 = 0.15$$

2. You have no use for either defective parts or resistors. What is the probability that a part is either defective and/or a resistor?
   Using (2.9) and the previous result we can write
   $$Pr[F + R] = Pr[F] + Pr[R] - Pr[FR] = 0.29 + 0.15 - 0.03 = 0.41$$

3. What is the probability that a part is useful to you?
   Let U represent the event that the part is useful. Then (see (2.6))
   $$Pr[U] = 1 - Pr[U^c] = 1 - 0.41 = 0.59$$

4. What is the probability of a bad diode?
   Observe that the events R, C, and G form a partition, since a component has to be one and only one type of part. Then using (2.11) we write
   $$Pr[F] = Pr[FR] + Pr[FC] + Pr[FD]$$
   Substituting the known numerical values and solving yields
   $$0.29 = 0.03 + 0.05 + Pr[FD] \quad \text{or} \quad Pr[FD] = 0.21$$

   □

It is worthwhile to consider what an appropriate representation of the sample space would be for this example. While the Venn diagram shown above represents the sample space in an abstract way, a more explicit representation is most useful. In this case, since a part can be a bad resistor, good resistor, bad capacitor, and so on, a suitable representation is the list of outcomes:

sample space: { FR   GR   FC   GC   FD   GD }

(You may want to check that this satisfies the requirements discussed in Section 2.1.2.) In later sections of this chapter, you will see that the answers to the four questions posed in this example can be easily obtained if the probabilities of these six outcomes were specified or could be computed. In this example however, the probabilities of all of these experimental outcomes are not known, i.e., only partial information is given.

2.2.2 Statistical independence

There is one more concept that is frequently used when solving basic problems in probability, but cannot be derived from either the algebra of events or any of the axioms. Because of its practical importance in solving problems, we introduce this concept early in our discussion of probability:

Two events A_1 and A_2 are said to be *statistically independent* if and only if

$$\Pr[A_1 A_2] = \Pr[A_1] \cdot \Pr[A_2] \qquad (2.12)$$

That is, for two independent events, the probability of both occuring is the product of the probabilities of the individual events. Independence of events is not generally something that you are asked to *prove* (although it may be). More frequently it is an assumption made when the conditions of the problem warrant it. The idea can be extended to multiple events. For example, if A_1, A_2 and A_3 are said to be *mutually independent* if and only if

$$\Pr[A_1 A_2 A_3] = \Pr[A_1] \Pr[A_2] \Pr[A_3]$$

Note also that for independent events, (2.9) becomes

$$\Pr[A_1 + A_2] = \Pr[A_1] + \Pr[A_2] - \Pr[A_1] \Pr[A_2]$$

so the computation of probability for the union of two events is also simplified.

The concept of statistical independence, as we have already said, cannot be derived from anything else presented so far, and does not have a convenient interpretation in terms of a Venn diagram. However it can be argued in terms of the relative frequency interpretation of probability. Suppose two events are "independent" in that they arise from two different experiments that have nothing to do with each other. Let it be known that in N_1 repetitions of the first experiment there are k_1 occurrences of the event A_1 and in N_2 repetitions of the second experiment there are k_2 occurrences of the event A_2. If N_1 and N_2 are sufficiently large, then the relative frequencies k_i/N_i remain approximately constant as N_i is increased. Let us now perform both experiments together a total of $N_1 N_2$ times. Consider the event A_1. Since it occurs k_1 times in N_1 repetitions of the experiment it will occur $N_2 k_1$ times in $N_1 N_2$ repetitions of the experiment. Now consider those $N_2 k_1$ cases where A_1 occured. Since event A_2 occurs k_2 times in N_2 repetitions, it will occur $k_1 k_2$ times in these $N_2 k_1$ cases where A_1 has occured. In other words the two events occur together $k_1 k_2$ times in all of these $N_1 N_2$ repetitions of the experiments. The relative frequency for the occurrence of the two events together is therefore

$$\frac{k_1 k_2}{N_1 N_2} = \frac{k_1}{N_1} \cdot \frac{k_2}{N_2}$$

which is the product of the relative frequencies of the individual events. So given the relative frequency interpretation of probability, the definition (2.12) makes good sense.

2.3 Some Applications

Let us continue with some examples in which many of the ideas discussed so far in this chapter are illustrated.

2.3.1 Repeated independent trials

Many problems involve a repetition of independent events. As a typical example of this, consider the experiment of tossing a coin three times in succession. The result of the second toss is independent of the result of the first toss; likewise the result of the third toss is independent of the result of the first two tosses. Let us denote the probability of a "head" (H) on any toss by p and the probability of a "tail" (T) by $q = 1 - p$. (For a fair coin, $p = q = 1/2$, but let us be more general.) Since the results of the tosses are independent, the probability of any experimental outcome such as HHT is simply the product of the probabilities: $p \cdot p \cdot q = p^2 q$. The sequence HTH has the same probability: $p \cdot q \cdot p = p^2 q$. Experiments of this type are said to involve *repeated independent trials*.

An application which is familiar to electrical and computer engineers is the transmission of a binary sequence over a communication channel. In many practical cases the bits (1 or 0) can be modeled as independent events. Thus the probability of a bit sequence of any length $1\,0\,1\,1\,0\,1\,\dots$ is simply equal to the product of probabilities: $p \cdot q \cdot p \cdot p \cdot q \cdot p \cdots$. This considerably simplifies the analysis of such systems.

An example is given below where the outcomes of the experiment are based on repeated independent trials. Once the sample space and probabilities of the outcomes have been specified, a number of other probabilistic questions can be answered.

Example 2.2:   Diskettes selected from the bins at Simon's Surplus are as likely to be good as to be bad. If three diskettes are selected independently and at random, what is the probability of getting exactly *three* good diskettes? Exactly *two* good diskettes? How about *one* good diskette?

Evidently buying a diskette at Simon's is like tossing a coin. The sample space is represented by the listing of outcomes shown below: where G represents a good diskette

BBB	BBG	BGB	BGG	GBB	GBG	GGB	GGG
A_1	A_2	A_3	A_4	A_5	A_6	A_7	A_8

and B represents a bad one. Each outcome is labeled as an event A_i; note that the events A_i are mutually exclusive and collectively exhaustive.

Three good diskettes is represented by only the last event (A_8) in the sample space. Since the probability of selecting a good diskette and the probability of selecting a bad diskette are both equal to $\frac{1}{2}$, and the selections are independent, we can write

$$\Pr[3 \text{ good diskettes}] = \Pr[A_8] = \Pr[G]\Pr[G]\Pr[G] = \left(\tfrac{1}{2}\right)^3 = \tfrac{1}{8}$$

(see Section 2.2.2).

The result of two good diskettes is represented by the events A_4, A_6, and A_7. By a procedure similar to the above, each of these events has probability $\frac{1}{8}$. Since these three events are mutually exclusive, their probabilities add (see Section 2.2.1). That is,

$$\Pr[2 \text{ good diskettes}] = \Pr[A_4 + A_6 + A_7] = \Pr[A_4] + \Pr[A_6] + \Pr[A_7] = \tfrac{1}{8} + \tfrac{1}{8} + \tfrac{1}{8} = \tfrac{3}{8}$$

Finally, a single good diskette is represented by the events A_2, A_3, and A_5. By an identical procedure it is found that this result also occurs with probability $\frac{3}{8}$.

□

To be sure you understand the steps in this example, you should repeat this example for the case where the probability of selecting a good diskette is increased to $\frac{5}{8}$. In this case the probability of all of the events A_i are not equal. For example, the probability of the event A_6 is given by $\Pr[G]\Pr[B]\Pr[G] = \frac{5}{8} \cdot \frac{3}{8} \cdot \frac{5}{8} = \frac{75}{512}$. When you work through the example you will find that the probabilities of three and two good diskettes is increased to $\frac{125}{512}$ and $\frac{225}{512}$ respectively while the probability of just one good diskette is decreased to $\frac{135}{512}$.

2.3.2 Problems involving counting

Many important problems involve adding up the probabilities of a number of equally-likely events. These problems involve some basic combinatorial analysis, i.e., counting the number of possible events, configurations, or outcomes in an experiment.

Some discussion of combinatorial methods is provided in Appendix A. All of these deal with the problem of counting the number of pairs, triplets, or k-tuples of elements that can be formed under various conditions. Let us review the main results here.

Rule of product. In the formation of k-tuples consisting of k elements where there are N_i choices for the ith element, the number of possible k-tuples is $\prod_{i=1}^{k} N_i$. An important special case is when there are the *same* number of choices N for each element. The number of k-tuples is then simply N^k.

Permutations. A *permutation* is a k-tuple formed by selecting from a set of N *distinct* elements, where each element can only be selected once. (Think of forming words from a finite alphabet where each letter can be used only once.) There are N choices for the first element, $N-1$ choices for the second element, ..., and $N-k+1$ choices for the kth element. The number of such permutations is given by

$$N \cdot (N-1) \cdots (N-k+1) = \frac{N!}{(N-k)!}$$

For $k = N$ the result is simply $N!$.

Combinations. A *combination* is a k-tuple formed by selecting from a set of N distinct elements where the *order* of selecting the elements makes no difference. For example, the sequences ACBED and ABDEC would represent two different permutations, but only a single *combination* of the letters A through E. The number of combinations k from a possible set of N is given by the binomial coeffient

$$\binom{N}{k} = \frac{N!}{k!(N-k)!}$$

This is frequently read as "N *choose* k," which provides a convenient mnemonic for its interpretation.

Counting principles provide a way to assign or compute probability in many cases. This is illustrated in a number of examples below.

The following example illustrates use of some basic counting ideas.

Example 2.3: In sequences of k binary digits, 1's and 0's are equally likely. What is the probability of encountering a sequence with a single '1' (in any position) and all other digits zero?

Imagine drawing the sample space for this experiment consisting of all possible sequences. Using the rule of product we see that there are 2^k events in the sample space and they are all equally likely. Thus we assign probability $1/2^k$ to each outcome in the sample space.

Now, there are just k of these sequences that have exactly one '1'. Thus the probability is $k/2^k$.

□

The next example illustrates the use of permutation.

Example 2.4: IT technician Chip Gizmo has a cable with four twisted pairs running from each of four offices to the service closet; but he has forgotten which pair goes to which office. If he connects one pair to each of four telephone lines arbitrarily, what is the probability that he will get it right on the first try?

The number of ways that four twisted pairs could be assigned to four telephone lines is $4! = 24$. Assuming that each arrangement is equally likely, the probability of getting it right on the first try is $1/24 = 0.0417$.

□

The following example illustrates the use of permutations versus combinations.

Example 2.5: Five surplus computers are available for adoption. One is an IBM, another is an HP, and the rest are nondescript. You can request two of the suplus computers but cannot specify which ones. What is the probability that you get the IBM and the HP?

Consider first the experiment of randomly selecting two computers. Let's call the computers A, B, C, D, and E. The sample space is represented by a listing of pairs

$$A,B \quad B,A \quad A,C \quad C,A \quad \cdots$$

representing the computers chosen. Each pair is a *permutation*, and there are $5!/(5 - 2)! = 5 \cdot 4 = 20$ such permutations that represent the outcomes in the sample space. Thus each outcome has a probability of $1/20$. We are interested in two of these outcomes, namely IBM,HP or HP,IBM. The probability is thus $2/20$ or $1/10$.

Another simpler approach is possible. Since we do not need to distinguish between ordering in the elements of a pair, we could choose our sample space to be

$$A,B \quad A,C \quad A,D \quad A,E \quad \cdots$$

where events such as B,A and C,A are not listed since they are equivalent to A,B and A,C. The number of pairs in this new sample space is the number of *combinations* of 5 objects taken 2 at a time:

$$\binom{5}{2} = \frac{5!}{2!(5-2)!} = 10$$

Thus each outcome in this sample space has probability $1/10$. We are interested only in the single outcome IBM,HP. Therefore this probability is again $1/10$.

□

The final example for this section illustrates a more advanced use of the combinatoric ideas.

Example 2.6: DEAL Computers Incorporated manufactures some of their computers in the US and others in Lower Slobbovia. The local DEAL factory store has a stock of 10 computers that are US made and 15 that are foreign made. You order five computers from the DEAL store which are randomly selected from this stock. What is the probability that two or more of them are US-made?

The number of ways to choose 5 computers from a stock of 25 is

$$\binom{25}{5} = \frac{25!}{5!(25-5)!} = 53130$$

This is the total number of possible outcomes in the sample space.

Now consider the number of outcomes where there are *exactly* 2 US-made computers in a selection of 5. Two US computers can be chosen from a stock of 10 in $\binom{10}{2}$ possible ways. For each such choice, three non-US computers can be chosen in $\binom{15}{3}$ possible ways. Thus the number of outcomes where there are exactly 2 US-made computers is given by

$$\binom{10}{2} \cdot \binom{15}{3}$$

Since the problem asks for "two or more" we can continue to count the number of ways there could be exactly 3, exactly 4, and exactly 5 out of a selection of five computers. Therefore the number of ways to choose 2 *or more* US-made computers is

$$\binom{10}{2}\binom{15}{3} + \binom{10}{3}\binom{15}{2} + \binom{10}{4}\binom{15}{1} + \binom{10}{5} = 36477$$

The probability of two or more US-made computers is thus the ratio $36477/53130 = 0.687$.

$\square$

2.3.3 Network reliability

Consider the set of communication links shown in Fig. 2.5. In both cases it is desired

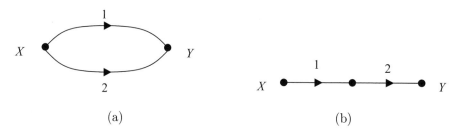

(a)   (b)

Figure 2.5 Connection of communication links. (a) Parallel. (b) Series.

to communicate between points X and Y. Let A_i represent the event that link i fails and F be the event that there is failure to communicate between X and Y. Further, assume that the link failures are *independent* events. Then for the parallel connection (Fig. 2.5(a))

$$\Pr[F] = \Pr[A_1 A_2] = \Pr[A_1]\Pr[A_2]$$

where the last equality follows from the fact that events A_1 and A_2 are independent. For the series connection (Fig. 2.5(b))

$$\Pr[F] = \Pr[A_1 + A_2] = \Pr[A_1] + \Pr[A_2] - \Pr[A_1 A_2] = \Pr[A_1] + \Pr[A_2] - \Pr[A_1]\Pr[A_2]$$

where we have applied (2.9) and again used the fact that the events are independent.

The algebra of events and the rules for probability can be used to solve some additional simple problems such as in the following example.

Example 2.7: In the simple communication network shown below, link failures occur in-

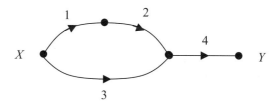

dependently with probability p. What is the largest value of p that can be tolerated if the overall probability of failure of communication between X and Y is to be kept less than 10^{-3}?

Let F represent the failure of communication; this event can be expressed as $F = (A_1 + A_2)A_3 + A_4$. The probability of this event can then be computed as follows:

$$
\begin{aligned}
\Pr[F] &= \Pr[(A_1 + A_2)A_3 + A_4] \\
&= \Pr[A_1 A_3 + A_2 A_3] + \Pr[A_4] - \Pr[A_1 A_3 A_4 + A_2 A_3 A_4] \\
&= \Pr[A_4] + \Pr[A_1 A_3] + \Pr[A_2 A_3] - \Pr[A_1 A_2 A_3] \\
&\quad - \Pr[A_1 A_3 A_4] - \Pr[A_2 A_3 A_4] + \Pr[A_1 A_2 A_3 A_4] \\
&= p + 2p^2 - 3p^3 + p^4
\end{aligned}
$$

To find the desired value of p, we set this expression equal to 0.001; thus we need to find the roots of the polynomial $p^4 - 3p^3 + 2p^2 + p - 0.001$. Using MATLAB, this polynomial is found to have two complex conjugate roots, one real negative root, and one real positive root $p = 0.001$, which is the desired value.

□

An alternative method can be used to compute the probability of failure in this example. You list the possible outcomes (the sample space) and their probabilities as shown in the table below and put a check ($\sqrt{}$) next to each outcome that results in failure to communicate.

outcome	probability	F
$A_1 A_2 A_3 A_4$	p^4	$\sqrt{}$
$A_1 A_2 A_3 A_4{}^c$	$p^3(1-p)$	$\sqrt{}$
$A_1 A_2 A_3{}^c A_4$	$p^3(1-p)$	$\sqrt{}$
$A_1 A_2 A_3{}^c A_4{}^c$	$p^2(1-p)^2$	
$\vdots$	$\vdots$	
$A_1{}^c A_2{}^c A_3{}^c A_4{}^c$	$(1-p)^4$	

Then you simply add the probabilities of the outcomes that comprise the event F. The procedure is straightforward but slightly tedious because of the algebraic simplification required to get to the answer (see Problem 2.19).

2.4 Conditional Probability and Bayes' Rule

2.4.1 Conditional probability

If A_1 and A_2 are two events, then the probability of the event A_1 when it is known that the event A_2 has occured is defined by the relation

$$\Pr[A_1|A_2] = \frac{\Pr[A_1A_2]}{\Pr[A_2]} \qquad (2.13)$$

$\Pr[A_1|A_2]$ is called the probability of "A_1 conditioned on A_2" or simply the probability of "A_1 *given* A_2." Note that in the special case that A_1 and A_2 are statistically independent, it follows from (2.13) and (2.12) that $\Pr[A_1|A_2] = \Pr[A_1]$. Thus when two events are independent, conditioning one upon the other has no effect.

The use of conditional probability is illustrated in the following simple example.

Example 2.8:   Remember Simon's Surplus and the diskettes? A diskette bought at Simon's is equally likely to be good or bad. Simon decides to sell them in packages of two and guarrantees that in each package, at least one will be good. What is the probability that when you buy a single package, you get two good diskettes?

Define the following events:

$$A_1 : \quad \text{Both diskettes are good.}$$
$$A_2 : \quad \text{At least one diskette is good.}$$

The sample space and these events are illustrated below:

The probability we are looking for is

$$\Pr[A_1|A_2] = \frac{\Pr[A_1A_2]}{\Pr[A_2]}$$

Recall that since all events in the sample space are equally likely, the probability of A_2 is $3/4$. Also, since A_1 is included in A_2, it folows that $\Pr[A_1A_2] = \Pr[A_1]$, which is equal to $1/4$. Therefore

$$\Pr[A_1|A_2] = \frac{1/4}{3/4} = \frac{1}{3}$$

□

It is meaningful to interpret conditional probability as the Venn diagram of Fig. 2.6. *Given* the event A_2, the only portion of A_1 that is of concern is the intersection that

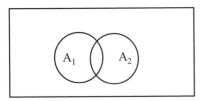

Figure 2.6 Venn diagram illustrating conditional probability.

A_1 has with A_2. It is as if A_2 becomes the new sample space. In defining conditional probability $\Pr[A_1|A_2]$, the probability of event $A_1 A_2$ is therefore "renormalized" by dividing by the probability of A_2.

Equation 2.13 can be rewritten as

$$\Pr[A_1 A_2] = \Pr[A_1|A_2]\Pr[A_2] = \Pr[A_2|A_1]\Pr[A_1] \qquad (2.14)$$

where the second equality follows from the fact that $\Pr[A_1 A_2] = \Pr[A_2 A_1]$. Thus *the joint probability of two events can always be written as the product of a conditional probability and an unconditional probability.* Now let $\{A_i\}$ be a (finite or countably infinite) set of mutually exclusive collectively exhaustive events, and B be some other event of interest. Recall the *principle of total probability* introduced in Section 2.2 and expressed by (2.11). This equation can be rewritten using (2.14) as

$$\boxed{\Pr[B] = \sum_i \Pr[B|A_i]\Pr[A_i]} \qquad (2.15)$$

Although both forms are equivalent, (2.15) is likely the more useful one to remember. This is because the information given in problems is more frequently in terms of conditional probabilities rather than joint probabilities. (And you must be able to recognize the difference!)

Let us consider one final fact about conditional probability before moving on. Again let $\{A_i\}$ be a (finite or countably infinite) set of mutually exclusive collectively exhaustive events. Then the probabilities of the A_i conditioned on *any* event B sum to one. That is,

$$\sum_i \Pr[A_i|B] = 1 \qquad (2.16)$$

The proof of this fact is straightforward. Using the definition (2.13), we have

$$\sum_i \Pr[A_i|B] = \sum_i \frac{\Pr[A_i B]}{\Pr[B]} = \frac{\sum_i \Pr[A_i B]}{\Pr[B]} = \frac{\Pr[B]}{\Pr[B]} = 1$$

where in the next to last step we used the principle of total probability in the form (2.11).

As an illustration of this result, consider the following brief example.

Example 2.9:   Consider the situation in Example 2.8. What is the probability that when you buy a package of two diskettes, only one is good?

Since Simon guarantees that there will be *at least* one good diskette in each package, we have the event A_2 defined in Example 2.8. Let A_3 represent the set of outcomes {BG GB} (only one good diskette). This event has probability $1/2$. The probability of only one good diskette given the event A_2 is thus

$$\Pr[A_3|A_2] = \frac{\Pr[A_3 A_2]}{\Pr[A_2]} = \frac{\Pr[A_3]}{\Pr[A_2]} = \frac{1/2}{3/4} = \frac{2}{3}$$

The events A_3 and A_1 are mutually exclusive and collectively exhaustive given the event A_2. Hence their probabilites, $2/3$ and $1/3$, sum to one.

$\square$

2.4.2 Event trees

As stated above, often the information for problems in probability is stated in terms of *conditional* probabilities. An important technique for solving some of these problems is to draw the sample space by constructing a tree of dependent events and to use the information in the problem to determine the probabilities of compound events.

The idea is illustrated in Fig. 2.7. In this figure, A is assumed to be an event whose

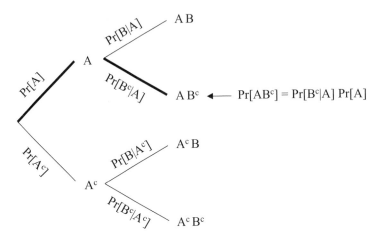

Figure 2.7 Sample space constructed using an event tree.

probability is known and does not depend on any other event. The probability of event B however depends on whether or not A occurred. These conditional probabilities are written on the branches of the tree. The endpoints of the tree comprise the sample space for the problem; they are a set of mutually exclusive collectively exhaustive events which represent the outcomes of the experiment. The probabilities of these events are computed using (2.14), which is equivalent to multiplying probabilities along the branches of the tree that form a path to the event (see figure). Once the probabilities of these elementary events are determined, you can find the probabilities for other compound events by adding the appropriate probabilities. The technique is best illustrated by an example.

Example 2.10:   You listen to the morning weather report. If the weather person forecasts rain, then the probability of rain is 0.75. If the weather person forecasts "no rain," then the probability of rain is 0.15. You have listened to this report for well over a year now and have determined that the weather person forecasts rain 1 out of every 5 days regardless of the season. What is the probability that the weather report is wrong? Suppose you take an umbrella if and only if the weather report forecasts rain. What is the probability that it rains and you are caught without an umbrella?

To solve this problem, define the following events:

$$F = \text{"rain is forecast"} \qquad R = \text{"it actually rains"}$$

From the problem statement, we have the following information:

$$\Pr[R|F] = 0.75 \quad \Pr[R|F^c] = 0.15$$

$$\Pr[F] = 1/5 \qquad \Pr[F^c] = 4/5$$

The events and conditional probabilities are depicted in the event tree shown below. The event that the weather report is wrong is represented by the two elementary events FR^c and F^cR. Since these events are mutually exclusive, their probabilities can be added to find

$$\Pr[\text{wrong report}] = \tfrac{1}{5}(0.25) + \tfrac{4}{5}(0.15) = 0.17$$

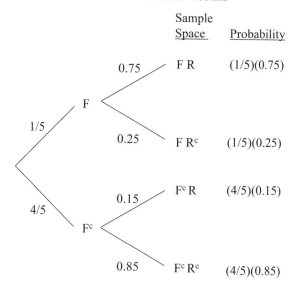

The probability that it rains and you are caught without an umbrella is the event $F^c R$. The probability of this event is $\frac{4}{5}(0.15) = 0.12$.

□

2.4.3 Bayes' rule

One of the most important uses of conditional probability was developed by Bayes in the late 1800's. It follows if (2.14) is rewritten as

$$\Pr[A_1|A_2] = \frac{\Pr[A_2|A_1] \cdot \Pr[A_1]}{\Pr[A_2]} \tag{2.17}$$

This allows one conditional probability to be computed from the other. A particularly important case arises when $\{A_j\}$ is a (finite or countably infinite) set of mutually exclusive collectively exhaustive events and B is some other event of interest. From (2.17) the probability of one of these events conditioned on B is given by

$$\Pr[A_i|B] = \frac{\Pr[B|A_i] \cdot \Pr[A_i]}{\Pr[B]} \tag{2.18}$$

But since the $\{A_j\}$ are a set of mutually exclusive collectively exhaustive events, the principle of total probability can be used to express the probability of the event B. Thus, substituting (2.15) into the last equation yields

$$\boxed{\Pr[A_i|B] = \frac{\Pr[B|A_i] \cdot \Pr[A_i]}{\sum_j \Pr[B|A_j] \Pr[A_j]}} \tag{2.19}$$

This result is known as *Bayes' theorem* or *Bayes' rule.* It is used in a number of problems that commonly arise in communications or other areas where decisions or inferences are to be made from some observed signal or data. Because (2.19) is a more complicated formula, it is sufficient in problems of this type to remember the simpler result (2.18) and to know how to compute $\Pr[B]$ using the principle of total probability.

As an illustration of Bayes' rule, consider the following example.

Example 2.11:   The US Navy is involved in a service-wide program to update memory in computers on-board ships. The Navy will buy memory modules only from American manufacturers known as A_1, A_2, and A_3. The probabilities of buying from A_1, A_2, and A_3 (based on availability and cost to the government) are given by 1/6, 1/3, and 1/2 repectively. The Navy doesn't realize, however, that the probability of failure for the modules from A_1, A_2, and A_3 is 0.006, 0.015, and 0.02 (respectively).

Back in the fleet, an enlisted technician upgrades the memory in a particular computer and finds that it fails. What is the probability that the failed module came from A_1? What is the probability that it came from A_3?

Let F represent the event that a memory module fails. Using (2.18) and (2.19) we can write

$$\Pr[A_1|F] = \frac{\Pr[F|A_1] \cdot \Pr[A_1]}{\Pr[F]} = \frac{\Pr[F|A_1] \cdot \Pr[A_1]}{\sum_{j=1}^{3} \Pr[F|A_j]\Pr[A_j]}$$

Then substituting the known probabilities yields

$$\Pr[A_1|F] = \frac{(0.006)1/6}{(0.006)1/6 + (0.015)1/3 + (0.02)1/2} = \frac{0.001}{0.016} = 0.0625$$

and likewise

$$\Pr[A_3|F] = \frac{(0.02)1/2}{(0.006)1/6 + (0.015)1/3 + (0.02)1/2} = \frac{0.01}{0.016} = 0.625$$

Thus in almost two-thirds of the cases the bad module comes from A_3.

□

2.5 More Applications

This section illustrates the use of probability as it occurs in two problems involving digital communication. A basic digital communication system is shown in Fig. 2.8. The system has three basic parts: a transmitter which codes the message into some

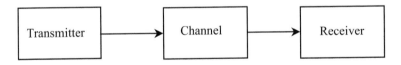

Figure 2.8 Digital communication system.

representation of binary data; a channel over which the binary data is transmitted; and a receiver, which decides whether a 0 or a 1 was sent and decodes the data. The simple binary communication channel discussed in Section 2.5.1 below is a probabilistic model for the binary communication system, which models the effect of sending one bit at a time. Section 2.5.2 discusses the digital communication system using the formal concept of "information," which is also a probabilistic idea.

2.5.1 The binary communication channel

A number of problems naturally involve the use of conditional probability and/or Bayes' rule. The binary communication channel, which is an abstraction for a communication system involving binary data, uses these concepts extensively. The idea is illustrated in the following example.

Example 2.12: The transmission of bits over a binary communication channel is represented in the drawing below, where we use notation like 0_S, 0_R ... to denote events "0

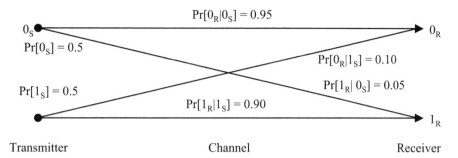

sent," "0 received," etc. When a 0 is transmitted, it is correctly received with proba-
bility 0.95 or incorrectly received with probability 0.05. That is $\Pr[0_R|0_S] = 0.95$ and
$\Pr[1_R|0_S] = 0.05$. When a 1 is transmitted, it it is correctly received with probability
0.90 and incorrectly received with probability 0.10. The probabilities of sending a 0
or a 1 are denoted by $\Pr[0_S]$ and $\Pr[1_S]$ and are known as the *prior* probabilities. It
is desired to compute the *probability of error* for the system.

This is an application of the principle of total probability. If two events A_1 and A_2
form a partition, then (2.15) can be written as

$$\Pr[B] = \Pr[B|A_1]\Pr[A_1] + \Pr[B|A_2]\Pr[A_2]$$

Since the two events 0_S and 1_S are mutually exclusive and collectively exhaustive, we
can identify them with the events A_1 and A_2 and take the event B to be the event
that an error occurs. It then follows that

$$
\begin{aligned}
\Pr[\text{error}] &= \Pr[\text{error}|0_S]\Pr[0_S] + \Pr[\text{error}|1_S]\Pr[1_S] \\
&= \Pr[1_R|0_S]\Pr[0_S] + \Pr[0_R|1_S]\Pr[1_S] \\
&= (0.05)(0.5) + (0.10)(0.5) = 0.075
\end{aligned}
$$

$\square$

The probability of error is an overall measure of performance that is frequently
used for a communication system. Notice that it involves not just the conditional
error probabilities $\Pr[1_R|0_S]$ and $\Pr[0_R|1_S]$ but also the prior probabilities $\Pr[0_S]$ and
$\Pr[1_S]$ for transmission of a 0 or a 1. One criterion for optimizing a communication
system is to minimize the probability of error.

If a communication system is correctly designed, then the probability that a 1 was
sent given that a 1 is received should be greater than the probability that a 0 was
sent given that a 1 is received. In fact, as will be shown later in the text, this condition applied to both 0's and 1's leads to minimizing the probability of error. The
computation of these "inverse" probabilities is illustrated in the next example.

Example 2.13: Assume for the communication system illustrated in the previous example
that a 1 has been received. What is the probability that a 1 was sent? What is the
probability that a 0 was sent?

This is an application of conditional probability and Bayes rule. For a 1, we have

$$\Pr[1_S|1_R] = \frac{\Pr[1_R|1_S]\Pr[1_S]}{\Pr[1_R]} = \frac{\Pr[1_R|1_S]\Pr[1_S]}{\Pr[1_R|1_S]\Pr[1_S] + \Pr[1_R|0_S]\Pr[0_S]}$$

Substituting the numerical values from the figure in Example 2.12 then yields

$$\Pr[1_S|1_R] = \frac{(0.9)(0.5)}{(0.9)(0.5) + (0.05)(0.5)} = 0.9474$$

For a 0, we have a similar analysis:

$$\Pr[0_S|1_R] = \frac{\Pr[1_R|0_S]\Pr[0_S]}{\Pr[1_R|1_S]\Pr[1_S] + \Pr[1_R|0_S]\Pr[0_S]}$$

$$= \frac{(0.05)(0.5)}{(0.9)(0.5) + (0.05)(0.5)} = 0.0526$$

Note that $\Pr[1_S|1_R] > \Pr[0_S|1_R]$ as would be expected, and also that $\Pr[1_S|1_R] + \Pr[0_S|1_R] = 1$.

□

2.5.2 *Measuring information and coding*

The study of *Information Theory* is fundamental to understanding the trade-offs in design of efficient communication systems. The basic theory was developed by Shannon [2, 3, 4] and others in the late 1940's and '50's and provides fundamental results about what a given communication system can or cannot do. This section provides just a taste of the results which are based on a knowledge of basic probability.

Consider the digital communication system depicted in Fig. 2.8 and let the events A_1 and A_2 represent the transmission of two codes representing the symbols 0 and 1. To be specific, assume that the transmitter, or source, outputs the symbol to the communication channel with the following probabilities: $\Pr[A_1] = \frac{1}{8}$ and $\Pr[A_2] = \frac{7}{8}$. The *information* associated with the event A_i is defined as

$$I(A_i) = -\log \Pr[A_i]$$

The logarithm here is taken with respect to the base 2 and the resulting information is expressed in *bits*.[2] The information for each of the two symbols is thus

$$I(A_1) = -\log \Pr[\tfrac{1}{8}] = 3 \text{ (bits)}$$
$$I(A_2) = -\log \Pr[\tfrac{7}{8}] = 0.193 \text{ (bits)}$$

Observe that events with *lower* probability have *higher* information. This corresponds to intuition. Someone telling you about an event that almost always happens provides little information. On the other hand, someone telling you about the occurrence of a very rare event provides you with much more information. The news media works on this principle in deciding what news to report and thus tries to maximize information.

The *average* information[3] H is given by the weighted sum

$$H = \sum_{i=1}^{2} \Pr[A_i] I(A_i) = \tfrac{1}{8} \cdot 3 + \tfrac{7}{8} \cdot 0.193 = 0.544$$

Notice that this average information is less than one bit, although it is not possible to transmit two symbols with less than one binary digit (bit).

Now consider the following scheme. Starting anywhere in the sequence, group together two consecutive bits and assign this pair to one of four possible codewords. Let

---

[2] Other less common choices for the base of the logarithm are 10, in which case the units of information are Hartleys, and e in which case the units are called nats.
[3] Average information is also known as the source *entropy* and is discussed further in Chapter 4.

the corresponding events (or codewords) be denoted by B_j for $j = 1, 2, 3, 4$ as shown in the table below and assume that two consecutive symbols are independent.

codeword	symbols	probability	information (bits)
B_1	00	$\frac{1}{8} \cdot \frac{1}{8} = \frac{1}{64}$	6
B_2	01	$\frac{1}{8} \cdot \frac{7}{8} = \frac{7}{64}$	3.193
B_3	10	$\frac{7}{8} \cdot \frac{1}{8} = \frac{7}{64}$	3.193
B_4	11	$\frac{7}{8} \cdot \frac{7}{8} = \frac{49}{64}$	0.386

Notice that since the probabilities of two consecutive symbols multiply, the corresponding information adds. For example, the symbol 0 by itself has information of 3 bits, while the pair of symbols 00 shown in the table has information of 6 bits. The average information for this scheme is given by the weighted sum

$$H = \frac{1}{64} \cdot 6 + \frac{7}{64} \cdot 3.193 + \frac{7}{64} \cdot 3.193 + \frac{49}{64} \cdot 0.386 = 1.087$$

The average information per bit is still $1.087/2 = 0.544$.

The previous analysis does not result in any practical savings since the average information H is still more than one bit and therefore in a binary communication system it will still require a minimum of two bits to send the four codewords. A continuation of this procedure using larger groups of binary symbols mapped to codewords however does lead to some efficiency. The table below lists the average information with increasing numbers of symbols per codeword.

No. symbols	Avg. Information	Avg. Inf. / bit
2	1.087	0.544
3	1.631	0.544
4	2.174	0.544
5	2.718	0.544
8	4.349	0.544
10	5.436	0.544
12	6.523	0.544

From this table, it is seen that when three symbols are grouped together, the average information is 1.631 bits. It would therefore seem that only two binary digits should be theoretically required to transmit the codewords, since the information $I = 1.631$ is less than 2 bits. Likewise, when 12 symbols are grouped together, it should require no more than 7 binary digits on average to code the message ($I = 6.523 < 7$). How to achieve such efficiency in practice has led to various coding algorithms such as Huffman coding and Shannon-Fano coding. The basic idea is to use variable length codes and assign fewer binary digits to codewords that occur more frequently. This reduces the average number of bits that are needed to transmit the message. The example below illustrates the technique using the Shannon-Fano algorithm.

Example 2.14: It is desired to code the message "ELECTRICAL ENGINEERING" in an efficient manner using Shannon-Fano coding. The probabilities of the letters (excluding the space) are represented by their relative frequency of occurrence in the message. The letters { E, L, C, T, ,R, I, A, N, G } are thus assigned the probabilities $\left\{ \dfrac{5}{21}, \dfrac{2}{21}, \dfrac{2}{21}, \dfrac{1}{21}, \dfrac{2}{21}, \dfrac{3}{21}, \dfrac{1}{21}, \dfrac{3}{21}, \dfrac{2}{21} \right\}$.

The codewords are assigned as illustrated in the steps below. The letters are arranged

Code assignment

E	5/21	0
I	3/21	0
N	3/21	0 (1)
L	2/21	1
C	2/21	1
R	2/21	1
G	2/21	1
T	1/21	1
A	1/21	1

in order of decreasing probability; any ties are broken arbitrarily. The letters are then partitioned into two groups of approximately equal probability (as closely as possible). This is indicated by the partition labeled 1. Those letters in the first group are assigned a codeword beginning with 0 while those in the second group are assigned a codeword beginning with 1.

Within each group, this procedure is repeated recursively to determine the second, third, and fourth digit of the codeword.

Code assignment

E	5/21	0	0 (2)
I	3/21	0	1
N	3/21	0 (1)	1
L	2/21	1	0
C	2/21	1	0
R	2/21	1	0 (2)
G	2/21	1	1
T	1/21	1	1
A	1/21	1	1

Code assignment

E	5/21	0	0 (2)	
I	3/21	0	1	0 (3)
N	3/21	0 (1)	1	1
L	2/21	1	0	0 (3)
C	2/21	1	0	1
R	2/21	1	0 (2)	1
G	2/21	1	1	0 (3)
T	1/21	1	1	1
A	1/21	1	1	1

The final result is as shown below:

		Code assignment						Codeword	Length
E	5/21	0	0 (2)					00	2
I	3/21	0	1	0 (3)				010	3
N	3/21	0 (1)	1	1				011	3
L	2/21	1	0	0 (3)				100	3
C	2/21	1	0	1	0 (4)			1010	4
R	2/21	1	0 (2)	1	1			1011	4
G	2/21	1	1	0 (3)				110	3
T	1/21	1	1	1	0 (4)			1110	4
A	1/21	1	1	1	1			1111	4

An inherent and necessary property (for decoding) of any variable-length coding scheme is that no codeword is a prefix of any longer codeword. Thus, for example, upon finding the sequence 011, we can uniquely determine that this sequence corresponds to the letter N, since there is no codeword of length 4 that has 011 as its first three binary digits.

Now consider the efficiency achieved by this coding scheme. The average number of bits used in coding of the message is the sum of the lengths of the codewords weighted by the probability of the codeword. For this example, the average length is given by (see final figure)

$$2 \cdot 5/21 + 3 \cdot 3/21 + \cdots + 4 \cdot 1/21 = 3.05 \text{ (bits)}$$

On the other hand, if a fixed-length coding scheme were used then the length of each codeword would be 4 bits. (Since there are nine letters, three bits are insufficient and four bits are needed to code each letter.) Thus the variable-length coding scheme, which is based on estimating the average information in the message, reduces the communication traffic by about 24%.

□

2.6 Summary

The study of probability deals with the occurrence of random "events." Such events occur as outcomes or collections of outcomes from an experiment. The complete set of outcomes from an experiment comprise the Sample Space. These outcomes must be mutually exclusive and collectively exhaustive and be of finest grain in representing the conditions of the experiment. Events are defined in the sample space.

The algebra of events is a form of set algebra that provides rules for describing arbitrarily complex combinations of events in an unambiguous way. Venn diagrams are useful as a complementary geometric method for depicting relationships among events.

Probability is a number between 0 and 1 assigned to an event. Several rules and formulae allow you to compute the probabilities of intersections, unions, and other more complicated expressions in the algebra of events when you know the probabilities of the other events in the expression. An important special rule applies to independent events: to compute the probability of the combined event, we simply multiply the individual probabilities.

Conditional probability is necessary when events are not independent. The formulas developed in this case provide means for computing joint probabilities of sets of events that depend on each other in some way. These conditional probabilities are also useful in developing a "tree diagram" from the facts given in a problem, and an associated representation of the sample space. Bayes' rule is an especially important use of conditional probability since it allows you to "work backward" and compute the probability of unknown events from related observed events. Bayes' rule forms the basis for methods of "statistical inference" and finds much use (as we shall see later) in engineering problems.

A number of examples and applications of the theory are discussed in this chapter. These are important to see how the theory applies in specific situations. The applications also illustrate some well-established models, such as the binary communication channel, which are important to electrical and computer engineering.

References

[1] Alvin W. Drake. *Fundamentals of Applied Probability Theory*. McGraw-Hill, New York, 1967.

[2] Claude E. Shannon. A mathematical theory of communication. *Bell System Technical Journal*, 27(3):379–422, July 1948. (See also [4].).

[3] Claude E. Shannon. A mathematical theory of communication (concluded). *Bell System Technical Journal*, 27(4):623–656, October 1948. (See also [4].).

[4] Claude E. Shannon and Warren Weaver. *The Mathematical Theory of Communication*. University of Illinois Press, Urbana, IL, 1963.

Problems

Algebra of events

2.1 Draw a set of Venn diagrams to illustrate each of the following identities in the algebra of events.

(a) $A \cdot (B + C) = AB + AC$

(b) $A + (BC) = (A + B)(A + C)$

(c) $(AB)^c = A^c + B^c$

(d) $(A + B)^c = A^c B^c$

(e) $A + A^c B = A + B$

(f) $AB + B = B$

(g) $A + AB + B = A + B$

2.2 A sample space S is given to be $\{a_1, a_2, a_3, a_4, a_5, a_6\}$. The following events are defined on this sample space: $A_1 = \{a_1, a_2, a_4\}$, $A_2 = \{a_2, a_3, a_6\}$, and $A_3 = \{a_1, a_3, a_5\}$.

(a) Find the following events: (i) $A_1 + A_2$, (ii) $A_1 A_2$, and (iii) $(A_1 + A_3^c)A_2$.

(b) Show the identities: (i) $A_1(A_2 + A_3) = A_1 A_2 + A_1 A_3$,
(ii) $A_1 + A_2 A_3 = (A_1 + A_2)(A_1 + A_3)$, and (iii) $(A_1 + A_2)^c = A_1^c A_2^c$.

Probability of events

2.3 By considering probabilities as represented by areas in a Venn diagram, show that the four axioms of probability and the results listed in Table 2.4 "make sense." If the result cannot be shown by Venn diagram, say so.

2.4 Starting with the expression for $\Pr[A_1 + A_2]$, show that for *three* events

$$\Pr[A_1 + A_2 + A_3] = \Pr[A_1] + \Pr[A_2] + \Pr[A_3] - \Pr[A_1 A_2] \\ - \Pr[A_1 A_3] - \Pr[A_2 A_3] + \Pr[A_1 A_2 A_3]$$

2.5 Consider the sample space in Problem 2.2 in which all the outcomes are assumed equally likely. Find the following probabilities: (i) $\Pr[A_1 A_2]$, (ii) $\Pr[A_1 + A_2]$, and (iii) $\Pr[(A_1 + A_3^c)A_2]$.

2.6 A signal has been sampled, quantized (8 levels), and encoded into 3 bits. These bits are sequentially transmitted over a wire.

(a) Draw the sample space corresponding to the received bits. Assume that no errors occur during transmission.

(b) Determine the probability that the received encoded sample has a value greater than 5.

(c) What is the probability that the value of the received sample is between 3 and 6 (inclusive)?

2.7 For purposes of efficient signal encoding, a given speech signal is subject to a companding operation. The μ-law companding used in commercial telephony in North America is given by

$$|y| = \frac{\log(1 + \mu|x|)}{\log(1 + \mu)}$$

where $\mu \approx 100$, x is the input signal value and y is the output signal value. Consider that the input is in the range of $0 \leq x \leq 5$.

(a) Draw the sample spaces that could be used to represent events involving the input and the output signals.

(b) What is the probability that the output value is less than 1?

(c) Find these probabilities: $\Pr[\frac{1}{2} \leq x \leq 1]$ and $\Pr[\frac{1}{2} \leq y \leq 1]$.

2.8 The following events and their probabilities are listed below.

event:	A	B	C	D
probability:	1/4	1/3	1/4	1/4

(a) Compute $\Pr[A + B]$ assuming $AB = \emptyset$.

(b) Compute $\Pr[A + B]$ assuming A and B are *independent*.

(c) Is it possible that all of A, B, C, and D are mutually exclusive? Tell why or why not.

2.9 Computers purchased from *Simon's Surplus* will experience hard drive failures with probability 0.3 and memory failures with probability 0.2 and will experience both types of failures simultaneously with probability 0.1

(a) What is the probability that there will be one type of failure but not the other?

(b) What is the probability that there will be no failures of either kind?

Use the algebra of events to write expressions for the desired events and the rules of probability to compute probabilities for these event expressions.

Applications

2.10 Consider the problem of purchasing diskettes described in Example 2.2.

(a) Assuming that the probabilities of good diskettes and bad diskettes are equal, what is the probability that you have one or more good diskettes?

(b) If the probability of a good disk is $\frac{5}{8}$, what is the probability that you have one or more good diskettes?

2.11 Simon has decided to improve the quality of the products he sells. Now only one out of five diskettes selected from the bins is defective (i.e., the probability that a diskette is bad is only 0.2). If three diskettes are chosen at random, what is the probability ...

(a) that all three diskettes are good?

(b) that all three diskettes are bad?

(c) that [exactly] one of the three diskettes is bad?

(d) of more good diskettes than bad ones?

2.12 In a certain digital control system, the control command is represented by a set of four hexadecimal characters.

(a) What is the total number of control commands that are possible?

(b) If each control command must have a unique prefix, i.e., starting from left to right no hex character must be repeated, how many control commands are possible?

2.13 A digital transmitter sends groups of 8 bits over a communication channel sequentially. The probability of a single bit error in the channel is p.

(a) How many different ways can two errors occur in 8 bits?

(b) Assume that a particular 2-bit error has occured: $\Phi x \Phi \Phi \Phi x \Phi \Phi$, where x indicates an error bit and Φ the correct bit (0 or 1). What is the probability of this particular event?

(c) Determine the probability of 2-bit errors in this system.

(d) Compute and plot the probability of 2-bit errors for $p = 0.1, 0.01, 0.001, 0.0001$.

2.14 In the Internet, the TCP/IP protocol suite is used for transmitting packets of information. The receiving TCP entity checks the packets for errors. If an error is detected, it requests retransmission of the packet that is found to be in error. Suppose that the packet error probability is 10^{-2}.

(a) What is the probability that the third packet requires retransmission?

(b) What is the probability that the tenth packet requires retransmission?

(c) What is the probability that there is a packet retransmission within the first 5 packets?

(d) Following on the lines of (c), how many packets are required in order for the probability of retransmission to be equal to or greater than 0.1?

2.15 In the serial transmission of a byte (such as over a modem) errors in each bit occur independently with probability Q. For every 8-bit byte sent a parity check bit is appended so that each byte transmitted actually consists of 9 bits. The parity bit is chosen so that the group of 9 bits has "even parity," i.e., the number of 1's, is even. Errors can be detected by checking the parity of the received 9-bit sequence.

(a) What is the probability that a *single* error occurs?

(b) What is the probability that errors occur but are not detected? Write your answer as an expression involving the error probability Q.

2.16 The diagram below represents a communication network where the source S communicates with the receiver R. Let A represent the event "link a fails," and B represent the event "link b fails," etc. Write an expression in the algebra of events for the event F = "S fails to communicate with R."

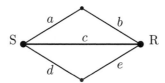

2.17 Repeat Problem 2.16 for the following network.

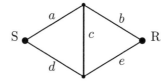

2.18 In Problem 2.16 assume that link failures are independent and that each link can fail with probability 0.1. Compute the probability of the event F = "S fails to communicate with R."

2.19 By using the alternate procedure described on page 21, compute the probability of failure for the network of Example 2.7.

2.20 (a) A communication network is shown below. Define the following events:

A   link a fails
B   link b fails
C   link c fails
F   S cannot communicate with R

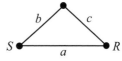

Assume link failures are independent events. Write an expression in the Algebra of Events for the event F. Your expression should be in terms of the events A, B, and C.

(b) Repeat part (a) for the following network, shown below

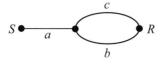

(c) If every link has the same probability of failure $\Pr[A] = \Pr[B] = \Pr[C] = p = 0.1$, then which network has the lowest $\Pr[F]$? Justify your answer.

2.21 The diagram below is meant to represent a communication network with links a, b, c, and d. Let A denote the event "Link a is OK." Events B, C, and D are defined similarly. Each link has a probability of failure of 0.5 (in other words, the probability that each link is OK is 0.5) and the failure of any link is *independent* of the failure of any other link.

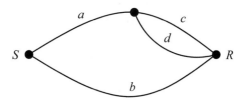

(a) Write an expression in the Algebra of Events that represents the event "S communicates with R." The expression should involve only the events A, B, C, and D. Do not use any probabilities here.

(b) What is the probability that c or d or both links are OK (i.e., what is the probability of the event CD)?

(c) What is the probability of the event "S communicates with R"?

(d) If S fails to communicate with R, which link has the highest probability of having failed? Which link has the next highest probability of having failed? What are these probabilities? In what order should we test the links (according to their probability of failure) to determine if each is functioning?

2.22 Consider the problem of failure of a communication network as described in Example 2.7.

(a) What is the largest value of p that can be tolerated if the probability of failure must be less than 0.001?

(b) What is the largest value of p if the probability of failure must be less than 0.05?

2.23 A schematic of Mincom's North American backbone network is shown below. The probability of failure of any link in the network is p, and the link failures are statistically independent.

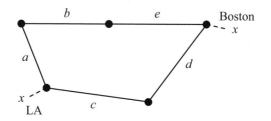

Let A, B, C, D, E be the events that the corresponding link fails.

(a) Write an algebraic expression for the failure to connect between subscribers in Los Angeles and Boston.

(b) Determine the probability of establishing a connection between these users.

(c) Let $p = 0.01$. Compute the probability of establishing the above connection.

Conditional probability and Bayes' rule

2.24 An analog to digital (A/D) converter generates 3-bit digital words corresponding to seven levels (0, 1, ..., 7) with the following probabilities: $\Pr[0] = \Pr[1] = \Pr[6] = \Pr[7] = \frac{1}{16}$, $\Pr[2] = \Pr[5] = \frac{1}{8}$, and $\Pr[3] = \Pr[4] = \frac{1}{4}$.

(a) Find the probability of receiving a value > 5 given that the first bit is a 1.

(b) Find the probability that the first bit is a 1 given that the value is greater than 5.

(c) What is the probability of a 1? Hint: Use the principle of total probability approach.

(d) Determine the probability of a 0.

2.25 In a certain computer, the probability of a memory failure is 0.01, while the probability of a hard disk failure is 0.02. If the probability that the memory and the hard disk fail simultaneously is 0.0014, then

(a) Are memory failures and hard disk failures independent events?

(b) What is the probability of a memory failure, given a hard disk failure?

2.26 Repeat Problem 2.25 if the probability of a memory failure is 0.02, the probability of a disk failure is 0.015, and the probability that both fail simultaneously is 0.0003.

2.27 In the process of transmitting binary data over a certain noisy communication channel, it is found that the errors occur in bursts. Given that a single bit error occurs, the probabilty that the next bit is in error is <u>twice</u> the probability of a single bit error. If it is known that two consecutive errors occur with probability 2×10^{-4}, what is the probability of a single bit error?

2.28 In the Navy's new advanced YK/2 (pronounced "yuk-2") computer, hardware problems are not necessarily catastrophic, but software problems will cause complete shutdown of ship's operations. Nothing but software failures can cause a complete shutdown.

It is known that the probability of a hardware failure is 0.001 and the probability of a software failure *given* a hardware failure is 0.02. Complete shutdowns (i.e., software failures) occur with probability 0.005. If a shutdown occurs, what is the probability that there was a hardware failure?

2.29 A random hexadecimal character in the form of four binary digits is read from a storage device.

(a) Draw the tree diagram and the sample space for this experiment.

(b) Given that the first bit is a zero, what is the probability of more zeros than ones?

(c) Given that the first two bits are 10, what is the probability of more zeros than ones?

(d) Given that there are more zeros than ones, what is the probability that the first bit is a zero?

2.30 Beetle Bailey has a date with Miss Buxley, but Beetle has an old jeep which will break down with probability 0.4. If his jeep breaks down he will be late with probability 0.9. If it does not break down he will be late with probability 0.2. What is the probability that Beetle will be late for his date?

2.31 Rudy is an astronaut and the engineer for Project Pluto. Rudy has determined that the mission's success or failure depends on only three major systems. Further, Rudy decides that the mission is a failure if and only if two or more of the major systems fail. The following is known about these systems. System I, the auxiliary generator, fails with probability 0.1 and does not depend on the other systems. System II, the fuel system, fails with probability 0.5 if at least one other system fails. If no other system fails, the probability that the fuel system fails is 0.1. System III, the beer cooler, fails with probability 0.5 if the generator system fails. Otherwise the beer cooler cannot fail.

(a) Draw the event space (sample space) for this problem by constructing a tree of possible events involving failure or nonfailure of the major systems. Use the notation G^c, B^c, F^c to represent failure of the generator, beer cooler, and fuel systems and G, B, F to represent nonfailure of those systems. The ends of your tree should be labeled with triples such as $G^c BF$ and their probabilities.

Hint: Start with the Generator. What system should go on the next level of the tree?

(b) Answer the following questions about the mission.

(i) What is the probability that the mission fails?
(ii) What is the probability that all three systems fail?

(c) Given that more than one system failed, what is the probability:

(i) that the generator did not fail?
(ii) that the beer cooler failed?
(iii) that both the generator and the fuel system failed?

(d) Given that the beer cooler failed, what is the probability that the mission succeeded?

[This problem is a modified version of a problem from Alvin W. Drake, *Fundamentals of Applied Probability Theory*, McGraw-Hill, New York, 1967. Reproduced by permission.]

2.32 Microsoft's Internet Explorer (IE) and Netscape's Navigator (NN) are two popular Web browsers. IE supports certain features that are present at a given site, such as security, that NN does not and vice versa. Assume that all features on all sites are exploitable by either IE or NN. The probability that IE supports a given feature is $1 - \epsilon$, and the probability that NN fails to support a feature is δ.

(a) A user selects a browser at random and attempts to access a site at random. What is the probability that the user will be forced to change the browsers?

(b) Given that the browser fails to access a site, what is the probability that it is IE?

More applications

2.33 In a certain binary communication channel, it is equally likely to send a 1 or a 0 (both probabilities are equal to 1/2). The probability of an error given that a 1 is sent is 2/9, while the probability of an error given a 0 is sent is 1/9.

(a) What is the probability that a 1 is received?
(b) What is the (unconditional) probability that an error occurs?
(c) What is the probability that a 1 was sent, given a 1 was received?

2.34 A binary symmetric communication channel is shown below.

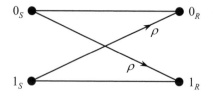

An "Error" occurs if 0 is sent and 1 is received or 1 is sent and 0 is received. Assume that $\Pr[1_R | 0_S] = \Pr[0_R | 1_S] = \rho$.

(a) Assume $\Pr[0_S] = \Pr[1_S] = \frac{1}{2}$. Find ρ such that $\Pr[\text{Error}] = 0.001$.

(b) Repeat part (a) for $\Pr[0_S] = 0.2$ and $\Pr[1_S] = 0.8$.

(c) What is the probability that a 0 was sent, given that a 0 is received? Assume the same conditions as part (a).

2.35 A binary communication channel is depicted below. Assume that the random experiment consists of transmitting a single binary digit and that the probability of transmitting a 0 or a 1 is the same.

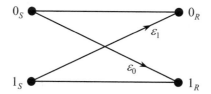

(a) Draw the sample space for the experiment and label each elementary event with its probability.

(b) What is the probability of an error?

(c) Given that an error occurred, what is the probability that a 1 was sent?

(d) What is the probability a 1 was sent given that a 1 was received?

2.36 In a digital communication channel, such as QPSK, one of the four symbols $\{A, B, C, D\}$ is transmitted at a time. The channel characteristics are indicated in the figure below. The *prior* probabilities of the input symbols are given to be: $\Pr[A_S] = \Pr[B_S] = \Pr[C_S] = \Pr[D_S] = \frac{1}{4}$.

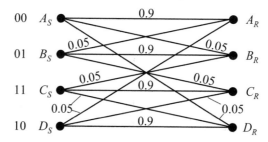

(a) What is the probability of receiving an A?

(b) Determine the probability of error in this channel.

(c) Given that a D is received, what is the probability that (i) an A was sent, (ii) a B was sent, (iii) a C was sent, and (iv) a D was sent.

2.37 Repeat Problem 2.36 for the channel shown in the figure below. The *prior* probabilities of the input symbols are given to be: $\Pr[A_S] = \frac{3}{16}$, $\Pr[B_S] = \frac{1}{4}$, $\Pr[C_S] = \frac{5}{16}$, $\Pr[D_S] = \frac{1}{4}$.

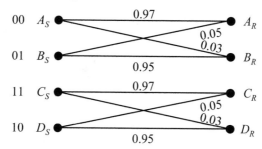

2.38 Never-Ready Wireless has just invented a new signaling system known as 3PSK. The system involves transmitting one of three symbols {0, 1, 2}. The conditional probabilities for their network are given in the following table:

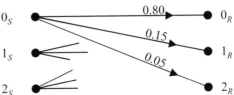

	0_R	1_R	2_R
0_S	0.80	0.15	0.05
1_S	0.05	0.80	0.15
2_S	0.15	0.05	0.8

For example, the probability that a 2 is received, given a 0 was transmitted is $\Pr[2_R|0_S] = 0.05$. A partial sketch of the communication channel is also provided above. The *prior* probabilities of transmitting each of the symbols are given by $\Pr[0_S] = 2/5$, $\Pr[1_S] = 2/5$ and $\Pr[2_S] = 1/5$.

(a) What is the probability of an error given that a 0 was sent?

(b) What is the (unconditional) probability of error?

(c) Given that a 1 is received, what is the probability that a 1 was sent?

(d) Given that a 1 is received, what is the probability that this is an error?

2.39 Two symbols A and B are transmitted over a binary communications channel shown below. The two received symbols are denoted by α and β. The numbers on the branches in the graph indicate *conditional probabilities* that α or β is received given that A or B is transmitted. For example, $\Pr[\alpha|A] = 0.8$.

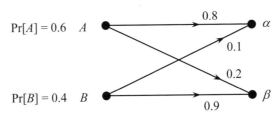

The symbols A and B are transmitted with probabilities indicated above.

(a) What is the probability that the received symbol is α (i.e., what is $\Pr[\alpha]$)?

(b) What decision rule should be used to minimize the probability of error? That is, tell whether A or B should be chosen when α is received and tell whether A or B should be chosen when β is received.

(c) Someone has "switched the wires" so now the channel looks like:

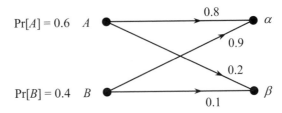

Answer the question in part (b) for this new situation. Also compute the *probability of error* corresponding to this decision rule.

2.40 Consider the following seemingly meaningless sentence:

MAT CAT MAN MITTEN MAM MEN EATEN.

(a) Estimate the probabilities of each of the symbols in the sentence by computing the relative frequency of the symbols. You may ignore spaces.

(b) Compute the average information for the sentence.

(c) Obtain the Huffman code for each of the symbols in the sentence.

(d) Determine the average length of the codewords.

HUFFMAN ALGORITHM

Step 1: Arrange the symbols in order of decreasing probability. If the probabilities are equal, break the tie randomly.

Step 2: Combine the two smallest values in the list to form a new entry whose probability is the sum of the two original entries.

Step 3: Continue combining two smallest values in the modified list (original untouched values and the new entry) until the root of the tree is reached.

Step 4: Assign bits from left to right on the tree: at each branch, assign a 0 to the higher valued branch and a 1 to the other.

Step 5: Form the codewords by reading from the root to the node of the symbol.

Computer Projects

Project 2.1

In this project, you are to simulate a simple binary communication channel characterized by appropriate conditional and prior probabilities and estimate the probability of error as well as the probability of receiving either a 1 or a 0.

We start out with a symmetric binary communication channel characterized by the conditional probabilities $\Pr[0_R|0_S] = \Pr[1_R|1_S] = 0.975$ and $\Pr[0_R|1_S] = \Pr[1_R|0_S] = 0.025$. The prior probabilities of a 0 or a 1 are given by $\Pr[0_S] = 0.512$ and $\Pr[1_S] = 0.488$. The input to the binary communication channel is to be a sequence of 0's and 1's. Each 0 or 1 in the sequence is statistically independent of the others and is generated according to the probabilities $\Pr[0_S]$ and $\Pr[1_S]$ given above.

1. Generate the data input sequence of 0's and 1's according to the required probabilities. The size of the sequence is your choice; however, to obtain meaningful results, it should be at least 5000 points long.

2. Simulate the channel by writing code to do the following:

(a) When a 0 is presented as input to the channel, the channel should generate an output of 0 with probability $\Pr[0_R|0_S]$ and an output of 1 with probability $\Pr[1_R|0_S]$ (where these numerical values are given above).

(b) When a 1 is presented to the channel, the channel should generate a 0 with probability $\Pr[0_R|1_S]$ and a 1 with probability $\Pr[1_R|1_S]$.

3. Compute the theoretical values for the following probabilities: $\Pr[0_S|0_R]$, $\Pr[1_S|1_R]$, and $\Pr[\text{error}]$ (see Section 2.5.1).

4. Apply the input data sequence generated in Step 1 to the channel in Step 2, estimate the probabilities in Step 3. To estimate the probability, use relative frequency; for example, to estimate $\Pr[0_S|0_R]$ you would compute

$$\frac{\#\text{ times 0 sent and 0 received}}{\#\text{ times 0 received}}$$

5. Compare the estimated values to the theoretical values. If necessary, repeat the experiment using a longer input sequence. You may also wish to compare the results for *various* length input sequences.

Repeat Steps 1 through 5 for a nonsymmetric binary communication channel with conditional probabilities $\Pr[0_R|0_S] = 0.975$, $\Pr[1_R|1_S] = 0.9579$, $\Pr[1_R|0_S] = 0.025$, $\Pr[0_R|1_S] = 0.0421$. Let the prior probabilities be $\Pr[0_S] = 0.5213$ and $\Pr[1_S] = 0.4787$.

MATLAB programming notes

You can generate a binary random number with the following statement:

```
x=rand(1)<P;
```

where P is the desired probability of a 1. Note that the expression to the right of the equal sign is a boolean expression; thus MATLAB assigns a 0 or 1 to the variable x. You can replace the argument '1' of the 'rand' function with the size of a vector or matrix and generate a whole vector or matrix of binary random numbers.

Project 2.2

The Huffman coding algorithm is described in Prob. 2.40 (above). In this project, you will measure the average information, code the given message using the Huffman algorithm, transmit the coded message over a simple binary channel, reconstruct the message, and compare the performance to that without coding. The following schematic diagram illustrates the sequence for implementing these operations.

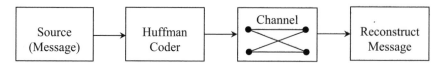

1. Consider the following message for source coding and transmission:

APPARENTLY NEUTRAL'S PROTEST IS THOROUGHLY DISCOUNTED AND IGNORED. ISMAN HARD HIT. BLOCKAGE ISSUE AFFECTS PRE-TEXT FOR EMBARGO ON BY-PRODUCTS, EJECTING SUETS AND VEGETABLE OILS.

(This message is included in the data package for this book as the file 'msg.txt'.)

2. Following the procedure of Section 2.5.2, determine the average information for this message. (You may ignore hyphens, spaces, and punctuation.)

3. Using the Huffman algorithm, code the above message.

4. Determine the average codeword length.

5. Transmit the codewords in binary form across the binary symmetric communication channel specified in Project 2.1.

6. Reconstruct the message from the received codewords.

7. Determine the average information of the received message. Would you expect it to be larger or smaller than that of the transmitted message?

3 Random Variables and Transformations

Situations involving probability do not always deal strictly with events. Frequently there are real-valued measurements or observations associated with a random experiment. Such measurements or observations are represented by *random variables*.

This chapter develops the necessary mathematical tools for the analysis of experiments involving random variables. We begin with discrete random variables, i.e., those random variables that take on only a discrete (but possibly countably infinite) set of possible values. Some common types of discrete random variables are described that are useful in practical applications. Moving from the discrete to the continuous, we discuss random variables that can take on an uncountably infinite set of possible values and some common types of these random variables.

The chapter also develops methods to deal with problems where one random variable is described in terms of another. This is the subject of "transformations."

The chapter concludes with two important practical applications. The first involves the detection of a random signal in noise. This problem, which is fundamental to every radar, sonar, and communication system, can be developed using just the information in this chapter. The second application involves the classification of objects on an assembly line. Although the problem may seem unrelated to the detection problem, some of the underlying principles are identical.

3.1 Discrete Random Variables

Formally, a random variable is defined as a function $X(\cdot)$ that assigns a real number to each elementary event in the sample space. In other words, it is a mapping from the sample space to the real line (see Fig 3.1). A random variable therefore takes on

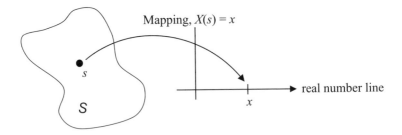

Figure 3.1 Illustration of a random variable.

a given numerical value with some specified probability. A simple example is useful to make this abstract idea more clear. Consider the experiment of rolling a pair of dice. The sample space showing the complete set of outcomes is illustrated in Chapter 2, Fig. 2.5. Let us define the random variable K as follows:

$$K(s) = \text{the number rolled}$$

Now, suppose s is the point in the sample space representing a roll of $(3,1)$ on the dice; then $K(s) = 4$. Likewise, if s represents the roll of $(2,2)$ or $(1,3)$ then again $K(s) = 4$. Thus the mapping provided by the random variable is in general many-to-one.

In discussing random variables it is common to drop the argument s and simply write (for example) $K = 4$. In this example K is a *discrete* random variable, meaning that it takes on only a discrete set of values, namely the integer values 2 through 12. The probability assigned to these values of K is determined by the probability of the outcomes s in the sample space. If each outcome in the sample space has probability 1/36, then the probability that $K = 4$ is 3/36 or 1/12. (This is the sum of the probabilities for the outcomes $(3,1)$, $(2,2)$, and $(1,3)$ all of which result in $K = 4$.) Similarly, the probability that $K = 2$ is 1/36. (There is only one experimental outcome which results in this value of K.)

The probability associated with random variables is conveniently represented by a plot similar to a bar chart and illustrated in Fig. 3.2. The function $f_K[k]$ depicted in

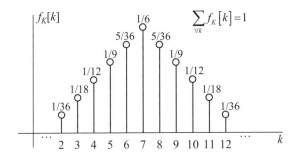

Figure 3.2 Probability mass function for rolling of dice.

the figure is called a *probability mass function* (PMF). The PMF shows the probability of the random variable for each possible value that it can take on. The sum of the probabilities represented in the PMF is therefore equal to 1.

Let us now establish some notation and a formal definition. It is common to represent a random variable by a capital letter such as K. The value that the random variable takes on is then denoted by the corresponding lower case letter k. The PMF is then a function $f_K[k]$ defined by

$$f_K[k] = \Pr[s \in \mathcal{S} : K(s) = k]$$

In other words, $f_K[k]$ is the probability that random variable K takes on the value k. Less formally, we can write

$$f_K[k] = \Pr[K = k] \tag{3.1}$$

where the precise meaning of the expression $\Pr[K = k]$ is given in the equation preceding (3.1).

The probability that the random variable takes on a set of values is obtained by summing the PMF over that whole set of values. For example, the probability that you roll a number greater than or equal to 10 is given by (see Fig. 3.2)

$$\Pr[K \geq 10] = \sum_{k=10}^{12} f_K[k] = \frac{1}{12} + \frac{1}{18} + \frac{1}{36} = \frac{1}{6}$$

Formally, the PMF must satisfy two conditions: First, since $f_K[k]$ represents a *probability*, it must be true that

$$0 \leq f_K[k] \leq 1 \qquad \forall k \tag{3.2}$$

($\forall k$ means "for all values of k"). Secondly, the sum of all the values is

$$\sum_{\forall k} f_K[k] = 1 \qquad (3.3)$$

as we observed earlier. Any discrete function satisfying these two conditions can be a PMF.

To further illustrate the PMF, let us consider another example.

Example 3.1:   The price of Simon's diskettes is three for $1. With the sales tax this comes to $1.08 or $.36 per diskette. Customers will always return bad diskettes and Simon will exchange any bad diskette for another one guaranteed to be good, but customers have determined that the cost of returning a disk in terms of trouble and aggravation is $.14. The net effect is that good diskettes cost $.36 while bad ones cost the customer $.50. What is the discrete probability density function for the random variable C, the cost of buying three diskettes at Simon's? What is the probability that you will end up paying more than $1.25?

The events of the sample space and the cost associated with each event are shown below:

Events:	BBB	BBG	BGB	BGG	GBB	GBG	GGB	GGG
Cost:	$1.50	$1.36	$1.36	$1.22	$1.36	$1.22	$1.22	$1.08
Probability:	1/8	1/8	1/8	1/8	1/8	1/8	1/8	1/8

Since all of the events in the sample space are equally likely, you can construct the PMF by counting events in the sample space that result in the same cost. The resulting PMF is shown below.

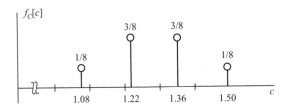

The probability of paying more than $1.25 is given by

$$f_C[1.36] + f_C[1.50] = \frac{3}{8} + \frac{1}{8} = \frac{1}{2}$$

□

This example illustrates that although the domain of the PMF is discrete, the random variable does not necessarily have to take on integer values. Integer values for a discrete random variable are the most common, however.

3.2 Common Discrete Probability Distributions

There are a number of common types of random variables that occur frequently in various problems. The PMFs that describe these random variables are sometimes referred to as "probability laws" or *distributions*. A few of these common distributions are discussed below. In all cases the random variable is assumed to take on values in the set of integers.

3.2.1 Bernoulli random variable

A Bernoulli random variable is a discrete random variable that takes on only one of two discrete values (usually 0 and 1). The Bernoulli PMF is defined by

$$f_K[k] = \begin{cases} p & k = 1 \\ 1 - p & k = 0 \\ 0 & \text{otherwise} \end{cases} \tag{3.4}$$

and is illustrated in Fig. 3.3. The parameter p is the probability of a one.

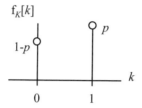

Figure 3.3 The Bernoulli PMF.

Example 3.2: A certain binary message $M[n]$ is represented by a sequence of zeros and ones. If we sample the sequence at a random time n_o the result $K = M[n_o]$ is a Bernoulli variable. Unless there is evidence to the contrary, you would expect the parameter of the distribution to be $p = 0.5$.

□

3.2.2 Binomial random variable

Consider a binary sequence of length n, where each element of the sequence is a Bernoulli random variable occurring independently. Let the random variable K be defined as the number of 1's in the sequence. Consider any such sequence, say

$$1\ 0\ 0\ 1\ 1\ \ldots\ 0\ 1$$

in which k 1's occur. The corresponding probability of this sequence is

$$p \cdot (1 - p) \cdot (1 - p) \cdot p \cdot p \cdots (1 - p) \cdot p = p^k (1 - p)^{(n-k)}$$

If you were to list all the possible sequences with k 1's, you would find that there are $C_k^n = \binom{n}{k}$ such sequences (see Appendix A). Therefore the probability that one of these sequences occurs is given by[1]

$$f_K[k] = \binom{n}{k} p^k (1 - p)^{(n-k)} \qquad 0 \le k \le n \tag{3.5}$$

This PMF is known as the binomial distribution. It is depicted in Fig. 3.4 for the parameter values $n = 20$ and $p = 0.25$.

---

[1] Throughout this section, $f_K[k]$ may be assumed to be 0 for values of k not explicitly represented in the formula.

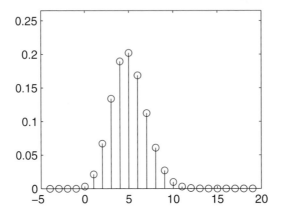

Figure 3.4 The binomial distribution.
$(n = 20, p = 0.25)$

Example 3.3:   A certain modem connection has a channel bit error rate of $p = 0.01$. Given that data are sent as packets of 100 bits, what is the probability that (a) just 1 bit is in error and (b) that 3 bits are in error?

The answers to the question are

$$\Pr[1 \text{ bit error}] = f_K[1] = \binom{100}{1} 0.01^1 \, 0.99^{99} = 0.3697$$

$$\Pr[3 \text{ bit errors}] = f_K[3] = \binom{100}{3} 0.01^3 \, 0.99^{97} = 0.0610$$

□

3.2.3 Geometric random variable

The geometric random variable is typically used to represent waiting times. Again, consider a sequence of binary digits being sent over some communication channel. Let us consider the number of bits that are observed before the first error occurs; this is a random variable K. (The *method* of error detection is irrelevant.) A tree diagram for this experiment is shown in Fig. 3.5. If the first error occurs at the kth bit, then $k - 1$

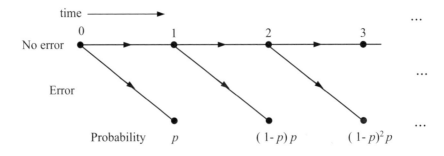

Figure 3.5 Tree diagram for error detection experiment.

bits were without error. The probability of this event is given by the geometric PMF:

$$f_K[k] = p(1-p)^{(k-1)} \qquad 1 \le k < \infty \tag{3.6}$$

Actually, there are two common variations of this distribution. The other form corresponds to an error occuring at the $k+1^{st}$ bit and is given by

$$f_K[k] = p(1-p)^k \qquad 0 \le k < \infty \tag{3.7}$$

We refer to (3.6) and (3.7) as Type 1 and Type 0 respectively. A plot of the Type 0 geometric PMF for $p = 0.25$ is shown in Fig. 3.6. The function is a simple decaying exponential.

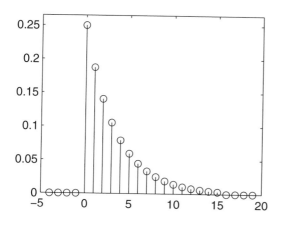

Figure 3.6 The geometric distribution (Type 0). $p = 0.25$)

Since this is our first example of a random variable that can take on an *infinite* number of possible values, it is interesting to check that the sum of the PMF is equal to 1 as required. This is most easily done by considering the second form of the distribution and defining $q = 1 - p$. Then since $q < 1$ we can write

$$\sum_{k=0}^{\infty} f_K[k] = \sum_{k=0}^{\infty}(1-q)q^k = (1-q)\sum_{k=0}^{\infty}q^k = (1-q)\frac{1}{1-q} = 1$$

where in the second to last step we have used the formula for the sum of an infinite geometric series.[2]

Example 3.4: Errors in the transmission of a random stream of bytes occur with probability ϵ. What is the probability that the first error will occur after 16 bytes?

Let K be the random variable representing the occurrence of the first error. So K is a geometric random variable. The probability that the first error occurs after 16 bytes is given by

$$\Pr[K > 16] = \sum_{k=16}^{\infty} \epsilon(1-\epsilon)^k = (1-\epsilon)^{16}$$

---

[2] See, e.g., [1].

The last step is left as a challenge to you to evaluate the sum using the formula for the summation of a infinite geometric series. You can check that the answer is correct by the following argument. The event that the first error occurs after the 16th byte is the same as the product or union of events:

$$\bigcup_{i=1}^{16}(\text{"No error on the }k^{\text{th}}\text{ byte"})$$

Since the events in the union are mutually independent and each has probability $1-\epsilon$, the probability of the overall event is $(1-\epsilon)^{16}$.

□

3.2.4 Poisson random variable

Consider the situation where events occur randomly in a certain time interval t. The "events" could be the arrival of print jobs at a network server, the number of "hits" on a particular Web page, the number of telephone calls occuring in a busy office, or any similar set of events. The time interval could be measured in microseconds or hours or any other set of units appropriate to the problem. Let λ be a variable that represents the *rate* of arrival of these events (e.g., $\lambda = 20$ hits/hour on a certain Web page). Then, a useful expression for the probability of k events happening in time t that has been experimentally verified is

$$\Pr[k \text{ events in } t] = \frac{(\lambda t)^k}{k!}e^{-\lambda t}$$

(This formula is derived later in the text.) The number of events K is said to be a Poisson random variable with corresponding PMF

$$f_K[k] = \frac{\alpha^k}{k!}e^{-\alpha} \qquad 0 \le k < \infty \tag{3.8}$$

The single parameter $\alpha = \lambda t$ can be interpreted as the average number of events arriving in the time interval t. For example, if $\lambda = 20$ events/hour and t=0.2 hour (12 minutes), then $\alpha = (20)(0.2) = 4$. A plot of the PMF for $\alpha = 4$ is given in Fig. 3.7.

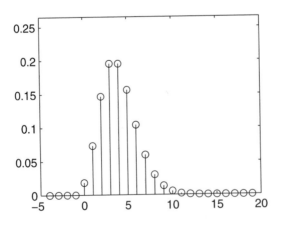

Figure 3.7 The Poisson PMF. ($\alpha = 4$)

Example 3.5: Packets at a certain node on the internet arrive with a rate of 100 packets per minute. What is the probability that no packets arrive in 6 seconds? What about the probability that 2 or more packets arrive in the first 6 seconds?

The parameter α is given by

$$\alpha = \lambda t = 100/\text{min} \cdot \frac{6}{60} \text{ min } = 10 \text{ (arrivals)}$$

The probability that no packets arrive is thus given by

$$f_K[0] = \frac{10^0}{0!}e^{-10} = e^{-10} = 4.540 \times 10^{-5}$$

To answer the second question, let us first compute the probability of a *single* arrival in the first 6 seconds. This is given by

$$f_K[1] = \frac{10^1}{1!}e^{-10} = 10e^{-10} = 4.540 \times 10^{-4}$$

The probability of two or more arrivals is thus given by

$$1 - f_K[0] - f_K[1] = 0.9995$$

□

The Poisson PMF is also useful as an approximation to the binomial PMF. This approximation is given by

$$\binom{n}{k}p^k(1-p)^{(n-k)} \approx \frac{\alpha^k}{k!}e^{-\alpha}$$

where $\alpha = np$ in the limit as $n \to \infty$ and $p \to 0$. This relation is discussed further in Section 8.1 of Chapter 8.

3.2.5 Discrete uniform random variable

The final distribution to be discussed is the discrete uniform distribution defined by

$$f_K[k] = \begin{cases} \dfrac{1}{n-m+1} & m \le k \le n \\ 0 & \text{otherwise} \end{cases} \tag{3.9}$$

and illustrated in Fig. 3.8 for $m = 0$, $n = 9$. This PMF is useful as the formal description of a discrete random variable all of whose values over some range are equally likely.

Example 3.6: Let K be a random variable corresponding to the outcome of the roll of a single die. K is a discrete uniform random variable with parameters $m = 1$ and $n = 6$. All of the nonzero values of the function are equal to $1/(n-m+1) = 1/6$.

□

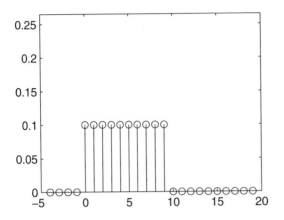

Figure 3.8 The discrete uniform distribution. ($m = 0$, $n = 9$)

3.3 Continuous Random Variables

Not all random variables take on a discrete set of values, even if the number of such values is infinite. For example suppose an experiment is to monitor the power line during an electrical storm. The peak voltage measured from a power surge on the line can be represented by a random variable that takes on a continuum of values (over some possible range). In order to deal with this kind of problem, we need tools to work with continuous random variables.

3.3.1 Probabilistic description of continuous random variables

Consider the example of the surge voltage cited above. The event that the peak voltage is less than, say 2000 volts has some reasonable likelihood of occurrence. The event that the voltage is *precisely equal* to 2000 volts (or any other specific value) is extremely unlikely however. A mathematical model involving continuous random variables should reflect these characteristics and assign probability to intervals of the real line. In order that these intervals have finite probability between 0 and 1, all (or at least most) of the infinite number of points within an interval should have probability approaching 0.

Let us begin with a random variable X and consider the probability

$$F_X(x) = \Pr[s \in \mathcal{S} : X(s) \leq x]$$

As with discrete random variables, we denote the random variable by a capital letter (X) and a value that it takes on by the corresponding lower case letter (x). To make the notation less cumbersome, reference to the sample space is frequently dropped and the previous equation is written simply as

$$F_X(x) = \Pr[X \leq x] \tag{3.10}$$

The quantity $F_X(x)$ is known as the *cumulative distribution function* (CDF) and has the general character depicted in Fig. 3.9. When plotted as a function of its argument x, this function starts at 0 and increases monotonically toward a maximum value of 1 (see Fig. 3.9). These properties, which can be shown rigorously, make sense intuitively:

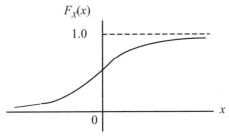

Figure 3.9 Cumulative distribution function.

1. Since no value of X can be less than minus infinity, $F_X(-\infty)$ should be 0.

2. If Δx is any positive value, then $\Pr[X \le x + \Delta x]$ should not be less than $\Pr[X \le x]$. Thus $F_X(x)$ should be monotonically increasing.

3. Since every value of X is less than infinity, $F_X(\infty)$ should be 1.

Next, consider the probability that X is in some more general interval[3] $a < X \le b$ where a and b are any two values with $a < b$. We can write

$$\Pr[X \le b] = \Pr[X \le a] + \Pr[a < X \le b]$$

This follows because the two intervals on the right are nonoverlapping and their union is the interval on the left ($X \le b$). Then rearranging this equation and using the definition (3.10) yields

$$\Pr[a < X \le b] = F_X(b) - F_X(a) \tag{3.11}$$

Now, *assume* for the moment that $F_X(x)$ is continuous and differentiable throughout its range, and let $f_X(x)$ represent the derivative

$$f_X(x) = \frac{dF_X(x)}{dx} \tag{3.12}$$

By using this definition, (3.11) can be written as

$$\Pr[a < X \le b] = \int_a^b f_X(x)dx \tag{3.13}$$

The function $f_X(x)$ is known as the *probability density function* (PDF). It will be our primary tool for dealing with continuous random variables.

3.3.2 More about the PDF

A typical generic PDF is shown in Fig. 3.10. The PDF for a continuous random variable

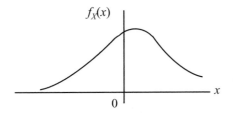

Figure 3.10 Typical probability density function for a continuous random variable.

has two fundamental properties. First, it is everywhere *nonnegative*:

$$f_X(x) \ge 0 \tag{3.14}$$

---

[3] The two different forms of inequality used here are deliberate and important in our development.

This follows directly from the monotonicity of $F_X(x)$ and (3.12). Secondly, the PDF must integrate to 1:

$$\int_{-\infty}^{\infty} f_X(x)dx = 1 \qquad (3.15)$$

This follows from (3.13) by taking the interval $(a, b]$ to be the entire real line. These two conditions correspond to the two conditions (3.2) and (3.3) for the discrete probability density function. Note however, that since $f_X(x)$ is *not* a probability, but rather a probability *density*, it is not required to be less than or equal to 1.

Let us explore this last point just a bit further. Consider (3.13) and let $b = a + \Delta_x$ where Δ_x is a small number. If $f_X(x)$ is continuous in the interval $(a, a + \Delta_x]$, then

$$\boxed{\Pr[a < X \le a + \Delta_x] = \int_a^{a+\Delta_x} f_X(x)dx \approx f_X(a) \cdot \Delta_x} \qquad (3.16)$$

The approximate equality follows from the mean value theorem of calculus and is consistent with our previous discussion. Equation 3.16 is a very important result because it shows that $f_X(x)$ by itself is *not* a probability, but it becomes a probability when multiplied by a small quantity Δ_x. *This interpretation of the probability density function for a continuous random variable will be used on several occasions in the text.*

3.3.3 A relation to discrete random variables

It is insightful to explore the relation between a continuous and discrete random variable in the context of a familiar application, namely that of analog-to-digital signal conversion. In the digital recording of audio signals such as speech or music, the analog signal has to be sampled and quantized. This operation is performed by an analog-to-digital coverter (ADC) as shown in Fig. 3.11. The ADC contains a quantizer which

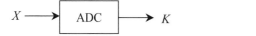

Figure 3.11 Analog-to-digital converter.

maps values of X into one of 2^N output values or "bins" which are represented by a binary number with N bits. The number N of bits in the output representation is known as the bit "depth." The mapping performed by the quantizer is such that the output is an integer k if the input x is in the interval $L_k < x \le L_{k+1}$ where $L_{k+1} = L_k + \Delta$. The input/output characteristic is shown in Fig. 3.12.

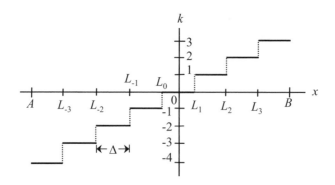

Figure 3.12 Quantization performed by ADC of Fig. 3.11.

Suppose that the input signal at a particular point in time is a continuous random variable X taking values in the interval $A < X \leq B$. The PDF for the random variable is depicted in Fig. 3.13 (a). The output K of the quantizer is a discrete random variable that takes on the values $k = 0, 1, 2, \ldots, 2^N - 1$. The PMF for the output K is depicted in Fig. 3.13 (b).

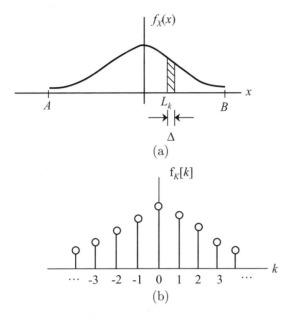

(a)

(b)

Figure 3.13 Discrete approximation for continuous random variable. (a) PDF for input random variable X. (b) PMF for output random variable K.

The probability that random variable K takes on a particular value k is given by the PMF $f_K[k]$. This probability is equal to the probability that X falls in the interval $L_k < X \leq L_k + \Delta$. Thus we have

$$f_K[k] = \int_{L_k}^{L_k + \Delta} f_X(x)\, dx$$

For any small value of Δ the PMF of the output is (approximately)

$$f_K[k] \approx f_X(L_k)\Delta$$

In a typical audio application, the bit depth N may be fairly large (e.g., $N = 16$ or $N = 20$). Therefore, except for the constant factor Δ, the PMF of the output is a very accurate discrete approximation to the PDF of the input.

3.3.4 Solving problems involving a continuous random variable

While solving problems for discrete random variables generally involves summing the PMF over various sets of points in its domain, solving problems involving continuous random variables usually involves integrating the PDF over some appropriate region or regions. In so doing, you must be careful to observe the limits over which various algebraic expressions may apply, and set up the limits of integration accordingly. In particular, the integral must *not* include regions where the PDF is 0. The following example illustrates these points.

Example 3.7: The time T (in seconds) for the completion of two sequential events is a continuous random variable with PDF given below:

$$f_T(t) = \begin{cases} Ct & 0 < t \le 5 \\ C(10 - t) & 5 < t \le 10 \\ 0 & \text{otherwise} \end{cases}$$

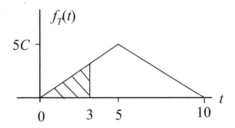

What is the value of the constant C so that $f_T(t)$ is a legitimate probability density function? Find the probability that both events are completed in 3 seconds or less. What is the probability that both events are completed in 6 seconds or less?

To answer the first question, use the fact that the PDF must integrate to 1:

$$\int_{-\infty}^{\infty} f_T(t)dt = \text{ area of triangle } = 25C = 1$$

Therefore the constant C is equal to $1/25$. Notice how the problem of formally evaluating the integral by calculus was replaced by a simple geometric computation. This is a useful technique whenever it is possible.

The probability that the events are completed in no more than 3 seconds is found by integrating the density over the region shown shaded below.

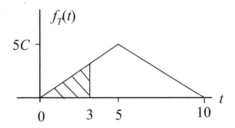

$$\Pr[T \le 3] = \int_{-\infty}^{3} f_T(t)dt = \int_{0}^{3} \frac{1}{25}t\, dt = \left. \frac{t^2}{50} \right|_{0}^{3} = \frac{9}{50} = 0.18$$

Observe that since the PDF is 0 for negative values of t, when the expression $\frac{1}{25}t$ is substituted for $f_T(t)$ the lower limit on the integral becomes 0 (not $-\infty$!).

The probability that the events are completed in no more than 6 seconds is found by integrating the density over the following region:

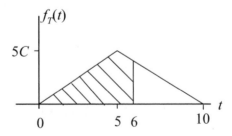

In this case the region of integration has to be broken up into two regions where different algebraic expressions apply. The computation is as follows:

$$\Pr[T \le 6] \quad = \quad \int_{-\infty}^{6} f_T(t)dt = \int_{0}^{5} \frac{1}{25}t \, dt + \int_{5}^{6} \frac{1}{25}(10 - t) \, dt$$

$$= \quad \left.\frac{t^2}{50}\right|_{0}^{5} + \left.\frac{-(10 - t)^2}{50}\right|_{5}^{6} = \frac{25}{50} + \frac{9}{50} = \frac{17}{25} = 0.68$$

☐

When the intervals of concern are very small, the interpretation of probability density provided by (3.16) is useful. The next example explores this point.

Example 3.8: Consider the time T (in seconds) for the completion of two sequential events as described by the PDF in Example 3.7. What is the most likely time of completion of the events? What is the (approximate) probability that the events are completed within 10 msec of the most likely time?

Since T is a continuous random variable, the probability for events to be completed at any *specific* time t is zero. We interpret the question however, to mean that the event is completed within a small interval of time $(t, t + \Delta_t]$. To find the most likely time of completion, notice that the probability that the event occurs in any such small interval is given by (see (3.16))

$$f_T(t)\Delta_t$$

From the sketch of $f_T(t)$ in Example 3.7, the time t that maximizes this probability is seen to be $t = 5$.

We interpret the second question to mean "What is the probability that T is between 4.99 and 5.01 seconds?" To compute this probability first use (3.16) to write

$$\Pr[5 < T \le 5 + 0.01] \approx f_T(5) \cdot 0.01 = \frac{1}{5}(0.01) = 0.002$$

Then, since the distribution is symmetric about the point $t = 5$, the probability that $4.99 < T \le 5$ is equal to the same value. Hence the probability that T is within 10 ms of the most likely value is approximately 0.004.

☐

3.4 Common Continuous Probability Density Functions

A number of common forms for the PDF apply in certain applications. These probability density functions have closed form expressions, which can be found listed in several places. A few of these PDFs are described here. Expressions for the CDF are generally *not* listed in tables however, and may not even exist in closed form at all. This is one reason why most engineering treatments of random variables focus on using the PDF.

3.4.1 Uniform random variable

A uniform random variable is defined by the density function (PDF)

$$f_X(x) = \begin{cases} \dfrac{1}{b - a} & a \le x \le b \\ \\ 0 & \text{otherwise} \end{cases} \tag{3.17}$$

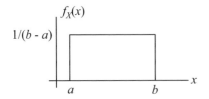

Figure 3.14 The uniform PDF.

and illustrated in Fig. 3.14.

Like its discrete counterpart, this PDF is useful as the formal description of a random variable all of whose values over some range are equally likely.

Example 3.9: An analog signal is hard limited to lie between -5 and $+5$ volts before digital-to-analog conversion. In the absence of any specific knowlege about the signal and its characteristics, we can model the signal value as a random variable having a uniform PDF with $a = -5$ and $b = +5$. We say the signal is "uniformly distributed" between -5 and $+5$ volts.

□

3.4.2 Exponential random variable

The exponential random variable has the negative exponential PDF given by

$$f_X(x) = \begin{cases} \lambda e^{-\lambda x} & x \geq 0 \\ 0 & x < 0 \end{cases} \tag{3.18}$$

and depicted in Fig. 3.15. The exponential density is used to describe waiting times

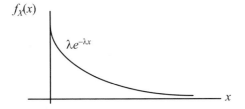

Figure 3.15 The exponential PDF.

for events, which usually follow this distribution in practice. The exponential density can be derived as a limiting form of the geometric distribution for discrete events (see e.g. [2]). It is also closely related to the Poisson random variable. In this context the parameter λ represents the arrival rate of any recurring events (electrons impinging on a cathode, jobs arriving at a printer server, buses arriving at a bus stop, etc.). While the Poisson PMF describes the *number* of arrivals in a fixed time interval t, the exponential density function $f_T(t)$ describes the waiting time T to the arrival of the *next* event.

Example 3.10: In a printer queue, the waiting time W for the next job is an exponential random variable. Jobs arrive at an average rate of $\lambda = 30$ jobs per hour. What is the probability that the waiting time to the next job is between 2 and 4 minutes?

This probability is easily calculated:

$$\Pr\left[\tfrac{2}{60} < W \leq \tfrac{4}{60}\right] = \int_{2/60}^{4/60} 30e^{-30w}\,dw = -e^{-30w}\Big|_{2/60}^{4/60} = e^{-1} - e^{-2} = 0.233$$

□

The exponential random variable is also a model for "time to failure" of a component or piece of equipment after it is put into service. In this context $1/\lambda$ is the average lifetime of the component, also called the "mean time to failure." Suppose that T represents the time to failure of a light bulb after it is installed. Then the probability that the bulb fails *after* some time t_o is given by

$$\Pr[T > t_0] = \int_{t_0}^{\infty} \lambda e^{-\lambda t} dt = e^{-\lambda t_0}$$

The exponential random variable has what is called the *memoryless property*, which states that if the bulb has *not* failed at a time t_0 then the probability that it fails after an additional time t_1 (i.e., at time $t_0 + t_1$) is just $e^{-\lambda t_1}$. In other words, what happens up to time t_0 is "forgotten." The proof of this fact is rather straightforward:

$$\Pr[T > t_0 + t_1 | T > t_0] = \frac{\Pr[[T > t_0 + t_1] \cdot [T > t_0]]}{\Pr[T > t_0]} = \frac{\Pr[T > t_0 + t_1]}{\Pr[T > t_0]}$$

$$= \frac{e^{-\lambda(t_0 + t_1)}}{e^{-\lambda t_0}} = e^{-\lambda t_1} = \Pr[T > t_1]$$

3.4.3 Gaussian random variable

The last continuous random variable to be discussed here is the Gaussian or "normal" random variable. It arises naturally in all kinds of engineering and real-world problems and plays a central role in probability and statistics.

The Gaussian random variable has the PDF

$$f_X(x) = \frac{1}{\sqrt{2\pi\sigma^2}} e^{-(x-\mu)^2/2\sigma^2} \quad -\infty < x < \infty \qquad (3.19)$$

which has the familiar "bell shaped" curve shown in Fig. 3.16. The parameter μ

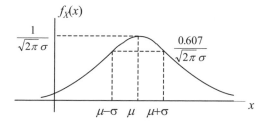

Figure 3.16 The Gaussian PDF.

represents the *mean* or average value of the random variable while the parameter σ^2 is called the *variance*. The square root of the variance (σ) is called the *standard deviation* and is a measure of the spread of the distribution. It can be shown that the probability that any Gaussian random variable lies between $\mu - \sigma$ and $\mu + \sigma$ is approximately 0.682.

While the Gaussian density is an extremely important PDF in applications, integrals of the form

$$\Pr[a < X \le b] = \frac{1}{\sqrt{2\pi\sigma^2}} \int_a^b e^{-(x-\mu)^2/2\sigma^2} dx \qquad (3.20)$$

cannot be evaluated analytically. Fortunately, numerical integration results are widely available. In particular, one of three normalized integrals listed in Table 3.1 are generally tabulated as a function of x. The function $\Phi(x)$ in the first row of the table is just the CDF of a Gaussian random variable with mean $\mu = 0$ and variance $\sigma^2 = 1$. The Q *function*, which is listed in the second row and is equal to $1 - \Phi(x)$, is frequently used

function	symmetry relation	$F_X(x) = \Pr[X \le x]$
$\Phi(x) = \dfrac{1}{\sqrt{2\pi}} \displaystyle\int_{-\infty}^{x} e^{-\frac{u^2}{2}}\,du$	$\Phi(-x) = 1 - \Phi(x)$	$\Phi\left(\dfrac{x-\mu}{\sigma}\right)$
$Q(x) = \dfrac{1}{\sqrt{2\pi}} \displaystyle\int_{x}^{\infty} e^{-\frac{u^2}{2}}\,du$	$Q(-x) = 1 - Q(x)$	$1 - Q\left(\dfrac{x-\mu}{\sigma}\right)$
$\mathrm{erf}(x) = \dfrac{2}{\sqrt{\pi}} \displaystyle\int_{0}^{x} e^{-u^2}\,du$	$\mathrm{erf}(-x) = -\mathrm{erf}(x)$	$\dfrac{1}{2}\left[1 + \mathrm{erf}\left(\dfrac{x-\mu}{\sqrt{2}\sigma}\right)\right]$

Table 3.1 Functions for evaluation of Gaussian density integrals.

in communication engineering applications. Since $Q\left(\frac{x-\mu}{\sigma}\right) = \Pr[X > x]$, this function provides the most direct way for computing probability of detection and false alarm (see Section 3.8.1). The *error function*, listed in the last row of the table is common in both statistics and engineering. This function is also available in MATLAB. Using these functions, integrals of the form (3.20) can be written as $F_X(b) - F_X(a)$ where $F_X(x)$ is the Gaussian cumulative distribution function, which can be evaluated according to the formula in the last column of Table 3.1. Where negative arguments for these functions are not tabulated, you can use the relations listed in the second column of the table, which derive from the symmetry of the Gaussian density function.

For the remainder of this text, the formulas based on the error function will be used. The most commonly encountered integrals occur when one of the limits a or b in (3.20) is infinite. The corresponding integrals and their formulas involving the error function are listed in Table 3.2.

$$\Pr[X \le b] = \frac{1}{\sqrt{2\pi\sigma^2}} \int_{-\infty}^{b} e^{-(x-\mu)^2/2\sigma^2}\,dx \quad \Longleftrightarrow \quad \frac{1}{2}\left[1 + \mathrm{erf}\left(\frac{b-\mu}{\sqrt{2}\sigma}\right)\right]$$

$$\Pr[X > a] = \frac{1}{\sqrt{2\pi\sigma^2}} \int_{a}^{\infty} e^{-(x-\mu)^2/2\sigma^2}\,dx \quad \Longleftrightarrow \quad \frac{1}{2}\left[1 - \mathrm{erf}\left(\frac{a-\mu}{\sqrt{2}\sigma}\right)\right]$$

Table 3.2 Computation of probability using the error function.

Example 3.11:   A radio receiver observes a voltage V which is equal to signal plus noise:

$$V = s + N$$

The signal s is a constant, 2 mv. The noise N is a Gaussian random variable with mean $\mu = 0$ and variance $\sigma^2 = 4$ (mv^2). The receiver will detect the signal s if the voltage V is greater than 1 mv. What is the probability that the signal is *not* detected?

The voltage V in this example is a Gaussian random variable with mean $S = 2$ and variance $\sigma^2 = 4$. (You can take this as a fact for now; it follows from the formula (3.27) discussed in Section 3.6.) The probability that the signal is not detected is therefore given by

$$\Pr[V \le 1] = \frac{1}{\sqrt{2\pi \cdot 4}} \int_{-\infty}^{1} e^{-(v-2)^2/(2 \cdot 4)}\,dv$$

The first result in Table 3.2 is then used to evaluate the integral:

$$\Pr[V \leq 1] \quad = \quad \tfrac{1}{2}\left[1 + \mathrm{erf}\left(\frac{1-2}{\sqrt{2}\cdot 2}\right)\right]$$

$$\approx \quad \tfrac{1}{2}\left[1 - \mathrm{erf}\left(0.35\right)\right] = \tfrac{1}{2}(1 - 0.38) = 0.31$$

where the table in the back cover of the book has been used for the error function.

□

3.5 CDF and PDF for Discrete and Mixed Random Variables

3.5.1 Discrete random variables

The cumulative distribution function (CDF) is introduced in Section 3.3.1 as a probabilistic description for continuous random variables. In that section it is assumed that the CDF is continuous and differentiable; its derivative is the PDF.

The CDF can apply to discrete random variables as well, although the function as defined by (3.10) will exhibit discontinuities. For example, consider the Bernoulli random variable whose PMF is shown in Fig. 3.4. The CDF for this random variable is defined by

$$F_K(k) = \Pr[K \leq k]$$

and is shown in Fig. 3.17.

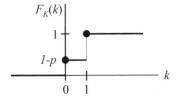

Figure 3.17 Cumulative distribution function for Bernoulli random variable.

Let us check out this plot according to the last equation, bearing in mind that K takes on only values of 0 and 1. Since the random variable K never takes on values less than 0, the function is 0 for $k < 0$. The probability that $K \leq k$ when k is a number between 0 and 1 is just the probability that $K = 0$. Hence the value of the CDF is a constant equal to $1 - p$ in the interval $0 \leq k < 1$. Finally, the probability that $K \leq k$ for $k \geq 1$ is equal to 1. Therefore, the CDF is 1 in the interval $k \geq 1$.

This function clearly does not satisfy the conditions of continuity and differentiability previously assumed for *continuous* random variables. If we tolerate the discontinuities however, the CDF is a useful analysis tool since it allows for description of both discrete and continuous random variables.

A complete mathematical description of the CDF is beyond the scope of this text. An interesting discussion can be found in [3], however. For most engineering applications, we can make the following assumption:

> The CDF is continuous and differentiable everywhere except at a countable number of points, where it may have discontinuities.

At points of discontinuity, the "$\leq$" in the definition (3.10) implies that the CDF is continuous *from the right*. In other words, the function is determined by its value on the right rather than on the left (hence the dot in the illustration). The definition (3.10) also implies that the CDF must be *bounded between 0 and 1 and be monotonically increasing*.

With this broader definition of the CDF, the corresponding PDF cannot be defined unless a certain type of singularity is allowed. The needed singularity however is well known in engineering applications as the unit impulse, denoted by $\delta(x)$. A brief discussion of the unit impulse is provided in Appendix B.

The unit impulse $\delta(x)$ satisfies the following condition:

$$\int_{-\infty}^{x} \delta(\xi)d\xi = u(x) = \begin{cases} 1 & x \geq 0 \\ 0 & x < 0 \end{cases} \tag{3.21}$$

The function $u(x)$ defined on the right is the *unit step function* (see Appendix B) and can be used to account for any discontinuities that may appear in the CDF. *The impulse $\delta(x)$ can be regarded as the formal derivative of the step $u(x)$.* If the upper limit x in (3.21) becomes infinite, then the equation becomes

$$\int_{-\infty}^{\infty} \delta(\xi)d\xi = 1$$

which shows that the "area" of the impulse is equal to 1.

With this "function" available, we can define the PDF for any random variable (continuous or discrete) as the derivative of the CDF. Fig. 3.18 shows the PDF for the

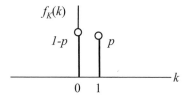

Figure 3.18 PDF for a Bernoulli random variable.

Bernoulli random variable whose CDF is depicted in 3.17. A comparison of this PDF to the PMF given in Fig. 3.3 shows that the discrete values have been replaced by impulses whos *areas* are equal to the corresponding probabilities. The areas are equal to the height of the discontinuities in the corresponding CDF. The general formula for expressing the PMF as a PDF with impulses is

$$f_K(k) = \sum_{\forall i} f_K[i]\delta(k - i) \tag{3.22}$$

Here the impulse occuring at $k = i$ is given the value (area) $f_K[i]$.

3.5.2 *Mixed random variables*

You may have noticed that while we have so far only given examples of continuous and discrete random variables, our definition for the CDF and PDF allows for random variables that may have aspects of both types. Such random variables are commonly referred to as *mixed* random variables. The CDF and PDF for a typical such random variable are illustrated in Fig. 3.19. While for most values of x the probability $\Pr[X = x]$ is equal to 0, at the point $x = x_o$ that probability is equal to a positive value P. (The overall area under the density function, including the impulse, must integrate to 1 however.)

Figure 3.19 CDF and PDF for a random variable with a continuous and discrete component. (a) CDF. (b) PDF.

To see how such mixed random variables may occur, consider the following example.

Example 3.12:   The input voltage X to the simple rectifier circuit shown below is a continuous Gaussian random variable with mean $m = 0$ and variance $\sigma^2 = 1$. The circuit contains an ideal diode which offers 0 resistance in the forward direction and infinite resistance in the backward direction. The circuit thus has the voltage input/output characteristic shown in the figure. It will be shown that the output voltage Y is a mixed random variable.

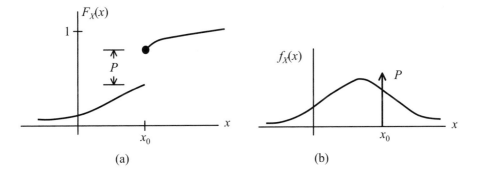

We will develop a more complete approach to problems of this type in Section 3.6. For now, we can use the following argument. When the input voltage is positive, the output is equal to the input. Therefore, for positive values of X, the input and output PDF are identical. For negative values of the input, the output is 0. The probability of negative values is given by

$$\int_{-\infty}^{0} f_X(x)dx = 0.5$$

where $f_X(x)$ represents the Gaussian density of (3.19) with parameter $m = 0$. Since the probability of a negative input is 0.5, the probability of a 0 output is correspondingly 0.5. Thus the output PDF has an impulse with area of 0.5 at the location $y = 0$. The sketch below shows the input and output density functions. The output Y is a mixed random variable.

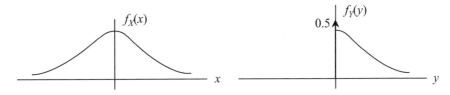

□

3.6 *Transformation of Random Variables*

In many important applications, one random variable is specified in terms of another. You can think of this as a *transformation* from a random variable X to a random variable Y as illustrated in Fig. 3.20.

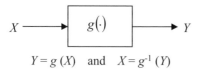

$$Y = g(X) \quad \text{and} \quad X = g^{-1}(Y)$$

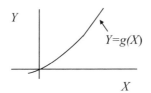

Figure 3.20 Transformation of a random variable.

Frequently this transformation is linear (or at least represented by a straight line). For instance, in Example 3.11, the output random variable V is given by $V = S + N$ where N is the random variable representing the noise and S is a constant. In some cases however, the transformation may be nonlinear. In the process of rectification, a voltage is applied to a diode or other nonlinear device to convert alternating current to direct current. In a neural network, a sum of inputs is applied to a sigmoidal or other nonlinearity before comparing to a threshold. In all of these cases it is important to be able to compute the output PDF ($f_Y(y)$) when you know the input PDF ($f_X(x)$).

In the following, we will consider three important cases for the transformation and develop some simple formulas for each. These three cases cover most of the situations you are likely to encounter.

3.6.1 *When the transformation is invertible*

Figure 3.21 shows a portion of the transfer characteristic when g is an invertible

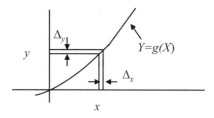

Figure 3.21 Section of invertible transfer characteristic.

function. Considering a point (x, y) on the diagram, we can write

$$\Pr[y < Y \leq y + \Delta_y] = \Pr[x < X \leq x + \Delta_x]$$

since both sides of the equation represent the same event. Thus

$$f_Y(y)\Delta_y = f_X(x)\Delta_x$$

Further, for any small Δ_x, we have to any desired degree of approximation

$$\Delta_y = \left| \frac{dy}{dx} \right| \Delta_x$$

(The absolute value is to allow for the fact that the slope of the curve might be negative.) Then combining these last two equations and cancelling the common term Δ_x produces

$$f_Y(y) = \frac{1}{|dy/dx|} f_X(x)\Big|_{x=g^{-1}(y)} \tag{3.23}$$

The entire expression on the right is evaluated at $x = g^{-1}(y)$ where g^{-1} represents the inverse transformation, because we want $f_Y(y)$ to be expressed in terms of y.

As an example of the use of this equation, suppose

$$Y = aX + b \tag{3.24}$$

Then $dy/dx = a$ and $g^{-1}(y) = (y - b)/a$. Therefore,

$$f_Y(y) = \frac{1}{|a|} f_X\left(\frac{y-b}{a}\right) \tag{3.25}$$

Notice that magnitude (absolute value) of a appears outside of the function but not inside. This is important in applying the formula.

Although (3.25) is a very useful result, it is probably easier to memorize it and apply it as two separate special cases. These are summarized in the two boxes below. For $b = 0$:

$$
\begin{aligned}
Y &= aX \\
f_Y(y) &= \frac{1}{|a|} f_X\left(\frac{y}{a}\right)
\end{aligned}
\tag{3.26}
$$

while for $a = 1$:

$$
\begin{aligned}
Y &= X + b \\
f_Y(y) &= f_X(y - b)
\end{aligned}
\tag{3.27}
$$

Example 3.13: Suppose random variables X and Y are related by $Y = 2X + 1$ and that X is uniformly distributed between -1 and $+1$. The PDF for X is shown in part (a) of the sketch below.

If we define the intermediate random variable $U = 2X$ then the PDF of U is given (from (3.26)) by $f_U(u) = \frac{1}{2} f_X\left(\frac{u}{2}\right)$. This PDF is shown in part (b) of the figure.

Now we can write $Y = U + 1$.

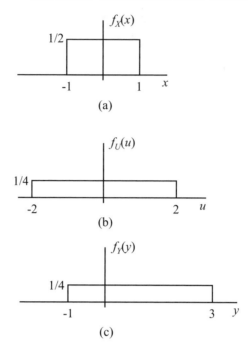

(a)

(b)

(c)

The PDF of Y is thus given (from (3.27)) by $f_Y(y) = f_U(y-1) = \frac{1}{2}f_X\left(\frac{y-1}{2}\right)$. This final PDF is shown in part (c).

$\square$

It is interesting to note in this example that both the original variable and the transformed variable turned out to be uniformly distributed. This is not a coincidence. Any time you have a straight-line transformation of the form (3.24), the *shape* of the PDF does not change. This is evident from (3.25) where it can be seen that f_Y is just a shifted and scaled version of f_X. Thus under a straight line transformation, a uniform random variable remains uniform (with different parameters), a Gaussian random variable remains Gaussian, and so on. Let us explore this point just a bit further for the Gaussian case.

Example 3.14:   Random variables X and Y are related by $Y = aX + b$. X is a Gaussian random variable with mean μ and variance σ^2. Demonstrate that Y is also a Gaussian random variable. What are the parameters describing the random variable Y?

To answer this question, simply apply the transformation formula (3.25) using f_X as given by (3.19):

$$f_Y(y) = \frac{1}{|a|}\frac{1}{\sqrt{2\pi\sigma^2}}e^{-[((y-b)/a)-\mu]^2/2\sigma^2} = \frac{1}{\sqrt{2\pi a^2\sigma^2}}e^{-[y-(a\mu+b)]^2/(2a^2\sigma^2)}$$

From this it is seen that Y has a Gaussian PDF with parameters $\mu' = a\mu + b$ and $\sigma'^2 = a^2\sigma^2$.

$\square$

3.6.2 *When the transformation is not invertible*

First of all, let us mention that the formula (3.23) derived above works just fine when the transformation provides a one-to-one mapping of X to Y but Y does not take on all possible values on the real line. For example, the transformation $Y = e^X$ results in only positive values of Y. The inverse transformation therefore does not exist for negative values of Y. (Mathematicians would say that the transformation is one-to-one but not *onto*.) Nevertheless you can use (3.23) to write

$$f_Y(y) = y f_X(\ln y) \qquad 0 \le y < \infty$$

The main problem arises when the transformation is not one-to-one. This case is illustrated in Fig. 3.22. In this illustration, $g(x_1)$, $g(x_2)$ and $g(x_3)$ all produce the same

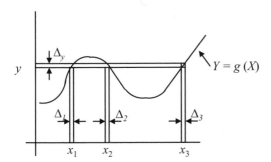

Figure 3.22 Section of noninvertible transfer characteristic.

value of y. In other words, the equation

$$g(x) = y$$

has three roots (x_1, x_2, x_3); thus there is no unique *inverse* for the function $g(x)$.

This case can be handled by a generalization of the technique applied in the previous subsection however. The event "$y < Y \le y + \Delta_y$" is the sum of three disjoint events involving X. Therefore we can write

$$\Pr[y < Y \le y + \Delta_y] = \Pr[x_1 < X \le x_1 + \Delta_1] + \Pr[x_2 < X \le x_2 + \Delta_2] + \Pr[x_3 < X \le x_3 + \Delta_3]$$

or

$$f_Y(y)\Delta_y = f_X(x_1)\Delta_1 + f_X(x_2)\Delta_2 + f_X(x_3)\Delta_3$$

Note that the increments in x are generally different from each other because they must correspond to the *same* value for Δ_y and the slope of the function may be different at each point. Specifically, the increments satisfy the relation

$$\Delta_y = \left| \frac{dy}{dx} \right|_{x_i} \Delta_i \qquad i = 1, 2, 3$$

Then by combining the last two equations and cancelling the common term Δ_y we obtain the desired result,

$$\boxed{f_Y(y) = \sum_{i=1}^{r} \frac{1}{|dy/dx|} f_X(x) \Bigg|_{x = g_i^{-1}(y)}} \qquad (3.28)$$

which has been generalized to an arbitrary number of roots r. The notation $g_i^{-1}(y)$ is not a true inverse, but represents an algebraic expression that holds in the region of the root x_i. The application of the formula (3.28) is best illustrated through an example.

Example 3.15: The noise X in an electronic system is frequently a Gaussian random variable with mean $\mu = 0$ and variance σ^2:

$$f_X(x) = \frac{1}{\sqrt{2\pi\sigma^2}} e^{-x^2/2\sigma^2} \qquad -\infty < x < \infty \qquad (1)$$

(see Fig. 3.16). The normalized instantaneous power of the noise is given by

$$Y = \frac{X^2}{\sigma^2}$$

where the parameter σ^2 can be interpreted as the average noise power.[4] Let us compute the PDF of the noise power.

The transformation given by the last equation is sketched below.

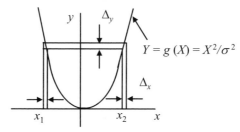

For a given value of $Y > 0$ the noise has two specific values, namely

$$x_1 = g_1^{-1}(y) = -\sigma\sqrt{y} \quad \text{and} \quad x_2 = g_2^{-1}(y) = +\sigma\sqrt{y}$$

The magnitude of the derivative is given by

$$\left|\frac{dy}{dx}\right| = |2x/\sigma^2| = 2\sqrt{y}/\sigma$$

The PDF of the noise power is then found by applying (3.28):

$$f_Y(y) = \frac{1}{2\sqrt{y}/\sigma} f_X(-\sigma\sqrt{y}) + \frac{1}{2\sqrt{y}/\sigma} f_X(+\sigma\sqrt{y})$$

Substituting the Gaussian density function (1) in this equation and simplifying yields

$$f_Y(y) = \frac{1}{\sqrt{2\pi y}} e^{-y/2} \qquad 0 \le x < \infty$$

(This PDF is known as a Chi-squared density with one degree of freedom.)
□

An alternative approach computing the density of a transformed random variable involves using the cumulative distribution function. This method, which works for both invertible and noninvertible transformations, can sometimes be algebraically simpler and thus is worth knowing about. The idea is to express the CDF of the new variable $F_Y(y)$ in terms of the CDF of the CDF of the original variable $F_X(x)$ and take the derivative. The procedure is best illustrated by an example.

Example 3.16: To compute the PDF of the random variable Y in Example 3.15 we could first compute the CDF of Y and then take the derivative.

Begin by observing from the figure in Example 3.15 that

$$\Pr[Y \le y] = \Pr[x_2 \le X \le x_1] = \Pr\left[\frac{-\sqrt{y}}{\sigma} \le X \le \frac{\sqrt{y}}{\sigma}\right]$$

---

[4] This fact is discussed further in Chapter 7.

This equation can be written in terms of cumulative distribution functions as

$$F_Y(y) = F_X\left(\frac{\sqrt{y}}{\sigma}\right) - F_X\left(\frac{-\sqrt{y}}{\sigma}\right)$$

Taking the derivative with respect to y then yields

$$f_Y(y) = \frac{1}{2\sqrt{y}/\sigma}f_X(\sigma\sqrt{y}) + \frac{1}{2\sqrt{y}/\sigma}f_X(-\sigma\sqrt{y})$$

which leads to the same result as in Example 3.15.

□

3.6.3 *When the transformation has discontinuities or flat regions*

An extreme case of a noninvertible transformation occurs when the transformation has step discontinuities, or regions over which it is constant. We will illustrate how to treat these cases using the method involving the cumulative distribution function.

Begin with the case where the transformation has a flat region as shown in Fig. 3.23 (a). For any value of y let us consider the event $Y \leq y$ and express this in terms of

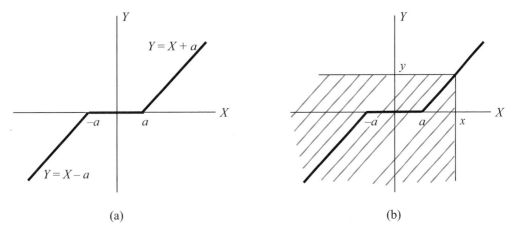

(a)   (b)

Figure 3.23 Transformation with flat region. (a) Transformation. (b) Region defined by $Y \leq y$ when y is positive.

events involving X. If we first consider first a positive value of y as illustrated in Fig. 3.23 (b), then we are on the rightmost branch of the curve and we can write

$$\Pr[Y \leq y] = \Pr[X \leq y - a]$$

Thus

$$F_Y(y) = F_X(y - a); \qquad y > 0$$

Similarly, we can observe that for $y = 0$

$$\Pr[Y \leq y] = \Pr[X \leq a]$$

while for $y < 0$

$$\Pr[Y \leq y] = \Pr[X \leq y + a]$$

and therefore,

$$F_Y(y) = \begin{cases} F_X(y - a) & y > 0 \\ F_X(a) & y = 0 \\ F_X(y + a) & y < 0 \end{cases}$$

An illustration of a typical CDF for X and the resulting CDF for Y is shown in Fig. 3.24 (a) and (b). We can then take the derivative to find the corresponding PDF:

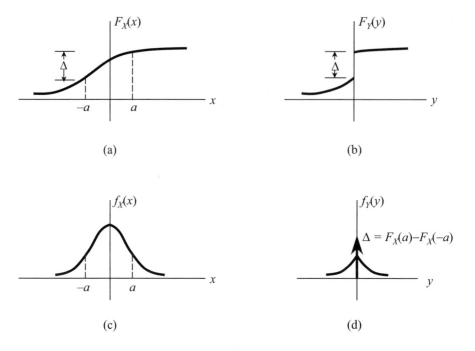

(a)

(b)

(c)

(d)

Figure 3.24 Distribution and density functions for the transformation of Fig. 3.23. (a) CDF of X. (b) CDF of Y. (c) PDF of X. (d) PDF of Y.

$$f_Y(y) = \begin{cases} f_X(y-a) & y > 0 \\ [F_X(a) - F_X(-a)]\,\delta(y) & y = 0 \\ f_X(y+a) & y < 0 \end{cases}$$

The results are shown in Fig. 3.24 (c) and (d).

A second case of transformations is one involving discontinuities, as in Fig. 3.25. In

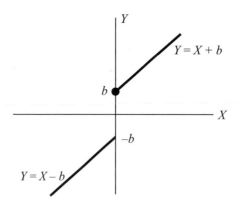

Figure 3.25 Transformation with discontinuity.

this case we can write

$$\Pr[Y \le y] = \begin{cases} \Pr[x \le y - b] & y > b \\ \Pr[x \le 0] & y = 0 \\ \Pr[x \le y + b] & y \le -b \end{cases}$$

Therefore,

$$F_Y(y) = \begin{cases} F_X(y-b) & y > b \\ F_X(0) & y = 0 \\ F_X(y+B) & y \leq -b \end{cases}$$

and

$$f_Y(y) = \begin{cases} f_X(y-b) & y > b \\ 0 & y = 0 \\ f_X(y+b) & y \leq -b \end{cases}$$

These various functions are illustrated in Fig. 3.26.

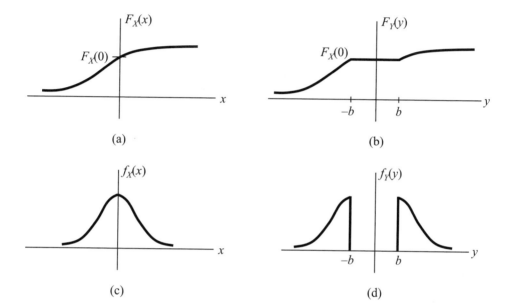

(a)   (b)

(c)   (d)

Figure 3.26 Distribution and density functions for the transformation of Fig. 3.25. (a) CDF of X. (b) CDF of Y. (c) PDF of X. (d) PDF of Y.

As a final example consider the transformation shown in Fig. 3.27, which has *only*

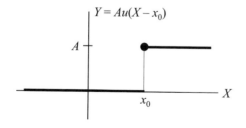

Figure 3.27 Transformation with only discontinuity and flat regions.

discontinuities and flat regions. Proceeding as before, we write

$$\Pr[Y \leq y] = \begin{cases} 0 & y < 0 \\ \Pr[X \leq x_0] & 0 \leq y < A \\ 1 & y \geq A \end{cases}$$

This leads to the CDF and PDF depicted in Fig. 3.28. (We'll let the reader fill in the last step and provide the analytic expressions.)

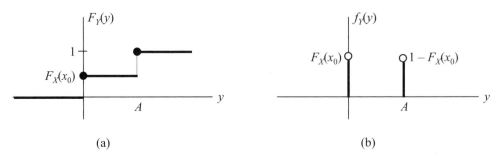

Figure 3.28 CDF and PDF corresponding to the transformation of Fig. 3.27.

You can easily generalize the preceding examples to other cases. In applying this method, it is important to observe whether or not each inequality is strict, especially if impulses are involved in the original density function!

3.7 Distributions Conditioned on an Event

Sometimes it is necessary to deal with a random variable in the context of a conditioning event. A good example of this is provided in the next section of this chapter.

For a discrete random variable, the *conditional* PMF is defined by

$$f_{K|A}[k|A] = \frac{\Pr[K = k, A]}{\Pr[A]} \tag{3.29}$$

where A is the conditioning event. In defining this quantity we have used a convenient shorthand notation on the right. In particular, the expression $\Pr[K = k, A]$ is interpreted as the joint probability of the event $\{s \in \mathcal{S} : K(s) = k\}$ and the event A. The notation $f_{K|A}[k|A]$ although redundant, is quite conventional.

The conditional PMF has all of the usual properties of a probability mass function. In particular, it satisfies the two conditions

$$0 \le f_K[k] \le 1 \qquad \forall k \tag{3.30}$$

and

$$\sum_{\forall k} f_K[k] = 1 \tag{3.31}$$

where in the last expression the sum is taken over all values of k that the event A allows.

For a discrete *or* continuous (or mixed) random variable, the conditional cumulative distribution function is defined by

$$F_{X|A}[x|A] = \frac{\Pr[X \le x, A]}{\Pr[A]} \tag{3.32}$$

This function has properties like any CDF; it is bounded between 0 and 1 and is a monotonically increasing function of x. The conditional PDF is then defined as the formal *derivative* of the conditional CDF,

$$f_{X|A}(x|A) = \frac{dF_{X|A}(x|A)}{dx} \tag{3.33}$$

where as before, the derivative is assumed to exist and can be represented by impulses at discontinuities. The conditional density is everywhere nonnegative and satisfies the condition

$$\int_{-\infty}^{\infty} f_{X|A}(x|A)dx = 1$$

For regions where there are no singularities, it is useful to think of the conditional PDF in terms of the following relation analogous to (3.16):

$$\Pr[x < X \le x + \Delta_x | A] \approx f_{X|A}(x|A) \cdot \Delta_x \qquad (3.34)$$

Equation 3.34 can be used to develop one more important result. Let us define the notation $\Pr[A|x]$ to mean

$$\Pr[A|x] = \lim_{\Delta_x \to 0} \Pr[A \mid x < X \le x + \Delta_x] \qquad (3.35)$$

This is the probability of the event A given that X is in an infinitesimally small increment near the value x. To show that this limit exists, we can use (2.17) of Chapter 2 and substitute (3.16) and (3.34) to write

$$\Pr[A|x < X \le x + \Delta_x] = \frac{\Pr[x < X \le x + \Delta_x \mid A] \cdot \Pr[A]}{\Pr[x < X \le x + \Delta_x]} \approx \frac{f_{X|A}(x|A)\Delta_x \cdot \Pr[A]}{f_X(x)\Delta_x}$$

Since the Δ_x cancels, the limit in (3.35) exists, and these last two equations combine to yield

$$\Pr[A|x] = \frac{f_{X|A}(x|A)\Pr[A]}{f_X(x)} \qquad (3.36)$$

The corresponding discrete form is easily shown to be

$$\Pr[A|k] = \frac{f_{K|A}[k|A]\Pr[A]}{f_K[k]} \qquad (3.37)$$

Equations 3.36 and 3.37 are useful in situations where the probability of an event is influenced by the observation of a random variable. In statistics, problems of this type occur in what is known as statistical inference. In electrical engineering, the same type of problem comes about in applications of detection and classification. The latter applications are addressed in the next section.

Example 3.17: Consider a random variable X whose PDF is given below.

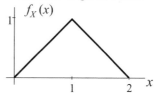

$$f_X(x) = \begin{cases} x & 0 \le x < 1 \\ 2 - x & 1 \le x \le 2 \\ 0 & \text{otherwise} \end{cases}$$

Let the event A be defined as the event "$X > 1$." Let us find the conditional PDF of the random variable.

The CDF of X can be found by integration. The result is sketched below. (The analytical expressions are not important in this example.)

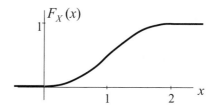

Next we find $F_{X|A}$ using (3.32). Observe that the numerator in (3.32) is just the original CDF restricted to the region where event A is true; that is,

$$\Pr[X \leq x, A] = \begin{cases} F_X(x) & x > 1 \\ 0 & \text{otherwise} \end{cases}$$

Additionally, it is easy to find that $\Pr[A] = 0.5$; thus from (3.32)

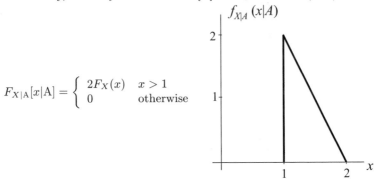

$$F_{X|A}[x|A] = \begin{cases} 2F_X(x) & x > 1 \\ 0 & \text{otherwise} \end{cases}$$

The conditional PDF is then found by taking the derivative (see (3.33)). The result is given below.

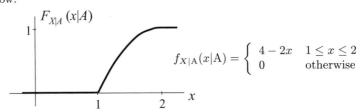

$$f_{X|A}(x|A) = \begin{cases} 4 - 2x & 1 \leq x \leq 2 \\ 0 & \text{otherwise} \end{cases}$$

Notice that both the conditional CDF and the conditional PDF are just scaled versions of the original (unconditional) CDF and PDF in the region where the condition applies, and are 0 elsewhere.

□

3.8 Applications Involving Random Variables

The tools developed in this chapter can be applied to two important engineering problems. The first is detection of a signal in noise; the second is automatic classification. Specific details in the design of systems to perform these tasks can of course be complex. The principles for an optimal design, however, can be discovered with a basic knowledge of random variables.

3.8.1 Optimal signal detection
Principles of optimal detection

The concepts associated with a random variable can be used to address a fundamental problem in electrical engineering, namely that of optimal signal detection. In this problem, it is assumed that a communications receiver is listening for transmissions which are subject to additive noise. For simplicity, it is assumed that processing in the receiver has reduced the observation to a single random variable X (see Fig. 3.29).[5]

---

[5] A more advanced version of this problem is discussed in Chapter 7, following the discussion of random signals.

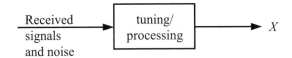

Figure 3.29 Receiver and processing prior to signal detection.

Let us further assume that the signal to be detected has a known constant value s and that the noise is represented by a continuous random variable N. There are two distinct possibilities for X. We denote these as *hypotheses* H_0 and H_1.

$$H_0 : \quad X = N \qquad \text{no signal is present (just noise)}$$

$$(3.38)$$

$$H_1 : \quad X = s + N \quad \text{signal is present (in additive noise)}$$

Given an observation x for the random variable X, we need to decide in an optimal way, between H_0 and H_1.

Note that this problem is very broad. The signal could represent the return from a target in a radar or sonar, an incoming call in a cellular telephone system, a radio or television broadcast, or some hidden signal in a military surveillance operation. In fact, the procedures about to be discussed are even more general, and pertain to situations requiring decision with random variables regardless of whether or not electronic systems are involved.

The hypotheses H_0 and H_1 can be regarded as *events*. Therefore, using the ideas developed in the previous section, we can define the following quantities for $i = 0, 1$:

term	name
$f_{X\|H_i}(x\|H_i)$	Conditional density function
$\Pr[H_i]$	Prior probability
$\Pr[H_i\|x]$	Posterior probability

The *conditional density function* is the PDF for the observation. When H_0 is true, it is just the PDF for the noise; when H_1 is true, it is the PDF for the signal plus noise. The *prior probabilities* reflect our knowledge of the likelihood of whether or not a signal was transmitted; these are assumed to be known. Finally, the *posterior probabilities* represent the likelihood that a signal is or is not present after having received the observation x. According to (3.36), these quantities are related as

$$\Pr[H_i|x] = \frac{f_{X|H_i}(x|H_i)\Pr[H_i]}{f_X(x)}; \qquad i = 0, 1 \qquad (3.39)$$

Now, let us return to the decision procedure and suppose that it is desired to minimize the probability of making an error. Consider the following logical argument. To minimize $\Pr[\text{error}]$ it is necessary to minimize $\Pr[\text{error}|x]$ for each possible value of x. This follows from a form of the principle of total probability written for random variables as

$$\Pr[\text{error}] = \int_{-\infty}^{\infty} \Pr[\text{error}|x]\, f_X(x)dx \qquad (3.40)$$

(Compare this formula with (2.15) of Chapter 2.) Now, suppose the curves for $\Pr[H_0|x]$ and $\Pr[H_1|x]$ and the observed value of x are as shown in Fig. 3.30. If H_0 is chosen, then an error is made if H_1 was actually the case. In other words, if H_0 is chosen, then

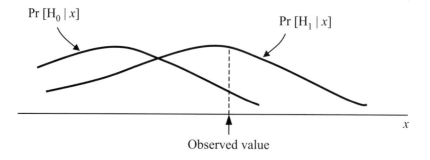

Figure 3.30 Typical plots of conditional probability.

$\Pr[\text{error}|x] = \Pr[H_1|x]$. Likewise, if H_1 is chosen, then $\Pr[\text{error}|x] = \Pr[H_0|x]$. For the situation shown in Fig. 3.30, $\Pr[H_1|x] > \Pr[H_0|x]$. Therefore to obtain the lower error for this particular value of x we should choose H_1.

A generalization of this argument shows that H_1 should be chosen *whenever* $\Pr[H_1|x] > \Pr[H_0|x]$ and H_0 should be chosen otherwise. The optimal decision rule, that minimizes the probability of error, is written as

$$\Pr[H_1|x] \underset{H_0}{\overset{H_1}{\gtrless}} \Pr[H_0|x] \tag{3.41}$$

This result is intuitively satisfying; the optimal decision rule is to *choose the hypothesis that is most likely* when conditioned on the observation.

To put the decision rule in a more workable form, we can substitute (3.39) in (3.41) and cancel the common denominator term to obtain

$$f_{X|H_1}(x|H_1)\Pr[H_1] \underset{H_0}{\overset{H_1}{\gtrless}} f_{X|H_0}(x|H_0)\Pr[H_0]$$

If we further define

$$P_i = \Pr[H_i]; \qquad i = 0, 1$$

then the decision rule can be written in the form

$$\boxed{\frac{f_{X|H_1}(x|H_1)}{f_{X|H_0}(x|H_0)} \underset{H_0}{\overset{H_1}{\gtrless}} \frac{P_0}{P_1}} \tag{3.42}$$

The term on the left side of this equation is called the *likelihood ratio* and is a fundamental quantity in statistical decision theory. The terms P_0 and P_1 are the *prior probabilities*.

The case of Gaussian noise

In many cases in the real world the noise has a Gaussian PDF with 0 mean:

$$f_N(n) = \frac{1}{\sqrt{2\pi\sigma^2}}e^{-n^2/2\sigma^2}$$

The conditional densities for the two hypotheses in (3.38) are thus given by

$$f_{X|H_0}(x|H_0) = f_N(x) = \frac{1}{\sqrt{2\pi\sigma^2}}e^{-x^2/2\sigma^2}$$

and

$$f_{X|H_1}(x|H_1) = f_N(x - s) = \frac{1}{\sqrt{2\pi\sigma^2}}e^{-(x-s)^2/2\sigma^2}$$

In this Gaussian case, substituting these equations into (3.42), taking the logarithm and simplifying leads to the decision rule (see Prob. 3.39)

$$x \underset{H_0}{\overset{H_1}{\gtrless}} \tau_o; \qquad \tau_o = \frac{s}{2} + \frac{\sigma^2}{s} \ln \frac{P_0}{P_1} \tag{3.43}$$

Here τ_o is an optimal threshold that minimizes the probability of error.

The performance of the decision rule can be characterized more completely using not only $\Pr[\text{error}]$ but also the two *conditional* errors

$$\epsilon_0 \overset{\text{def}}{=} \Pr[\text{error}|H_0]; \qquad \epsilon_1 \overset{\text{def}}{=} \Pr[\text{error}|H_1] \tag{3.44}$$

For any decision rule of the form

$$x \underset{H_0}{\overset{H_1}{\gtrless}} \tau$$

where τ is an *arbitrary* threshold, these latter two probabilities are given by

$$\begin{aligned} \epsilon_0 &= \Pr[X > \tau | H_0] \\ \epsilon_1 &= \Pr[X \le \tau | H_1] \end{aligned}$$

These probabilities are represented by the areas under the density functions depicted in Fig. 3.31. Moving the threshold permits trading off one type of error for the other.

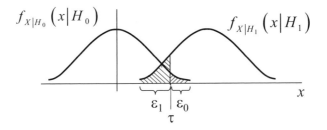

Figure 3.31 Illustration of errors ϵ_0 and ϵ_1.

The term ϵ_0 is usually called the *false alarm* probability and denoted as p_{FA}. This is the probability that the decision rule proclaims the existence of a signal when no signal is actually present. The term ϵ_1 is called the probability of a "miss"; a condition that occurs when the decision rule fails to detect a signal that is actually present. The complementary probability $p_D = 1 - \epsilon_1$ is called the probability of *detection* and is the more commonly used measure in engineering applications. A plot of p_D versus p_{FA} as the threshold τ is varied is called the *receiver operating characteristic* (ROC) and is illustrated in the example below. The ROC shows in a quantitative sense how high probability of detection can be traded off for low probability of false alarm.

For any threshold τ the unconditional probability of error can be related to the two conditional errors by using the principle of total probability. In particular,

$$\Pr[\text{error}] = \Pr[\text{error}|H_0] \Pr[H_0] + \Pr[\text{error}|H_1] \Pr[H_1]$$

or

$$\Pr[\text{error}] = \epsilon_0 P_0 + \epsilon_1 P_1 \tag{3.45}$$

The minimum value of $\Pr[\text{error}]$ occurs when the threshold τ is set to $\tau = \tau_o$ where τ_o is given by the expression in (3.43).

Example 3.18: Consider a detection problem where the signal S in (3.38) is equal to 2 mv and the noise is a zero-mean Gaussian random variable with variance $\sigma^2 = 4\,\text{mv}^2$. The conditional densities for the two hypotheses are then given by

$$H_0\,(X = N): \qquad f_{X|H_0}(x|H_0) = f_N(x) = \frac{1}{\sqrt{2\pi \cdot 4}}e^{-x^2/(2\cdot 4)}$$

$$H_1\,(X = s + N): \quad f_{X|H_1}(x|H_1) = f_N(x-2) = \frac{1}{\sqrt{2\pi \cdot 4}}e^{-(x-2)^2/(2\cdot 4)}$$

Substituting these equations into (3.42) and simplifying leads to

$$x \underset{H_0}{\overset{H_1}{\gtrless}} \tau_0 ; \qquad \tau_0 = 1 + 2\ln\frac{P_0}{P_1}$$

which corresponds to (3.43).

If the prior probabilities are equal ($P_0 = P_1 = 0.5$), then the threshold τ_0 is equal to 1 and the error probabilities are given by (see Table 3.2)

$$\epsilon_0 = \Pr[X > 1|H_0] \quad = \quad \int_1^\infty f_{X|H_0}(x|H_0)dx = \tfrac{1}{2}\left[1 - \text{erf}\left(\frac{1-0}{\sqrt{2}\cdot 2}\right)\right] = 0.3085$$

$$\epsilon_1 = \Pr[X \le 1|H_1] \quad = \quad \int_{-\infty}^1 f_{X|H_1}(x|H_1)dx = \tfrac{1}{2}\left[1 + \text{erf}\left(\frac{1-2}{\sqrt{2}\cdot 2}\right)\right] = 0.3085$$

The probabilities of detection and false alarm are thus given by

$$p_D = 1 - \epsilon_1 = 0.6915 \qquad p_{FA} = \epsilon_0 = 0.3085$$

If we now let the threshold τ take on a range of values, we can plot the receiver operating characteristic (p_D versus p_{FA}) as shown below.

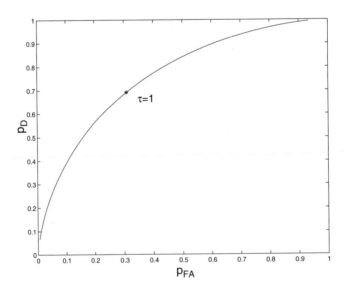

The point corresponding to $\tau = 1$ is shown on the curve. The total error given by (3.45) is also plotted below.

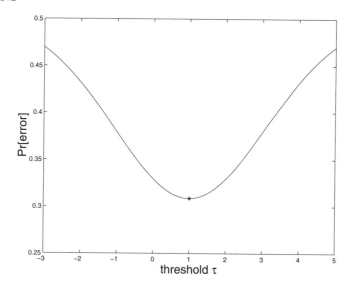

Observe that the minimum error occurs, as expected, at $\tau = 1$ and is given by

$$\Pr[\text{error}] = \epsilon_0\,P_0 + \epsilon_1\,P_1 = 0.3085$$

Note that τ_0, ϵ_0, ϵ_1 and $\Pr[\text{error}]$ in the above curve would change if P_1 and P_0 were chosen to be some other values. This point is explored in the computer project at the end of this chapter.

$\square$

3.8.2 Object classification

In a large number of problems, it is desired to provide automatic classification of objects based on some type of measurements. Examples of this include optical text recognition, military target identification, sorting of recyclable materials, and others. This is typically studied under the topic of Pattern Recognition [4, 5] and the statistical form of pattern recognition can be thought of as a generalization of the detection problem described above. The approach can be illustrated through the simple example presented below.

Example 3.19: It is desired to develop an automatic system for sorting jellybeans on a production line. Assume there are three colors of jellybeans (red, green, and blue) and because of variations in food coloring, the color of each jellybean is random. In particular, let the color be described by its wavelength X in microns and assume that X is a Gaussian random variable with mean μ and variance σ^2 given in the table below. A hypothesis H_i is also associated with each color for the purpose of making decisions.

	color	μ	σ^2
H_1:	red	700	10,000
H_2:	green	546	10,000
H_3:	blue	436	10,000

It can be shown by an argument similar to that in the previous subsection, that to minimize the error in classification, one should use the following decision rule:

$$\text{choose } H_k = \text{argmax } \Pr[H_i|x]$$

This is the generalization of (3.41).

By applying (3.39) the decision rule can be written in terms of the conditional densities and the prior probabilities. Further, if the prior probabilities $\Pr[H_i]$ are all equal then the decision rule becomes just a comparison of the conditional density functions:

$$\text{choose } H_k = \text{argmax } f_{X|H_i}(x|H_i)$$

For this case where the prior probabilities are all equal, the decision regions are defined by whichever density function is larger. This is illustrated in the figure below.

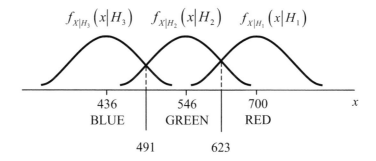

For example, if the color X is measured to be $x = 625$ microns, the density function $f_{X|H_1}$ evaluated at this point is larger than $f_{X|H_2}$ or $f_{X|H_3}$ (see figure). Therefore, the object would be classified as red. The decision region for 'red' is thus the region where the red density function $f_{X|H_1}$ is highest.

The decision region boundaries shown here are centered between the means of the densities. This occurs not only because the priors are equal, but also because the variance of the density under each hypothesis is the same. If the variances were different (as they well could be) a more complex set of decision regions would result.

Given the above decision regions, we may now consider the following questions:

What is the probability that a red jellybean is misclassified as a blue one?

This probability is given by

$$\Pr[X \le 491|H_1] = \tfrac{1}{2}\left[1 + \text{erf}\left(\frac{491 - 700}{100\sqrt{2}}\right)\right] \approx \tfrac{1}{2}[1 - \text{erf}(1.48)] \approx 0.02$$

What is the probability that a red jellybean is mistaken for a green one?

This probability is

$$\Pr[491 < X \le 623|H_1] \quad = \quad \Pr[X \le 623|H_1] - \Pr[X \le 491|H_1]$$

$$= \quad \tfrac{1}{2}\left[1 + \text{erf}\left(\frac{623 - 700}{100\sqrt{2}}\right)\right] - \tfrac{1}{2}\left[1 + \text{erf}\left(\frac{491 - 700}{100\sqrt{2}}\right)\right]$$

$$\approx \quad \tfrac{1}{2}[\text{erf}(1.48) - \text{erf}(0.54)] \approx 0.20$$

What is the probability that a green jellybean is misclassified?

This probability is for the union of two non-overlapping events

$$\Pr[\{X \leq 491\} + \{X > 623\}|H_2] = \Pr[X \leq 491|H_2] + \Pr[X > 623|H_2]$$

Again, using the results of Table 3.2, we can evaluate these probabilities as

$$\frac{1}{2}\left[1 + \text{erf}\left(\frac{491 - 546}{100\sqrt{2}}\right)\right] \quad + \quad \frac{1}{2}\left[1 - \text{erf}\left(\frac{623 - 546}{100\sqrt{2}}\right)\right]$$

$$\approx \quad 1 - \tfrac{1}{2}\text{erf}\,(0.39) - \tfrac{1}{2}\text{erf}\,(0.54) \approx 0.51$$

□

In problems involving classification, it is typical to have more than one type of measurement. The principles of classification are the same, however, but involve the joint density functions of the combined set of measurements. Problems of this sort are discussed briefly in Chapter 5, which deals with multiple random variables.

3.9 *Summary*

Many probabilistic experiments result in measurements or outcomes that take on numerical values. Such measurements are modeled as random variables.

Random variables may be discrete, continuous, or a combination of both. Purely discrete random variables take on only a countable number of values with nonzero probability. Such random variables can be characterized by the probability mass function (PMF). Continuous random variables can take on an uncountably infinite number of possible values and are characterized by the cumulative distribution function (CDF) and its derivative called the probability density function (PDF). For a continuous random variable, the PDF $f_X(x)$ has the following interpretation. The probability that X is in the interval $x < X \leq x + \Delta_x$ for an arbitrarily small increment Δ_x is given by $f_X(x)\Delta_x$. This probability goes to 0 as $\Delta_x \to 0$.

Discrete and mixed random variables can be also be characterized by the CDF; this function then has discontinuities at discrete points where the probability of the random variable is non-zero. If impulses are allowed, then the PDF can be considered as the formal derivative of the CDF. Thus PDFs containing impulses can represent both continuous and discrete random variables and are especially useful when the random variable has both continuous and discrete components.

A number of useful discrete and continuous distributions for random variables are described in this chapter. One of the most important and ubiquitous is the Gaussian. Since integrals of the Gaussian density function cannot be expressed in closed form, several related normalized quantities have been computed and listed in tables. We focus on the use of the "error function" to compute probabilities involving Gaussian random variables. The error function is available in MATLAB and other scientific computing languages.

When one random variable is expressed in terms of another, i.e., $Y = g(X)$, it may be necessary to compute the PDF of the new random variable (Y) from the PDF of the original (X). Some basic methods for dealing with transformations of random variables are discussed in this chapter. An especially important case is when the transformation can be represented on a graph as a straight line. In this case the form of the density does not change; only a shift and scaling occurs.

Distributions may be conditioned on an *event* and the probability of an event can be conditioned on the observance of a random variable. These concepts are developed and applied in the application areas of signal detection and object classification. In these areas, we demonstrate how to form optimal decison rules that minimize the probability of error, and compute the probability of various types of errors that occur when the decision rules are applied.

References

[1] Ross L. Finney, Maurice D. Weir, and Frank R. Giordano. *Thomas' Calculus*. Addison-Wesley, Reading, Massachusetts, tenth edition, 2001.

[2] Alberto Leon-Garcia. *Probability and Random Processes for Electrical Engineering*. Addison-Wesley, Reading, Massachusetts, second edition, 1994.

[3] Wilbur B. Davenport, Jr. *Probability and Random Processes*. McGraw-Hill, New York, 1970.

[4] Keinosuke Fukunaga. *Introduction to Statistical Pattern Recognition*. Academic Press, New York, second edition, 1990.

[5] Richard O. Duda, Peter E. Hart, and David G. Stork. *Pattern Classification*. John Wiley & Sons, New York, second edition, 2001.

Problems

Discrete random variables

3.1   In a set of independent trials, a certain discrete random variable K takes on the values

$$1,\ 0,\ 0,\ 0,\ 1,\ 1,\ 0,\ 1,\ 1,\ 1,\ 0,\ 1$$

(a) Plot a histogram for the random variable.

(b) Normalize the histogram to provide an estimate of the PMF of K.

3.2   Repeat Problem 3.1 for the discrete random variable J that takes on the values

$$1,\ 3,\ 0,\ 2,\ 1,\ 2,\ 0,\ 3,\ 1,\ 1,\ 3,\ 1,\ 0,\ 1,\ 2,\ 3,\ 0,\ 2,\ 1,\ 3,\ 3,\ 2$$

3.3   A random variable that takes on values of a sequence of 4QAM symbols, I, is transmitted over a communication channel: 3, 0, 1, 3, 2, 0, 2, 1, 3, 0, 0, 2, 3, 1, 3, 0, 2, 1, 0, 3. The channel introduces additive noise, J, equivalent to adding the following values to the above symbol sequence: 0, 0, 1, 0, 0, 1, 1, 0, 0, 0, 0, 0, 1, 0, 1, 0, 0, 0, 1, 0, 1. The addition is performed in modulo form.

(a) Form the resulting sequence: $I+J$ modulo 4. Let K be the random variable that takes on these values.

(b) Plot histograms for random variables I and K.

(c) Normalize the histograms to provide an estimate of the respective discrete probability density functions.

3.4 Consider transmission of ASCII characters, 8 bits long. Define a random variable as the number of 0's in a given ASCII character.

    (a) Determine the range of values that this random variable takes.

    (b) Determine and plot the probability mass function.

Common discrete probability distributions

3.5 Sketch the PMF $f_K[k]$ for the following random variables and label the values.

    (a) Uniform. $m = 0, n = 9$

    (b) Bernoulli. $p = 1/10$

    (c) Binomial. $n = 10, p = 1/10$

    (d) Geometric. $p = 1/10$ (plot $0 \le k \le 9$ only)

    For each of these distributions, compute the probability of the event $1 \le k \le 4$.

3.6 In the transmission of a packet of 128 bytes, byte errors occur independently with probability $1/9$. What is the probability that less than five byte errors occur in the packet of 128 bytes?

3.7 In transmission of a very long (you may read "infinite") set of characters, byte errors occur independently with probability $1/9$. What is the probability that the first error occurs within the first five bytes transmitted?

3.8 The Poisson PMF has the form

$$f_K[k] = \frac{\alpha^k}{k!} e^{-\alpha} \qquad k = 0, 1, 2, \ldots$$

where α is a parameter.

    (a) Show that this PMF sums to 1.

    (b) What are the restrictions on the parameter α, that is, what are the allowable values?

3.9 (a) Sketch the PMF of the following random variables:

        (i) geometric, $p = 1/8$

        (ii) uniform, $m = 0$, $n = 7$

        (iii) Poisson, $\alpha = 2$

    (b) For each of these, compute the probability that the random variable is greater than 5.

    (c) For each of these, if you desire that $\Pr[K > 5] \le 0.5$, what should be the respective values of the parameters?

Continuous random variables

3.10 The PDF for a random variable X is given by

$$f_X(x) = \begin{cases} 2x/9 & 0 \le x \le 3 \\ 0 & \text{otherwise} \end{cases}$$

    (a) Sketch $f_X(x)$ versus x.

    (b) What is the probability of the following events?

        (i) $X \le 1$

(ii) $X > 2$

(iii) $1 < X \le 2$

(c) Find and sketch the CDF $F_X(x)$.

(d) Use the CDF to check your answers to part (b).

3.11 The PDF of a random variable X is given by:

$$f_X(x) = \begin{cases} 0.4 + Cx, & 0 \le x \le 5 \\ 0 & \text{otherwise.} \end{cases}$$

(a) Find C that makes it a valid PDF.

(b) Determine: $\Pr[X > 3]$, $\Pr[1 < X \le 4]$.

(c) Find the CDF $F_X(x)$.

3.12 A PDF for a continuous random varaiable X is defined by

$$f_X(x) = \begin{cases} C & 0 < x \le 2 \\ 2C & 4 < x \le 6 \\ C & 7 < x \le 9 \\ 0 & \text{otherwise} \end{cases}$$

where C is a constant.

(a) Find the numerical value of the constant C.

(b) Compute $\Pr[1 < X \le 8]$.

(c) Find the value of M for which

$$\int_{-\infty}^{M} f_X(x)dx = \int_{M}^{\infty} f_X(x)dx = \tfrac{1}{2}$$

M is known as the *median* of the random variable.

3.13 For what values or ranges of values of the parameters would the following PDFs be valid? If there are no possible values that would make the PDF valid, state that.

(a)
$$f_X(x) = \begin{cases} B + A\sin x & 0 \le x \le 2\pi \\ 0 & \text{otherwise.} \end{cases}$$

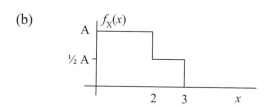

(b)

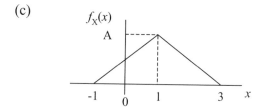

(c)

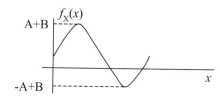

3.14 The CDF for a random variable X is given by
$$F_X(x) = \begin{cases} A\arctan(x) & x > 0 \\ 0 & x \le 0 \end{cases}$$

(a) Find the constant A.
(b) Find and sketch the PDF.

3.15 A random variable X has the CDF specified below.
$$F_X(x) = \begin{cases} 0 & x \le -\dfrac{\pi}{2\alpha} \\ \frac{1}{2}(1+\sin \alpha x) & -\dfrac{\pi}{2\alpha} < x \le \dfrac{\pi}{2\alpha} \\ 1 & x \ge \dfrac{\pi}{2\alpha} \end{cases}$$

(a) Sketch $F_X(x)$.
(b) For what values of the real parameter α is this expression valid?
(c) What is the PDF $f_X(x)$? Write a formula and sketch it.
(d) Determine the probability of the following events:
   (i) $X \le 0$
   (ii) $X \le -1$
   (iii) $|X| \le .001$ (approximate)

3.16 The PDF of a random variable X is shown below:

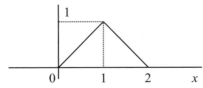

(a) Obtain the CDF.

(b) Find $\Pr[0.5 \leq X \leq 1.5]$, $\Pr[1 \leq X \leq 5]$, $\Pr[X < 0.5]$, and $\Pr[0.75 \leq X \leq 0.7501]$.

3.17 Given the CDF,

$$F_X(x) = \begin{cases} 0 & x < 0 \\ Cx^2 & 0 \leq x \leq 2 \\ 1 & x > 2 \end{cases}$$

(a) Find C.

(b) Sketch this function.

(c) Find the PDF and sketch it.

(d) Find $\Pr[0.5 \leq X \leq 1.5]$.

3.18 The PDF of a random variable X is of the form

$$f_X(x) = \begin{cases} Cxe^{-x} & x > 0 \\ 0 & x \leq 0 \end{cases}$$

Where C is a normalization constant.

(a) Find the constant C and sketch $f_X(x)$.

(b) Compute the probability of the following events:

(i) $X \leq 1$

(ii) $X > 2$

(c) Find the probability of the event $x_0 < X \leq x_0 + 0.001$ for the following values of x_0:

(i) $x_0 = 1/2$

(ii) $x_0 = 2$

What value of x_0 is "most likely," i.e., what value of x_0 would give the *largest* probability for the event
$x_0 < X \leq x_0 + 0.001$?

(d) Find and sketch the CDF.

Common probability density functions

3.19 Sketch the PDFs of the following random variables:

(a) Uniform:

(i) $a = 2, b = 6$;

(ii) $a = -4, b = -1$;

(iii) $a = -7, b = 1$;

(iv) $a = 0, b = a + \delta$, where $\delta \to 0$.

  (b) Gaussian:

    (i) $\mu = 0$, $\sigma = 1$;
    (ii) $\mu = 1.5$, $\sigma = 1$;
    (iii) $\mu = 0$, $\sigma = 2$;
    (iv) $\mu = 1.5$, $\sigma = 2$;
    (v) $\mu = -2$, $\sigma = 1$;
    (vi) $\mu = -2$, $\sigma = 2$;
    (vii) $\mu = 0$, $\sigma \to 0$.

  (c) Exponential:

    (i) $\lambda = 1$;
    (ii) $\lambda = 2$;
    (iii) $\lambda \to 0$;
    (iv) $\lambda \to \infty$;
    (v) Can λ be a negative value? Explain.

3.20 The *median* M of a random variable X is defined by the condition

$$\int_{-\infty}^{M} f_X(u)\,du = \int_{M}^{\infty} f_X(u)\,du$$

Find the median for:

  (a) The uniform PDF.

  (b) The exponential PDF.

  (c) The Gaussian PDF.

3.21 A certain random variable X is described by a Gaussian PDF with parameters $\mu = -1$ and $\sigma^2 = 2$. Use the error function (erf) or the Q function to compute the probability of the following events:

  (a) $X \leq 0$

  (b) $X > +1$

  (c) $-2 < X \leq +2$

3.22 Given a Gaussian random variable with mean $\mu = 1$ and variance $\sigma^2 = 3$, find the following probabilities:

  (a) $\Pr[X > 1]$ and $\Pr\left[X > \sqrt{3}\right]$.

  (b) $\Pr\left[|X - 1| > \sqrt{12}\right]$ and $\Pr\left[|X - 1| < 6\right]$.

  (c) $\Pr\left[X > 1 + \sqrt{3}\right]$ and $\Pr\left[X > 1 + 3\sqrt{3}\right]$.

CDF and PDF for discrete and mixed random variables

3.23 Consider the random variable described in Prob. 3.3.

  (a) Sketch the CDF for I and K.

  (b) Using the CDF, compute the probability that the transmitted signal is between 1 and 2 (inclusive).

  (c) Using the CDF, compute the probability that the received signal is between 1 and 2 (inclusive).

3.24 Determine and plot the CDF for the random variable described in Prob. 3.4.

3.25 The PDF of a random variable X is of the form
$$f_X(x) = \begin{cases} Ce^{-x} + \frac{1}{3}\delta(x-2) & x > 0 \\ 0 & x \le 0 \end{cases}$$

Where C is an unknown constant.

(a) Find the constant C and sketch $f_X(x)$.
(b) Find and sketch the CDF.
(c) What is the probability of the event $1.5 < X \le 2.5$?

3.26 Given the PDF of a random variable X:
$$f_x(x) = \begin{cases} Cx + 0.5\delta(x-1), & 0 \le x \le 3 \\ 0 & \text{otherwise.} \end{cases}$$

(a) Find C and sketch the PDF.
(b) Determine CDF of X.
(c) What are the probabilities for the following events?
   (i) $0.5 \le X \le 2$
   (ii) $X > 2$.

3.27 The CDF of a random variable X is
$$F_X(x) = \begin{cases} 0 & x \le 0 \\ x & 0 < x \le \frac{1}{2} \\ 1 & \frac{1}{2} \le x \end{cases}$$

(a) Sketch $F_X(x)$.
(b) Determine and sketch the PDF of X.

Transformation of random variables

3.28 A voltage X is applied to the circuit shown below and the output voltage Y is measured.

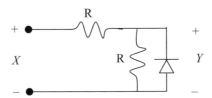

The diode is assumed to be ideal, acting like a short circuit in the forward direction and an open circuit in the backward direction. If the voltage X is a Gaussian random variable with parameters $m = 0$ and $\sigma^2 = 1$, find and sketch the PDF of the output voltage Y.

3.29 A random variable X is *uniformly distributed* (i.e., it has a uniform PDF) between -1 and $+1$. A random variable Y is defined by $Y = 3X + 2$.

(a) Sketch the density function for X and the function describing the transformation from X to Y.
(b) Find the PDF and CDF for the random variable Y and sketch these functions.

3.30 The Laplace PDF is defined by

$$f_X(x) = \frac{\alpha}{2} e^{-\alpha|x-\mu|} \qquad -\infty < x < \infty$$

Consider the transformation $Y = aX + b$. Demonstrate that if X is a Laplace random variable, then Y is also a Laplace random variable. What are the parameters α' and μ' corresponding to the PDF of Y?

3.31 Consider a Laplace random variable with PDF

$$f_X(x) = e^{-2|x|}, \qquad -\infty < x < \infty.$$

This random variable is passed through a 4-level uniform quantizer with output levels $\{-1.5d, -0.5d, 0.5d, 1.5d\}$.

(a) Determine the PMF of the quantizer output.

(b) Find $\Pr[X > 2d]$ and $\Pr[X < -2d]$.

3.32 Consider a half-wave rectifier with a Gaussian input having mean 0 and standard deviation 2.

(a) Find the CDF of the output.

(b) Find the PDF of output by differentiating the CDF.

3.33 Consider a clipper with the following transfer characteristic:

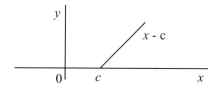

The random variable X is exponential with $\lambda = 2$.

(a) Determine the CDF of the output.

(b) Find the PDF of output by differentiating the CDF.

Distributions conditioned on an event

3.34 For the PDF given in Example 3.17, find and sketch the conditional density $f_{X|A}(x|A)$ using the following choices for the event A:

(a) $X < 1$

(b) $\frac{1}{3} < X < \frac{5}{3}$

(c) $|\sin(X)| < \frac{1}{2}$

3.35 Consider the exponential density function

$$f_X(x) = \begin{cases} \lambda e^{-\lambda x} & x \geq 0 \\ 0 & \text{otherwise} \end{cases}$$

Find and sketch the conditional density function $f_{X|A}(x|A)$ where A is the event $X \geq 2$.

3.36 Let K be a discrete uniform random variable with parameters $n = -2$, $m = 2$. Find the conditional PMF $f_{K|A}[k|A]$ for the following choices of the event A:

(a) $K < 0$

(b) $K \geq 0$

Applications

3.37 In a certain communication problem, the received signal X is a random variable given by

$$X = 2 + W$$

where W is a random variable representing the noise in the system. The noise W has a PDF which is sketched below:

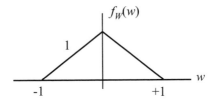

(a) Find and sketch the PDF of X.

(b) Using your answer to Part (a), find $\Pr[X > 2]$.

3.38 A signal is sent over a channel where there is additive noise. The noise N is uniformly distributed between -10 and 10 as shown below:

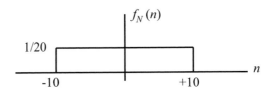

The receiver wishes to decide between two possible events based on the value of a random variable X. These two possible events are:

$$H_0 : \quad X = N \qquad \text{(no signal sent, only noise received)}$$
$$H_1 : \quad X = S + N \quad \text{(signal was sent, signal plus noise received)}$$

Here S is a fixed constant set equal to 18. It is *not* a random variable.

(a) Draw the density functions $f_{X|H_0}(x|H_0)$ and $f_{X|H_1}(x|H_1)$ in the same picture. Be sure to label your plot carefully.

(b) What decision rule, based on the value of X, would guarantee perfect detection of the signal (i.e., $\Pr[\text{Error}|H_1] = 0$), but minimize the probability of false alarm (i.e., $\Pr[\text{Error}|H_0]$)? What is this probability of false alarm?

(c) What decision rule would make the probability of a false alarm go to 0 but maximize the probability of detection? What is the probability of detection for this decision rule?

(d) If we could choose another value of S rather than S=18, what is the *minimum* value of S that would make it possible to have perfect detection and 0 false alarm probability?

3.39 Carry out the algebraic steps involved to show that the decision rule of (3.42) reduces to the simpler form given by (3.43).

Computer Projects

Project 3.1

This project requires writing computer code to generate random variables of various kinds and measure their probability distributions.

1. Generate a sequence of each of the following types of random variables; each sequence should be at least 10,000 points long.

   (a) A binomial random variable. Let the number of Bernoulli trials be $n = 12$. Recall that the binomial random variable is defined as the number of 1's in n trials for a Bernoulli (binary) random variable. Let the parameter p in the Bernoulli trials be $p = 0.5109$.

   (b) A Poisson random variable as a limiting case of the binomial random variable with $p = 0.0125$ or less and $n = 80$ or more while maintaining $\alpha = np = 1$.

   (c) A type 1 geometric random variable with parameter $p = 0.09$.

   (d) A (continuous) uniform random variable in the range $[-2, 5]$.

   (e) A Gaussian random variable with mean $\mu = 1.3172$ and variance $\sigma^2 = 1.9236$.

   (f) An exponential random variable with parameter $\lambda = 1.37$.

2. Estimate the CDF and PDF or PMF as appropriate and plot these functions next to their theoretical counterparts. Compare and comment on the results obtained.

MATLAB programming notes

To generate uniform and Gaussian random variables in Step 1(d) and 1(e), use the MATLAB functions 'rand' and 'randn' respectively with an appropriate transformation.

To generate an exponential random variable, you may use the function 'expon' from the software package for this book. This function uses the "transformation" method to generate the random variable.

In Step 2 you may use the functions 'pmfcdf' and 'pdfcdf' from the software package.

Project 3.2

In this project you will explore the problem of detection of a simple signal in Gaussian noise.

1. Consider a binary communication scheme where a logical 1 is represented by a constant voltage of value $A = 5$ volts, and a logical 0 is represented by zero volts. Additive Gaussian noise is present, so during each bit interval the received signal is a random variable X described by

$$X = A + N \quad \text{if a 1 was sent}$$
$$X = N \quad \text{if a 0 was sent}$$

The noise N is a Gaussian random variable with mean zero and variance σ^2. The signal-to-noise ratio in decibels is defined by

$$\text{SNR} = 10 \log_{10} \frac{A^2}{\sigma^2} = 20 \log_{10} \frac{A}{\sigma}$$

For this study the SNR will be taken to be 10 dB. This formula allows you to compute a value for σ^2.

(a) Let H_0 be the event that a logical 0 was sent and H_1 be the event that a 1 was sent. Show that the decision rule for the receiver to minimize the probability of error is of the form

$$x \underset{H_0}{\overset{H_1}{\gtrless}} \tau$$

Evaluate the threshold for the case where $P_0 = P_1 = 0.5$ where P_0 and P_1 are the prior probabilities of a 0 and 1, i.e., $P_0 = \Pr[H_0]$ and $P_1 = \Pr[H_1]$.

(b) Now consider a decision rule of the above form where τ is any arbitrary threshold. Evaluate ϵ_0 and ϵ_1 and plot the probability of error (3.45) as a function of τ for $1 \le \tau \le 4$. (Continue to use $P_0 = P_1 = 0.5$ in this formula.) Find the minimum error, and verify that it occurs for the value of τ computed in Step 1(a).

(c) Repeat Step 1(b) using $P_0 = 1/3$ and $P_1 = 2/3$. Does the minimum error occur where you would expect it to occur? Justify your answer.

2. The detection of targets (by a radar, for example) is an analogous problem. The signal, which is observed in additive noise, will have a mean value of A if a target is present (event H_1) , and mean 0 if only noise (no target) is present (event H_0). Let p_D represent the probability of detecting the target and p_{FA} represent the probability of a false alarm.

(a) Plot the ROC for this problem for the range of values of τ used in Step 1 (above). Indicate the direction of increasing τ to show how the probability of detection can be traded off for the probability of false alarm by adjusting the threshold.

(b) Show where the two threshold values used in Step 1(b) and 1(c) lie on the ROC curve, and find values of p_D and p_{FA} corresponding to each threshold value.

3. In this last part of the project you will simulate data to see how experimental results compare to the theory.

(a) i. Generate 1000 samples of a Gaussian random variable with mean 0 and variance σ^2 as determined in Step 1 for the 10 dB SNR. Generate 1000 samples of another random variable with mean 5 and variance σ^2. These two sets of random variables represent the outcomes in 1000 trials of the events H_0 and H_1. For each of these 2000 random variables make a decision according to the decision rule in Step 1(a) using the values of τ that you found in Steps 1(a) and 1(b). Form estimates for the errors ϵ_1 and ϵ_0 by computing the relative frequency, that is:

$$\hat{\epsilon}_i = \frac{\#\text{misclassified samples for } H_i}{1000} \quad \text{for } i = 0, 1$$

ii. Also compute p_D, p_{FA}, and the probability of error (3.45) using these estimates. Plot the values on the graphs generated in Steps 1 and 2 for comparison.

(b) Repeat Step (a)(i) above using two other threshold values of your choice for τ. You can use the *same* 2000 random variables generated in Step (i); just make your decision according to the new thresholds. Plot these points on the ROC of Step 2.

MATLAB programming notes

The MATLAB function 'erf' is available to help evaluate the integral of the Gaussian density.

To generate uniform and Gaussian random variables in Step 1(d) and 1(e), use the MATLAB functions 'rand' and 'randn' respectively with an appropriate transformation.

4 Expectation, Moments, and Generating Functions

One of the cornerstones in the fields of probability and statistics is the use of averages. The most important type of average for random variables involves a weighting by the PDF or PMF and is known as *expectation*. This chapter discusses expectation and how it applies to the analysis of random variables. It is shown that some of the most important descriptors of random variable X are not just its expectation, but also the expectation of products of the random variable such as X^2, X^3, and so on. These averages are called *moments* and collectively the complete set of moments for a random variable is equivalent to knowledge of the density function itself. This last idea is manifested through what are known as "generating functions," which are also discussed in this chapter. Generating functions are (Laplace, Fourier, z) transformations of the density function and thus provide a different but equivalent probabilistic description of a random variable. The name "generating function" comes from the fact that moments are easily derived from these functions by taking derivatives.

As an application involving expectation, we discuss the concept of entropy. Entropy is the expected value of information in an experiment, where 'information' is defined as in Chapter 2. We cite a theorem due to Shannon that provides a lower bound on the average number of bits needed to code a message in terms of the entropy of the code words. This result, though developed in the first half of the twentieth century, has clear relevance to modern problems of data compression and coding of digital information on computers and computer networks.

4.1 Expectation of a Random Variable

In the previous chapter, a random variable is defined as a function $X(s)$ that assigns a real number to each outcome s in the sample space of a random experiment. The *expectation* or *expected value* of a random variable can be thought of as the "average value" of the random variable, where the average is computed weighted by the probabilities. More specifically, if X is a random variable defined on a discrete sample space $\mathcal{S}$, then the expectation $\mathcal{E}\{X\}$ is given by

$$\mathcal{E}\{X\} = \sum_{\mathcal{S}} X(s) \Pr[s] \tag{4.1}$$

where the sum is over all elements of the sample space. The following example, based on Example 3.1 of the previous chapter, illustrates the idea of expectation of a random variable.

Example 4.1: Recall from Example 3.1 that the cost of buying 3 diskettes from Simon depends on the number of good and bad diskettes that result (good diskettes cost $.36 while bad diskettes cost the customer $.50). What is the expected value of the cost C of buying three diskettes at Simon's?

The events of the sample space and the cost associated with each event are listed below:

Events:	BBB	BBG	BGB	BGG	GBB	GBG	GGB	GGG
Cost:	$1.50	$1.36	$1.36	$1.22	$1.36	$1.22	$1.22	$1.08
Probability:	1/8	1/8	1/8	1/8	1/8	1/8	1/8	1/8

All of the events in the sample space are equally likely and have probability 1/8. The expected value of the cost is the weighted sum:

$$\mathcal{E}\{C\} = (1/8)1.50 \; + \; (1/8)1.36 + (1/8)1.36 + (1/8)1.22$$
$$+(1/8)1.36 + (1/8)1.22 + (1/8)1.22 + (1/8)1.08 = \$1.29$$

□

4.1.1 Discrete random variable

The expectation for a discrete random variable can be computed in a more convenient way. We *define* the expectation of a discrete random variable K to be

$$\mathcal{E}\{K\} = \sum_{\forall k} k\, \mathrm{f}_K[k] \tag{4.2}$$

Thus the expected value is the sum of all the values of the random variable weighted by their probabilities. Since the sum may be infinite, the expectation is defined only if the sum converges absolutely[1] [1]. Let us compute the expectation of the previous example using this formula.

Example 4.2: The PMF for the cost C of buying Simon's diskettes was computed in Example 3.1 and is shown in Fig. 3.4 (Chapter 3). Using (4.2), the expected value of the cost is given by

$$\mathcal{E}\{C\} = 1.08 \cdot \mathrm{f}_C[1.08] + 1.22 \cdot \mathrm{f}_C[1.22] + 1.36 \cdot \mathrm{f}_C[1.36] + 1.50 \cdot \mathrm{f}_C[1.50]$$

Then substituting the values of $\mathrm{f}_C[c]$ shown in Fig. 3.4 yields

$$\mathcal{E}\{C\} = 1.08 \cdot (1/8) + 1.22 \cdot (3/8) + 1.36 \cdot (3/8) + 1.50 \cdot (1/8) = 1.29$$

□

The fact that both examples produce the same result shouldn't be surprising. After all, we are just slicing up the probability in different ways.

4.1.2 Continuous random variable

The expectation of a continuous random variable can also be defined in terms of the sample space. Conceptually the procedure is similar to (4.1); however the result is written as

$$\mathcal{E}\{X\} = \int_{\mathcal{S}} X(s)dP(s) \tag{4.3}$$

where the expression on the right is a Lebesgue-Stieltjes integral and $P(s)$ is an appropriately defined probability measure for the sample space.[2]

---

[1] Absolute convergence is defined by: $\sum_{\forall k} |k|\, \mathrm{f}_K[k] < \infty$.
[2] Some further discussion of this idea can be found in [2], [3], and especially in [4], which provides an excellent treatment for engineers.

A more convenient definition however, is given in terms of the probability density function (PDF) when it exists, as we have assumed. For a continuous random variable we define the expectation to be

$$\mathcal{E}\{X\} = \int_{-\infty}^{\infty} x f_X(x) dx \qquad (4.4)$$

whenever the infinite integral converges absolutely [5].[3]

This formula can also be applied to discrete and mixed random variables as long as impulses are allowed in the PDF. In particular, let the PDF for a discrete random variable be represented as

$$f_K(k) = \sum_{\forall i} f_K[i] \delta(k - i) \qquad (4.5)$$

where $f_K[k]$ denotes the PMF and δ is the unit impulse. In other words, the PDF is represented as a series of impulses with areas equal to the probability of the random variable. Then substitution of this expression into (4.4) leads directly to (4.2). Alternatively, if a continuous random variable X is related to a discrete random variable K through quantization as in Section 3.3.3, then (4.4) can be thought of as the limiting version of (4.2).

In light of the preceding discussion, you can henceforth consider (4.4) to be the definition of expectation. For most analyses, it is convenient to regard $\mathcal{E}\{\cdot\}$ as a linear *operator* defined by (4.4). The definition of this operator will have to be expanded slightly for the treatment of multiple random variables in Chapter 5.

4.1.3 Invariance of expectation

An extremely important property of expectation is *invariance under a transformation*. This property is really grounded in the interpretation of expectation in the sample space as provided by (4.1) and (4.3). However, it can be explained using the definition (4.4).

Let X be a random variable and let $Y = g(X)$. Then (4.4) states that

$$\mathcal{E}\{Y\} = \int_{-\infty}^{\infty} y f_Y(y) dy \qquad (4.6)$$

What is *not* so obvious is that $\mathcal{E}\{Y\}$ can also be computed from

$$\mathcal{E}\{Y\} = \int_{-\infty}^{\infty} g(x) f_X(x) dx \qquad (4.7)$$

In other words, to compute the expectation of $Y = g(X)$ you do not have to transform and find the PDF $f_Y(y)$. You can simply place the function $g(x)$ under the integral along with the probability density f_X for X and integrate! Thus for any function $g(X)$, we can define expectation of that quantity to mean:

$$\mathcal{E}\{g(X)\} = \int_{-\infty}^{\infty} g(x) f_X(x) dx \qquad (4.8)$$

A simple argument can be given here to convince you of this result.

---

[3] Absolute convergence is defined as: $\int_{-\infty}^{\infty} |x| f_X(x) dx < \infty$.

Let (4.6) be represented by the discrete approximation

$$\mathcal{E}\{Y\} \approx \sum_{\forall k} y_k f_Y(y_k)\Delta_y$$

The term $f_Y(y_k)\Delta_y$ in the summation is the probability that Y is in a small interval $y_k < y \le y_k + \Delta_y$. Now refer to Fig 4.1. The probability that Y is in that small interval

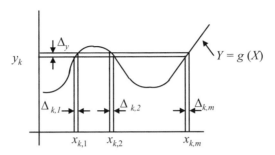

Figure 4.1 Transformation $Y = g(X)$.

is the same as the sum of the probabilities that X is in each of the corresponding small intervals shown in the figure. In other words,

$$f_Y(y_k)\Delta_y = f_X(x_{k,1})\Delta_{k,1} + f_X(x_{k,2})\Delta_{k,2} + \cdots + f_X(x_{k,m})\Delta_{k,m}$$

where m is the number of intervals in x that map into the interval $(y_k, y_k + \Delta_y]$ in y. Substituting this expression in the previous equation yields

$$\mathcal{E}\{Y\} \approx \sum_{\forall k}\sum_{i=1}^{m} y_k f_X(x_{k,i})\Delta_{k,i} = \sum_{\forall k}\sum_{i=1}^{m} g(x_{k,i})f_X(x_{k,i})\Delta_{k,i}$$

Since the double sum merely covers all possible regions of x, we can reindex the variables and replace it with the single summation

$$\mathcal{E}\{Y\} = \sum_{\forall l} g(x_l)f_X(x_l)\Delta_l$$

which in the limit becomes the integral (4.7).

Although Fig. 4.1 depicts a continuous function, the result (4.8) is actually quite general and applies for almost any function $g(X)$ that you are likely to encounter. Our argument was based on the usual Riemann integral; however, in advanced treatments of random variables, it is possible to state results in terms of Lebesgue integration for which invariance of expectation is almost a direct consequence. In most cases, the Lebesgue and the Riemann integrals produce identical results, so operations of integration are carried out in the usual way.

Invariance of expectation can be expressed for discrete random variables as

$$\mathcal{E}\{g(K)\} = \sum_{\forall k} g(k)\,f_K[k] \tag{4.9}$$

Although this equation is implied by (4.8) whenever the PDF contains impulses, it is sometimes useful to have an explicit formula for the discrete case.

Let us close this section with one further example on the use of expectation.

Example 4.3: The continuous random variable Θ is uniformly distributed between $-\pi$ and π. The expected value of this random variable is found using (4.4) as

$$E\{\Theta\} = \int_{-\infty}^{\infty} \theta f_\Theta(\theta) d\theta = \int_{-\pi}^{\pi} \theta \frac{1}{2\pi} d\theta = \left. \frac{\theta^2}{4\pi} \right|_{-\pi}^{\pi} = 0$$

Now consider the random variable $Y = A \cos^2(\Theta)$ where A is assumed to be a constant (not a random variable). Then using (4.8), we have

$$
\begin{aligned}
E\{Y\} &= E\{A \cos^2(\Theta)\} = \int_{-\pi}^{\pi} A \cos^2(\theta) \frac{1}{2\pi} d\theta \\
&= \frac{A}{2\pi} \int_{-\pi}^{\pi} \tfrac{1}{2}(1 + \cos 2\theta) d\theta = \left. \frac{A}{4\pi} (\theta + \tfrac{1}{2} \sin 2\theta) \right|_{-\pi}^{\pi} = \frac{A}{2}
\end{aligned}
$$

$\square$

The same result for $E\{Y\}$ can be obtained by finding $f_Y(y)$ and using (4.6). This would require much more work, however.

4.1.4 Properties of expectation

As we proceed in the study of random variables, you will find that we often use a number of properties of expectation, without ever needing to carry out the integral or summation involved in the definition of expectation. Some of the most important properties result from viewing the expectation $E\{\cdot\}$ as a linear operator.

Consider any two functions of X, say $g(X)$ and $h(X)$ and form a linear combination $ag(X) + bh(X)$ where a and b are any constants. Using 4.8, we can write

$$
\begin{aligned}
E\{ag(X) + bh(X)\} &= \int_{-\infty}^{\infty} (ag(x) + bh(x)) f_X(x) dx \\
&= a \int_{-\infty}^{\infty} g(x) f_X(x) dx + b \int_{-\infty}^{\infty} h(x) f_X(x) dx \\
&- aE\{g(X)\} + bE\{h(X)\}
\end{aligned}
$$

If we let $Y = g(X)$ and $Z = h(X)$, then

$$\boxed{E\{aY + bZ\} = aE\{Y\} + bE\{Z\}}$$

(4.10)

which defines expectation as a *linear operator*.

A number of properties of expectation are now apparent, which are listed in Table 4.1. The first property follows from the fact the PDF integrates to 1. The second and

$$
\boxed{
\begin{aligned}
E\{c\} &= c \quad (c \text{ is a constant}) \\
E\{X + c\} &= E\{X\} + c \\
E\{cX\} &= cE\{X\} \\
E\{E\{X\}\} &= E\{X\}
\end{aligned}
}
$$

Table 4.1 Some simple properties of expectation.

third properties follow from the linearity (4.10), and the last follows from the first and the observance that $E\{X\} = m_X$ is a constant.

4.1.5 Expectation conditioned on an event

Occasionally there is a need to compute expectation conditioned on events. If A is an event, then the expectation conditioned on the event A is defined by

$$\mathcal{E}\{X|A\} = \int_{-\infty}^{\infty} x f_{X|A}(x|A)dx \tag{4.11}$$

Here $f_{X|A}$ denotes a density function for X conditioned on the event A, which usually controls one or more of the parameters.

Many applications and variations of this idea are possible. For example, since A and A^c are mutually exclusive and collectively exhaustive events, it is possible to write

$$f_X(x) = f_{X|A}(x|A)\Pr[A] + f_{X|A^c}(x|A^c)\Pr[A^c]$$

This is just a form of the principle of total probability. It follows from this last expression that

$$\mathcal{E}\{X\} = \mathcal{E}\{X|A\}\Pr[A] + \mathcal{E}\{X|A^c\}\Pr[A^c]$$

4.2 Moments of a Distribution

Expectation can be used to describe properties of the PDF or PMF of random variables. If you let $g(X) = X^n$ for $n = 1, 2, 3, \ldots$ in (4.8) and (4.9), the quantity

$$\mathcal{E}\{X^n\} = \int_{-\infty}^{\infty} x^n f_X(x)dx \tag{4.12}$$

for a continuous random variable or

$$\mathcal{E}\{X^n\} = \sum_{\forall k} k^n \, \mathrm{f}_K[k] \tag{4.13}$$

for a discrete random variable is called the nth *moment* of the distribution. The first moment $(n = 1)$ is called the *mean*, and represents the average value of the random variable for the given PDF. We will denote the mean by the variable m with an appropriate subscript, i.e.,

$$m_X = \mathcal{E}\{X\} \qquad \text{or} \qquad m_K = \mathcal{E}\{K\} \tag{4.14}$$

Second and higher order moments further characterize the PDF in a way that is explained below. In the special case where X is a random electrical signal (voltage), X^2 represents the "power" associated with the signal; therefore, the second moment $\mathcal{E}\{X^2\}$ represents *average signal power*.

4.2.1 Central moments

When the mean of a random variable is not 0, then the moments centered about the mean are generally more useful than the moments defined above. These *central moments* are defined by

$$\boxed{\begin{aligned} \mathcal{E}\{(X - m_X)^n)\} &= \int_{-\infty}^{\infty} (x - m_X)^n f_X(x)dx \quad (a) \text{ CONTINUOUS RV} \\[2mm] \mathcal{E}\{(K - m_K)^n\} &= \sum_{\forall k} (k - m_K)^n \, \mathrm{f}_K[k] \qquad (b) \text{ DISCRETE RV} \end{aligned}} \tag{4.15}$$

The second central moment is called the *variance* and is given by[4]

$$\sigma_X^2 = \text{Var}\,[X] = \mathcal{E}\left\{(X - m_X)^2\right\} \tag{4.16}$$

(Both the symbol σ_X^2 and the notation $\text{Var}\,[X]$ are commonly used to denote the variance of X.) The square root of the variance (σ_X) is called the *standard deviation* and has the same units as the random variable X. The quantity σ_X is important because it is a measure of the *width* or *spread* of the density function.

For the Gaussian PDF, the variance appears explicitly in the formula (i.e., $\sigma_X^2 = \sigma^2$ in (3.19)), while for other PDFs the variance and standard deviation must be expressed in terms of the other parameters. For example, the variance of a uniform random variable (see Prob. 4.15) is given by

$$\sigma_X^2 = \frac{(b-a)^2}{12}$$

and thus the standard deviation σ_X is approximately $0.289(b-a)$. The integral

$$\int_{m_X-\sigma_X}^{m_X+\sigma_X} f_X(x)dx$$

of the PDF from one standard deviation below the mean to one standard deviation above the mean provides a rough indication of how the probability is distributed. For the Gaussian PDF this integral is approximately 0.683 while for the uniform PDF it is $1/\sqrt{3}$ or 0.577. This indicates that for the Gaussian random variable, the probability is distributed more densely in the region closer to the mean.

The following example shows how to compute the mean, variance, and standard deviation for an exponential random variable.

Example 4.4: Suppose X is an exponential random variable:

$$f_X(x) = \begin{cases} \lambda e^{-\lambda x} & x \ge 0 \\ 0 & x < 0 \end{cases}$$

To help in the evaluation of moments, we can make use of the formula

$$\int_0^\infty u^n e^{-u} du = n! \qquad n = 0, 1, 2, 3, \ldots$$

(See e.g., [6].)

The mean is given by

$$m_X = \int_{-\infty}^\infty x f_X(x)dx = \int_0^\infty x\lambda e^{-\lambda x} dx$$

Making the change of variables $u = \lambda x$ yields

$$m_X = \frac{1}{\lambda} \int_0^\infty u e^{-u} du = \frac{1}{\lambda} \cdot 1! = \frac{1}{\lambda}$$

The variance is computed from

---

[4] To avoid unnecessary redundancy, we assume in the following that the random variable X can be either continuous or discrete. The expectation is then carried out using whichever formula (i.e., (4.15) (a) or (b)) applies.

$$\sigma_X^2 = \int_{-\infty}^{\infty} (x - m_X)^2 f_X(x)dx = \int_0^{\infty} (x - \frac{1}{\lambda})^2 \lambda e^{-\lambda x} dx$$

$$= \int_0^{\infty} x^2 \lambda e^{-\lambda x} dx - \frac{2}{\lambda} \int_0^{\infty} x\lambda e^{-\lambda x} dx + \frac{1}{\lambda^2} \int_0^{\infty} \lambda e^{-\lambda x} dx$$

$$= \frac{1}{\lambda^2} \int_0^{\infty} u^2 e^{-u} du - \frac{2}{\lambda^2} \int_0^{\infty} u e^{-u} du + \frac{1}{\lambda^2}$$

$$= \frac{1}{\lambda^2} \cdot 2! - \frac{2}{\lambda^2} \cdot 1! + \frac{1}{\lambda^2} = \frac{1}{\lambda^2}$$

From this you can see that the mean and standard deviation (σ_X) of the exponential random variable are the same; both are equal to $1/\lambda$.

□

4.2.2 Properties of variance

The variance is such an important descriptor for a random variable, that it is appropriate to list some of its properties. This has been done in Table 4.2. Although the

$$\text{Var}[c] = 0 \quad (c \text{ is a constant})$$
$$\text{Var}[X + c] = \text{Var}[X]$$
$$\text{Var}[cX] = c^2 \text{Var}[X]$$

Table 4.2 Some properties of variance.

properties are stated without proof, you should be able to easily show them. Note especially the last property and the fact that the constant in front of the variance on the right hand side is squared. This occurs because the variance is the expectation of a *squared* quantity. Check yourself out on the use of these properties. Suppose the variance of X is equal to 1 and $Y = 2X + 3$. What is the variance of Y?[5]

The last formula involving variance that will be mentioned is given below:

$$\text{Var}[X] = \mathcal{E}\{X^2\} - m_X^2 \tag{4.17}$$

This states that the variance can be found by computing the second moment and subtracting the mean squared. Frequently, this method for computing the variance is much easier than using the formula (4.16) directly.

The proof of (4.17) is a good example of considering $\mathcal{E}\{\cdot\}$ as a linear *operator* and using the properties of expectation listed in Table 4.1. We begin with (4.16) and expand the argument algebraically:

$$\mathcal{E}\{(X - m_X)^2\} = \mathcal{E}\{X^2 - 2m_X X + m_X^2\}$$

Since the expectation is a linear operator, we can distribute the expectation over the three terms and write

$$\mathcal{E}\{(X - m_X)^2\} = \mathcal{E}\{X^2\} - \mathcal{E}\{2m_X X\} + \mathcal{E}\{m_X^2\} = \mathcal{E}\{X^2\} - 2m_X \mathcal{E}\{X\} + m_X^2$$

[5] Answer: Var$[Y] = 4$.

where in the last step we have used some of the properties in Table 4.1. Finally, since $\mathcal{E}\{X\} = m_X$, the middle term on the far right is equal to $-2m_X^2$. Therefore, by adding up terms, (4.17) follows.

An example, involving a discrete random variable, illustrates the use of (4.17).

Example 4.5: A Bernoulli random variable has the PMF shown in Fig. 3.5 of Chapter 3. The mean is computed as follows:

$$m_K = \mathcal{E}\{K\} = \sum_{k=0}^{1} k\, f_K[k] = (1-p) \cdot 0 + p \cdot 1 = p$$

The variance is computed by first computing the second moment:

$$\mathcal{E}\{K^2\} = \sum_{k=0}^{1} k^2\, f_K[k] = (1-p) \cdot 0^2 + p \cdot 1^2 = p$$

Then (4.17) is applied to compute the variance:

$$\sigma_K^2 = \mathcal{E}\{K^2\} - m_K^2 = p - p^2 = p(1-p)$$

(You can check this result by computing σ_K^2 from (4.16) directly.)

$\square$

4.2.3 Some higher-order moments

Some normalized quantities related to the third and fourth order moments are sometimes used in the analysis of non-Gaussian random variables. These quantities are called the *skewness* and *kurtosis* and are defined by

$$\alpha_3 = \frac{\mathcal{E}\{(X - m_X)^3\}}{\sigma_X^3} \qquad \text{(a) SKEWNESS}$$

$$\alpha_4 = \frac{\mathcal{E}\{(X - m_X)^4\}}{\sigma_X^4} - 3 \quad \text{(b) KURTOSIS}$$

(4.18)

The skewness and kurtosis for a Gaussian random variable are identically 0.

The skewness can be used to measure the asymmetry of a distribution. PDFs or PMFs which are symmetric about their mean have *zero* skewness. The kurtosis is sometimes used to measure the deviation of a distribution from the Gaussian PDF. In other words distributions that are in this sense "closer to Gaussian" have lower values of kurtosis. The two quantities α_3 and α_4 belong to a more general family of higher-order statistics known as *cumulants* [7, 8]. These quantities have recently found use in various areas of signal processing (see e.g., [9]). Some problems on computing skewness and kurtosis are included at the end of this chapter (see Prob. 4.20).

4.3 Generating Functions

In the analysis of signals and linear systems, Fourier, Laplace, and z-transforms are essential to show properties of the signal or system that are not readily apparent in the time domain. For similar reasons formal transforms applied to the PDF or PMF can be useful tools in problems involving probability and random variables. One especially important use of the transforms is in generating *moments* of the distribution; hence the name "generating functions." This section provides a short introduction to these transforms or "generating functions" for continuous and discrete random variables.

4.3.1 The moment generating function

The moment generating function (MGF) corresponding to a random variable X is defined by

$$M_X(s) = \mathcal{E}\left\{e^{sX}\right\} = \int_{-\infty}^{\infty} f_X(x)e^{sx}dx \qquad (4.19)$$

where s is a complex variable taking on values such that the integral converges.[6] These values of s define a region of the complex plane called the "region of convergence." The definition (4.19) provides two interpretations, *both of which are useful*. The MGF can be thought of as either: (1) the expected value of e^{sX}, *or* (2) as the Laplace transform of the PDF.[7] The MGF is most commonly used for continuous random variables; however, it can be applied to mixed or discrete random variables as long as impulses are allowed in the PDF.

When the MGF is evaluated at $s = \jmath\omega$ (i.e., on the imaginary axis of the complex plane), the result is the Fourier tranform:

$$M_X(\jmath\omega) = \mathcal{E}\left\{e^{\jmath\omega X}\right\} = \int_{-\infty}^{\infty} f_X(x)e^{\jmath\omega x}dx$$

This quantity, which is called the characteristic function, is often used instead of the MGF to avoid discussion of region of convergence and some subtle mathematical difficulties that can arise when we stray off the $s = \jmath\omega$ axis. Notice that on the $\jmath\omega$ axis, there is absolute convergence for the integral:

$$\int_{-\infty}^{\infty} |f_X(x)||e^{\jmath\omega x}|dx = \int_{-\infty}^{\infty} f_X(x)dx = 1 < \infty$$

Therefore the characteristic function always exists and the MGF always converges *at least* on the $s = \jmath\omega$ axis.

The ability to derive moments from the MGF is apparent when you expand the term e^{sX} in a power series and take the expectation[8]

$$\begin{aligned} M_X(s) = \mathcal{E}\left\{e^{sX}\right\} &= 1 + \mathcal{E}\left\{sX\right\} + \mathcal{E}\left\{\tfrac{1}{2}s^2X^2\right\} + \mathcal{E}\left\{\tfrac{1}{3!}s^3X^3\right\} + \cdots \\ &= 1 + \mathcal{E}\left\{X\right\}s + \tfrac{1}{2}\mathcal{E}\left\{X^2\right\}s^2 + \tfrac{1}{3!}\mathcal{E}\left\{X^3\right\}s^3 + \cdots \end{aligned}$$

The moments $\mathcal{E}\left\{X^n\right\}$ appear as coefficients of the expansion in s.

Given the moment generating function $M_X(s)$, the moments can be derived by taking derivatives with respect to s and evaluating the result at $s = 0$.[9] To see how this works, we can use the last equation to write

$$\left.\frac{dM_X(s)}{ds}\right|_{s=0} = \mathcal{E}\left\{X\right\} + 2\cdot\tfrac{1}{2}\mathcal{E}\left\{X^2\right\}s + 3\cdot\tfrac{1}{3!}\mathcal{E}\left\{X^3\right\}s^2 + \cdots \left.\right|_{s=0} = \mathcal{E}\left\{X\right\}$$

Then by taking the derivative once again:

$$\left.\frac{d^2M_X(s)}{ds^2}\right|_{s=0} = \mathcal{E}\left\{X^2\right\} + 3\cdot2\cdot\tfrac{1}{3!}\mathcal{E}\left\{X^3\right\}s + \cdots \left.\right|_{s=0} = \mathcal{E}\left\{X^2\right\}$$

---

[6] Lately it is common to define the MGF for only *real* values of s (e.g., [10]). Although this simplifies the mathematical discussion, it leads to cases where the MGF may not exist.

[7] Notice that the transform definition (4.19) uses s instead of $-s$ which is common in electrical engineering. Hence any arguments involving the left and right half planes have to be reversed.

[8] This expansion is possible in the region of convergence since there are no poles at $s = 0$. [11]

[9] While the derivative of a function of a complex variable may not always exist, we assume here that they exist in our region of interest.

Generalizing this result produces the formula

$$\boxed{\mathcal{E}\{X^n\} = \frac{d^n M_X(s)}{ds^n}\bigg|_{s=0}}$$

(4.20)

The following examples illustrate the calculation of the MGF and the use of (4.20) to compute the mean and variance.

Example 4.6:   Refer to the exponential random variable defined in Example 4.4. The PDF is given by

$$f_X(x) = \begin{cases} \lambda e^{-\lambda x} & x \geq 0 \\ 0 & x < 0 \end{cases}$$

The MGF is computed by applying (4.19)

$$
\begin{aligned}
M_X(s) &= \int_{-\infty}^{\infty} f_X(x)e^{sx}dx = \int_0^{\infty} \lambda e^{-\lambda x}e^{sx}dx \\
&= \lambda \int_0^{\infty} e^{-(\lambda-s)x}dx = \frac{\lambda}{-(\lambda-s)}e^{-(\lambda-s)x}\bigg|_0^{\infty} = \frac{\lambda}{\lambda-s}
\end{aligned}
$$

The integral exists as long as $\mathrm{Re}[s] < \lambda$. This inequality defines the region of convergence.

□

Example 4.7:   To compute the mean of the exponential random variable from its MGF, apply (4.20) with $n = 1$:

$$m_X = \mathcal{E}\{X\} = \frac{dM_X(s)}{ds}\bigg|_{s=0} = \frac{d}{ds}\left(\frac{\lambda}{\lambda-s}\right) = \frac{\lambda}{(\lambda-s)^2}\bigg|_{s=0} = \frac{1}{\lambda}$$

To compute the variance, first compute the second moment using (4.20):

$$\mathcal{E}\{X^2\} = \frac{d^2 M_X(s)}{ds^2}\bigg|_{s=0} = \frac{2\lambda}{(\lambda-s)^3}\bigg|_{s=0} = \frac{2}{\lambda^2}$$

Then use (4.17) to write

$$\sigma_X^2 = \mathcal{E}\{X^2\} - m_X^2 = \frac{2}{\lambda^2} - \left(\frac{1}{\lambda}\right)^2 = \frac{1}{\lambda^2}$$

The results agree with the results of Example 4.4.

□

4.3.2 The probability generating function

The moments for a discrete integer-valued random variable K are more easily dealt with by using the probability generating function (PGF) defined by

$$G_K(z) = \mathcal{E}\{z^K\} = \sum_{k=-\infty}^{\infty} f_K[k]z^k$$

(4.21)

where z is a complex variable in the *region of convergence* (i.e., the region where the infinite sum converges). Again, two interpretations are equally valid; the PGF can be thought of as either the expectation of z^K or the z-transform of the PMF.[10] The name probability generating function comes from the fact that if $f_K[k] = 0$ for $k < 0$ then

$$G_K(z) = f_K[0] + zf_K[1] + z^2f_K[2] + z^3f_K[3] + \cdots$$

From this expansion it is easy to show that

$$\frac{1}{k!}\frac{d^k G_K(z)}{dz^k}\bigg|_{z=0} = f_K[k] = \Pr[K=k] \tag{4.22}$$

Our interest in the PGF, however, is more in generating moments than it is in generating probabilities. For this, it is not necessary to require that $f_K[k] = 0$ for $k < 0$. Rather we can deal with the full two-sided transform defined in (4.21).

The method for generating moments can be seen clearly by using the first form of the definintion in (4.21), i.e.,

$$G_K(z) = \mathcal{E}\{z^K\}$$

The derivative of this expression is[11]

$$\frac{dG_K(z)}{dz} = \mathcal{E}\{Kz^{K-1}\}$$

If this is evaluated at $z = 1$, the term z^{K-1} goes away and leaves the formula

$$\mathcal{E}\{K\} = \frac{dG_K(z)}{dz}\bigg|_{z=1} \tag{4.23}$$

This is the mean of the discrete random variable. To generate higher order moments, we repeat the process. For example,

$$\frac{d^2 G_K(z)}{dz^2}\bigg|_{z=1} = \mathcal{E}\{K(K-1)z^{K-2}\}\big|_{z=1} = \mathcal{E}\{K^2\} - \mathcal{E}\{K\}$$

While this result is not as "clean" as the corresponding result for the MGF, we can use the last two equations to express the second moment as

$$\mathcal{E}\{K^2\} = \frac{d^2 G_K(z)}{dz^2} + \frac{dG_K(z)}{dz}\bigg|_{z=1} \tag{4.24}$$

Table 4.3 summarizes the results for computing the first four moments of a discrete random variable using the PGF. The primes in the table denote derivatives.

$$\boxed{\begin{aligned}
\mathcal{E}\{K\} &= G'(z)|_{z=1} \\[4pt]
\mathcal{E}\{K^2\} &= G''(z) + G'(z)|_{z=1} \\[4pt]
\mathcal{E}\{K^3\} &= G'''(z) + 3G''(z) + G'(z)|_{z=1} \\[4pt]
\mathcal{E}\{K^4\} &= G''''(z) + 6G'''(z) + 7G''(z) + G'(z)|_{z=1}
\end{aligned}}$$

Table 4.3 Formulas for moments computed from the PGF. G', G'' etc. denote derivatives of $G(z)$.

---

[10] Once again, there is a difference with electrical engineering, where the convention is to define the z-transform in terms of negative powers of z.

[11] It is assumed valid to interchange the operations of differentiation and expectation. This is true in most cases.

Before closing this section, let us just mention a relationship between the MGF and the PGF. If K is an integer-valued discrete random variable, then a PDF for K can be defined as in (4.5). If (4.5) is then substituted in (4.19) we are left with

$$M_K(s) = \sum_{k=-\infty}^{\infty} f_K[k]e^{sk}$$

This is the same as (4.21) evaluated at $z = e^s$. Hence for an integer-valued discrete random variable, we have the relationship

$$M_K(s) = G_K(e^s) \tag{4.25}$$

We end this section with an example of using the PGF to compute the mean and variance of a discrete random variable.

Example 4.8:   Consider the Poisson random variable whose PMF is given by

$$f_K[k] = \begin{cases} \dfrac{\alpha^k}{k!}e^{-\alpha} & k \geq 0 \\ 0 & k < 0 \end{cases}$$

The PGF for this random variable can be computed from the second form in (4.21), namely

$$G_K(z) = \sum_{k=-\infty}^{\infty} f_K[k]z^k = \sum_{k=0}^{\infty} \frac{\alpha^k}{k!}e^{-\alpha}z^k = e^{-\alpha}\sum_{k=0}^{\infty} \frac{(\alpha z)^k}{k!} = e^{-\alpha}e^{\alpha z} = e^{\alpha(z-1)}$$

Given the above PMF, the mean can then be computed using (4.23):

$$m_K = \mathcal{E}\{K\} = \left.\frac{dG_K(z)}{dz}\right|_{z=1} = \left.e^{-\alpha}\alpha e^{\alpha z}\right|_{z=1} = \alpha$$

The second moment can be computed from (4.24):

$$\mathcal{E}\{K^2\} = \left.\frac{d^2G_K(z)}{dz^2} + \frac{dG_K(z)}{dz}\right|_{z=1} = \left.e^{-\alpha}\alpha^2 e^{\alpha z} + e^{-\alpha}\alpha e^{\alpha z}\right|_{z=1} = \alpha^2 + \alpha$$

Finally, the variance is computed from

$$\sigma_K^2 = \mathcal{E}\{K^2\} - m_K^2 = \alpha^2 + \alpha - (\alpha)^2 = \alpha$$

□

4.4 Application: Entropy and Source Coding

In Chapter 2 we discussed the topics of information theory and coding. Recall that the *information* associated with an event A_i is defined as

$$I(A_i) = -\log \Pr[A_i]$$

where the logarithm is taken with base 2 and the information is therefore measured in 'bits.' Now reconsider the problem where the 'events' represent the transmission of symbols over a communication channel. As such, the events A_i are mutually exclusive and collectively exhaustive.

With the concept of expectation developed in this chapter, we are in a position to formally define *entropy* H for the set of symbols as the average information, that is,

$$H = \mathcal{E}\{I\} = -\sum_{\forall i} \Pr[A_i]\log \Pr[A_i] \tag{4.26}$$

Now suppose that a binary code is associated with each symbol to be transmitted. Let the discrete random variable L represent the length of a code word and define the average length of the code as

$$m_L = \mathcal{E}\{L\} = \sum_{\forall l} l\, f_L[l] \tag{4.27}$$

Shannon's source coding theorem can be stated as follows [12].

Theorem. *Given any discrete memoryless source with entropy H, the average length m_L of any lossless code for encoding the source satisfies*

$$m_L \geq H \tag{4.28}$$

Here the term "lossless" refers to the property that the output of the source can be completely reconstructed without any errors or substitutions. The theorem can be illustrated by a continuation of Example 2.10 of Chapter 2.

Example 4.9: Recall that in the coding of the message "ELECTRICAL ENGINEERING" the probability of the letters (excluding the space) are represented by their relative frequency of occurrence and thus are assigned the probabilities listed in Example 2.14. The codeword and length of the codeword associated with each letter by the Shannon-Fano algorithm are also shown in the figure.

It can be seen that the random variable L representing the length of the codeword can take on only three possible values. The PMF describing the random variable can be contructed by adding up the probabilities of the various length codewords. This PMF is shown below:

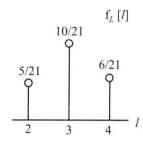

The average length of the codewords is computed from (4.27) and is given by[12]

$$m_L = \mathcal{E}\{L\} = 2 \cdot \tfrac{5}{21} + 3 \cdot \tfrac{10}{21} + 4 \cdot \tfrac{6}{21} = 3.047 \text{ (bits)}$$

Now, let us compute the source entropy. From (4.26) and Fig. 2.16, the entropy is given by

$$H = -\tfrac{5}{21}\log\left(\tfrac{5}{21}\right) + 3\left(-\tfrac{3}{21}\log\left(\tfrac{3}{21}\right)\right) + 4\left(-\tfrac{2}{21}\log\left(\tfrac{2}{21}\right)\right) + 2\left(-\tfrac{1}{21}\log\left(\tfrac{1}{21}\right)\right)$$

$$= 3.006 \text{ bits}$$

The result shows that the bound (4.28) in Shannon's theorem is satisfied. The average code length produced by the Shannon-Fano procedure is very close to the bound however. This shows that the Shannon-Fano code is very close to an optimal code for this message.

□

---

[12] This could also be computed by multiplying the length of each codeword in Fig. 2.16 by its probability and adding.

The original definition of entropy is in terms of source and channel coding. The basic definition extends to any random variables however (not just codewords). The entropy for any discrete random variable K is defined as

$$H_K = -\mathcal{E}\{\log(f_K[K])\} = -\sum_{\forall k} f_K[k] \log(f_K[k]) \tag{4.29}$$

while the entropy for a continuous random variable X is given by

$$H_X = -\mathcal{E}\{\log(f_X(X))\} = -\int_{-\infty}^{\infty} f_X(x) \log(f_X(x)) dx \tag{4.30}$$

These formulas have many uses in engineering problems and a number of optimal design procedures are based on minimizing or maximizing entropy.

4.5 Summary

The expectation of a random variable $\mathcal{E}\{X\}$ is a sum or integral over all possible values of the random variable weighted by their associated probability. Further, for any function $g(X)$, the expected value $\mathcal{E}\{g(X)\}$ can be computed by summing or integrating $g(X)$ with the PMF or PDF of X.

The moments of a random variable are expected values of products of X. Thus $m_X = \mathcal{E}\{X\}$ is the first moment (also called the *mean*), $\mathcal{E}\{X^2\}$ is the second moment, and so on. *Central* moments are expectations of products of the term $(X - m_X)$. The quantity $\sigma_X^2 = \mathcal{E}\{(X - m_X)^2\}$ is known as the *variance* of the distribution and measures how the random variable is spread about the mean.

The moment generating function is defined as $\mathcal{E}\{e^{sX}\}$ where s is a complex-valued parameter. The MGF can also be interpreted as the Laplace transform of the PDF. The power series expansion of the MGF reveals the moments of the distribution as coefficients. Thus, knowing all the moments of a distribution is equivalent to knowing the distribution. Some results and properties for random variables are discovered more easily from the MGF than by using the PDF.

For discrete random variables, a number of operations are carried out more conveniently using the probability generating function. The PGF is defined as $\mathcal{E}\{z^K\}$ and is the z-transform of the PMF.

The last section of this chapter discusses the concept of entropy as average information. Shannon's theorem states that entropy provides a lower bound on the average number of bits needed to code a message without loss.

References

[1] William Feller. *An Introduction to Probability Theory and Its Applications - Volume I.* John Wiley & Sons, New York, second edition, 1957.

[2] Athanasios Papoulis and S. Unnikrishna Pillai. *Probability, Random Variables, and Stochastic Processes.* McGraw-Hill, New York, fourth edition, 2002.

[3] Henry Stark and John W. Woods. *Probability, Random Processes, and Estimation Theory for Engineers.* Prentice Hall, Inc., Upper Saddle River, New Jersey, third edition, 2002.

[4] Wilbur B. Davenport, Jr. *Probability and Random Processes.* McGraw-Hill, New York, 1970.

[5] William Feller. *An Introduction to Probability Theory and Its Applications - Volume II.* John Wiley & Sons, New York, second edition, 1971.

[6] Daniel Zwillinger. *CRC Standard Mathematical Tables and Formulae*. CRC Press, Boca Raton, 2003.

[7] David R. Brillinger. *Time Series: Data Analysis and Theory*. Holden-Day, Oakland, California, expanded edition, 1981.

[8] Murray Rosenblatt. *Stationary Sequences and Random Fields*. Birkhauser, Boston, 1985.

[9] Chrysostomos L. Nikias and Athina P. Petropulu. *Higher-Order Spectra Analysis: A Nonlinear Signal Processing Framework*. Prentice Hall, Inc., Upper Saddle River, New Jersey, 1993.

[10] Sheldon M. Ross. *A First Course in Probability*. Prentice Hall, Inc., Upper Saddle River, New Jersey, sixth edition, 2002.

[11] Ruel V. Churchill and James Ward Brown. *Complex Variables and Applications*. McGraw-Hill Book Company, New York, fourth edition, 1984.

[12] Claude E. Shannon and Warren Weaver. *The Mathematical Theory of Communication*. University of Illinois Press, Urbana, IL, 1963.

[13] Alvin W. Drake. *Fundamentals of Applied Probability Theory*. McGraw-Hill, New York, 1967.

Problems

Expectation of a random variable

4.1 Rolf is an engineering student at the Technical University. Rolf plans to ask one of two women students for a Saturday night date: Claudia Schönstück or Ursula von Doppeldoof. Rolf is more attracted to Claudia but Ursula is more available. Further Claudia likes the finer things in life and a night with her is more expensive. All things considered, Rolf estimates that landing a date with Claudia is a "50/50 chance" and Rolf will end up spending 200 euros. On the other hand, Ursula is about twice as likely to accept a date with Rolf as not, and Ursula is happy with beer and pretzels (40 euros).

Rolf has the following procedure to determine whom he will ask first. He tosses a 1-euro coin. If it comes up "heads" he asks Claudia. If it comes up "tails," he flips it again (just to be sure). If it comes up "heads" this time he asks Claudia; otherwise he asks Ursula.

(a) What are the probabilities that Rolf will initially ask Claudia or Ursula?

According to the outcome of his toss(es) of the coin, Rolf asks either Claudia or Ursula. If he asks Claudia and she turns him down, then he asks Ursula. If he asks Ursula (first) and she turns him down, he is so despondent that he does not have the courage to ask Claudia. If he does not have a date, he spends the night drinking beer and eating pretzels by himself, so the evening still costs 40 euros.

(b) What is the probability that Rolf has a Saturday night date with Claudia?

(c) What is the expected value of the money that Rolf spends on a Saturday night?

(d) If C is a random variable representing Rolf's expenditure on a Saturday night, sketch the PMF for C. Compute $\mathcal{E}\{C\}$ using the PMF.

4.2 Consider the rolling of a pair of fair dice. The PMF for the number K that is rolled is given in Chapter 3. What is $\mathcal{E}\{K\}$?

4.3 Eduardo de Raton, known to his business partners as *Ed the Rat*, has invited you to play a "friendly" game of dice, which you dare not refuse. Ed always uses tetrahedral dice and plays on a glass coffee table with a mirror mounted below.[13] This table is useful to Ed in some of his other games. This situation is useful to you as well in that there are fewer cases you need to consider than with six-sided dice.

The basic dice game is as follows. Both you and Ed have one die; he rolls first then you roll. If you match him, you win; if you don't match him, you lose. Ed's die happens to be loaded. The probability that Ed rolls a 1, 2, or 3 is each $1/5$ while the probability that he rolls a 4 is $2/5$. Your die is a fair one; the probability that you roll a 1, 2, 3, or 4 is each equal to $1/4$. If you win, Ed pays you what was rolled on the dice (Example: Ed rolls 3 and you roll 3. Ed pays you \$6.) If you lose you pay Ed \$3 (i.e., you "win" \$−3).

(a) Draw the sample space for the game and label each elementary event with its probability.

(b) What is the expected value of your winnings? (This value may be positive or negative.)

(c) At your option, you can play another game with Ed. This game is called the "high stakes" game. This game is the same as the basic game except that if "doubles" are rolled you win twice the amount of money shown on the dice. (Example: Ed rolls 3 and you roll 3. You win $2 \times 6 = \$12$.) Except if a double 4 is rolled, *you pay Ed*. (Ed rolls 4 and you roll 4, you *lose* \$16.) Which game should you play (the basic game or the high stakes game) if you want to maximize the money you win or minimize the money you lose (expected value)?

4.4 Let K be a discrete uniform random variable with parameters $n = -2$ and $m = 4$. Define the random variable $I = 5K + 2$. Determine the following expectations:

(a) $\mathcal{E}\{K\}$

(b) $\mathcal{E}\{I\}$

(c) $\mathcal{E}\{K + I\}$

(d) $\mathcal{E}\{K \cdot I\}$

Hint: Don't forget to use $\mathcal{E}\{\cdot\}$ as a linear operator.

4.5 Consider the continuous uniform random variable with PDF

$$f_X(x) = \begin{cases} \dfrac{1}{2b} & -b \le x \le b \\ \\ 0 & \text{otherwise} \end{cases}$$

(a) What is $\mathcal{E}\{X\}$?

(b) What is $\mathcal{E}\{X|X > 0\}$?

---

[13] Tetrahedral dice have four sides with 1, 2, 3, and 4 dots; they were apparently invented by Alvin Drake [13]. The outcome of the roll is determined by which side is facing *down*.

4.6 Consider the discrete random variable I, with PMF shown below.

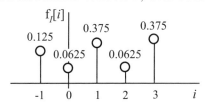

(a) What is $\mathcal{E}\{I|I > 0\}$?

(b) What is $\mathcal{E}\{I|I \leq 0\}$?

4.7 Claudia is a student at the Technical University majoring in statistics. On any particular day of the week, her friend Rolf may ask her for a Saturday night date with probability 3/4. Claudia, however, is more attracted to Roberto de la Dolce, who is very handsome, drives an expensive Italian car, and really knows how to treat women! Roberto has other women he likes however, so the probability that he asks Claudia out in any particular week is only 2/3. Roberto is also very self-confident and does not plan his activities early in the week. Let D represent the event that Roberto asks Claudia for a Saturday night date. Then the day of the week on which he asks her (beginning on Monday) is a random variable K with PMF $f_{K|D}[k|D] = k/15$ for $1 \leq k \leq 5$ and 0 otherwise. Claudia is aware of this formula and the probabilities.

Claudia needs to plan whether or not to accept if Rolf asks her for a date before Roberto; thus she decides to rate her emotional state (α) for the week on a scale of 0 to 10. A date with Roberto is actually way off the scale but she assigns it a 10. She further determines that a date with with Rolf is worth 4 points, and a Saturday night without a date is worth -5 points. She decides that if Rolf asks her out on the kth day of the week she will compute the expected value of α given that she accepts and the expected value of α given that she does not accept. Then she will make a choice according to which expected value is larger.

(a) Make a plot of the probability that Roberto does *not* ask Claudia for a date given that he has not asked her by the end of the lth day $1 \leq l \leq 5$.

(b) Sketch the conditional PMF for K given that Roberto asks Claudia out but has not done so by the end of the second day. Given this situation, what is the probability that Roberto asks her out

    (i) on the third day $(k = 3)$?

    (ii) on the fifth day $(k = 5)$?

(c) By the middle of the third day of the week Roberto has not asked Claudia for a date; but Rolf decides to ask her. Will she accept Rolf or not?

(d) Rolf has not studied statistics (not even probability), and thinks that his chances for a date with Claudia will be better if he asks her earlier in the week. Is he right or wrong?

(e) What is the optimal strategy for Rolf (i.e., when should he ask Claudia) in order to maximize the probability that Claudia will accept if he asks her for a Saturday night date?

<u>Hint</u>: It is useful to draw a sample space for this problem.

Moments of a distribution

4.8 Let K be a random variable representing the number shown at the roll of a die. The PMF is given by

$$f_K[k] = \begin{cases} 1/6 & 1 \le k \le 6 \\ 0 & \text{otherwise} \end{cases}$$

Determine the mean and variance of the random variable K.

4.9 A discrete random variable K has the PMF shown below.

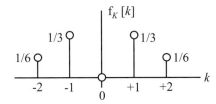

Find the mean and variance of this random variable.

4.10 In a certain digital control system a 0 is represented by a negative 1 volt level and a 1 is represented by a positive 1 volt level. The voltage level is modeled by a discrete random variable K with PMF

$$f_K[k] = \begin{cases} p & k = +1 \\ 1 - p & k = -1 \\ 0 & \text{otherwise} \end{cases}$$

(a) Sketch the PMF.

(b) Find $\mathcal{E}\{K\}$, $\mathcal{E}\{K^2\}$, and $\text{Var}[K]$.

(c) For what value of p is the variance maximized and what is this maximum value?

4.11 Find the mean and variance of a discrete uniform random variable in terms of the parameters m and n. <u>Hint:</u> To find the variance, use (4.17).

4.12 Consider the discrete random variable I, defined in Prob. 4.6.

(a) Find the expected value of random variable I.

(b) Determine the second moment of I.

(c) Compute the variance of I

    (i) starting with the definition of the second central moment,

    (ii) using the relationship among mean, second moment, and variance.

    Compare the results.

(d) What is the $\Pr[I > m_I]$ where m_I is the mean?

4.13 A type 0 geometric random variable takes on the values 0, 1, 2, 3, ...

(a) What is the mean of this random variable?

(b) Determine the second moment and variance.

(c) Given that the parameter $p = 1/3$, compute the mean and the variance.

4.14 A random variable X has the PDF shown below.

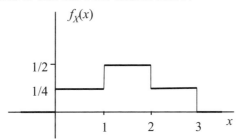

(a) Sketch the cumulative distribution function.

(b) Find the mean $E\{X\}$.

(c) Determine the variance of X.

4.15 Find the mean and variance of a continuous uniform random variable in terms of the parameters a and b. <u>Hint:</u> To find the variance, use (4.17).

4.16 A random variable Y is defined in terms of another random variable X as

$$Y = 2X + 3$$

X is known to be a Gaussian random variable with mean $\mu = 1$ and variance $\sigma^2 = 1$.

(a) What is the mean of Y?

(b) What is the variance of Y?

(c) Is Y a Gaussian random variable?

4.17 The PDF of a random variable X is shown below.

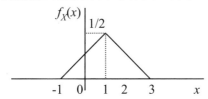

(a) Find the mean m_X and the second moment of this random variable.

(b) Determine the variance σ_X^2.

(c) Determine $\Pr[m_X - \sigma_X \leq X \leq m_X + \sigma_X]$.

4.18 The probability density function of a Laplace random variable is given by

$$f_X(x) = \frac{\alpha}{2} e^{-\alpha|x-\mu|}, \quad -\infty < x < \infty.$$

(a) Find the mean m_X, second moment, and variance σ_X^2.

(b) Determine the following probabilities for $\alpha = 4$:

    (i) $\Pr[|X - m_X| > 2\sigma_X]$

    (ii) $\Pr[|X - m_X| < \sigma_X]$

4.19 Find the mean and variance of the random variable in Prob. 3.12 of Chapter 3.

4.20 Find the *skewness* and *kurtosis* of a continuous uniform random variable with the PDF defined in Prob. 4.5. How would the results change if the PDF were uniform with a lower limit $a \neq -b$?

4.21 Show that the mean and variance of a Gaussian random variable are given by μ and σ^2 respectively, where these are the parameters that appear in the PDF. Hint: Make a change of variables and integrate "by parts."

Generating functions

4.22 A Gaussian random variable has mean μ and variance σ^2.

(a) Show that the moment generating function (MGF) for the Gaussian random variable is given by

$$M_X(s) = e^{(\mu s + \frac{1}{2}\sigma^2 s^2)}$$

Hint: Use the technique of "completing the square."

(b) Assume that $\mu = 0$ and use the MGF to compute the first four moments of X as well as the variance, skewness, and kurtosis.

(c) What are the mean, variance, skewness, and kurtosis for $\mu \neq 0$?

4.23 A random variable X is described by the Laplace density:

$$f_X(x) = \tfrac{1}{2}e^{-|x|}$$

(a) Find the MGF $M_X(s) = \mathcal{E}\{e^{sX}\}$. Assume that $-1 < \mathrm{Re}[s] < +1$ in order to evaluate the integral.

(b) What is the mean of X?

(c) What is the variance of X?

(d) Find a transformation such that the new random variable Y has a mean of $+1$. Find and sketch the density function $f_Y(y)$.

(e) Find a transformation such that the new random variable Y has a variance equal to 4. What is the density function $f_Y(y)$?

4.24 Find the PGF for the random variable described in Prob. 4.10. Use it to compute the first four moments of the random variable.

4.25 Consider the geometric random variable as defined in Prob. 4.13.

(a) Determine the PGF.

(b) From the PGF, find the mean and the variance.

Application: entropy

4.26 Let K be a Bernoulli random variable with parameter p.

(a) Find an expression for the entropy H_K of the Bernoulli random variable.

(b) Sketch H_K as a function of p for $0 < p < 1$. For what value of p is the entropy maximized?

4.27 Let X be a Gaussian random variable with mean μ and variance σ^2. The entropy in units of "nats" is given by (4.30) where we use the natural (Naperian) logarithm instead of logarithm with base 2. Find a simple expression for the entropy of the Gaussian random variable. Hint: Don't get hung up on the integration. Use the first form of the expression in (4.30) and recognize that the expectation of certain terms produces the variance σ^2.

4.28 Consider the situation described in Prob. 4.3.

    (a) Is the entropy for the first (basic) game greater than, less than, or equal to the entropy for the second (high stakes) game?

    (b) If the second game were played with a fair set of dice (both dice are fair dice) is the entropy for the first game greater than, less than, or equal to the entropy for the second game?

Computer Projects

Project 4.1

The objective of this project is to estimate the mean and variance of random variables of various kinds. This project follows Computer Project 3.1 of Chapter 3.

1. For each type of random variable described in Project 3.1 Step 1, calculate the theoretical mean and variance.
2. Generate the random sequences as specified in Project 3.1 Step 1.
3. Estimate the mean and the variance of these random variables using the following formulas:

$$\mathcal{E}\{X\} \approx M_n = \frac{1}{n}\sum_{i=1}^{n} X_i \qquad \text{Var}\,[X] \approx \Sigma_n^2 = \frac{1}{n}\sum_{i=1}^{n}(X_i - M_n)^2$$

(These formulas are discussed in Chapter 6.)

4. Repeat Steps 2 and 3 by increasing the sequence length n to (a) 20,000, and (b) 40,000.
5. Compare the estimated values in each case with the corresponding theoretical values.

Project 4.2

This project serves as an extension to Computer Project 2.2 of Chapter 2 to explore the formal definition of entropy and the source coding theorem presented in this chapter.

1. Follow Steps 1, 2, and 3 of Project 2.2 to code the given message using the Huffman algorithm.
2. Let L be a random variable representing the codeword length. Find an estimate of the PMF of L by plotting a histogram of the individual codeword lengths and normalizing it so the probabilities sum to 1. Use this estimated PMF $\hat{f}_L[l]$ in (4.27) to compute the mean length of the codewords, $\hat{m}_L$. Compare this mean length to the average codeword length computed in Step 4 of Project 2.2. (They should be the same.)
3. Consider the average information for the set of symbols computed in Step 2 of Project 2.2. This is an estimate of the source entropy H. Compare this estimate of the entropy to the mean codeword length $\hat{m}_L$ computed above, and verify that Shannon's source coding theorem (4.28) is satisfied experimentally.

5 Two and More Random Variables

This chapter expands the scope of discussion to two and more random variables. Among the topics introduced is that of dependency among random variables; specifically the case where the value of one random variable is *conditioned* upon that of another. You will also learn what it means for random variables to be *independent*.

We begin the chapter with the case of two discrete random variables to motivate the idea of dependency of one random variable upon another and develop the concepts of joint and conditional probability mass functions. We move on to continuous random variables and carry on a parallel development there. We then discuss expectation for two random variables and define the notion of "correlation."

Carrying on, the chapter discusses several properties and formulas involving sums of multiple random variables. These properties and formulas are needed sooner or later in any course in probability or statistics, and they fit well with the developments in this chapter. Although an extensive treatment of multiple random variables is an advanced topic, and not the intent of this book, we nevertheless provide an introduction to multiple random variables through the topic of random vectors. This topic can be skipped if there is not a need for it in the course of study or if readers do not have the necessary background in matrix algebra.

5.1 Two Discrete Random Variables

Recall that a random variable is a mapping from the sample space for an experiment to the set of real numbers. Many problems involve two or more random variables. If these random variables pertain to the same experiment, then these random variables may be interrelated; thus it is necessary to have a *joint* characterization of these random variables. To make the discussion more specific, consider the following example involving two discrete random variables.

Example 5.1: In the secure communication link provided by British Cryptologic Telecommunication Ltd (popularly known as "BRITTEL"), the overall probability of bit errors is 0.2. But errors tend to occur in bursts, so once an error occurs, the probability that the next bit has an error is 0.6, while if an error does not occur, the probability that the next bit is in error is only 0.1.

We start observing the transmission at some random time and observe two successive bits. An event tree and sample space corresponding to this experiment is shown in the figure on the next page, where E represents the occurrence of an error. Recall that the probabilities listed for the the four outcomes in the sample space are found by multiplying the probabilities along the branches of the tree leading to the outcome. Let the random variables K_i represent the total number of errors that have occured after the ith bit. The values of random variables K_1 and K_2 after observing two bits are also shown in the figure. These random variables are clearly interrelated.

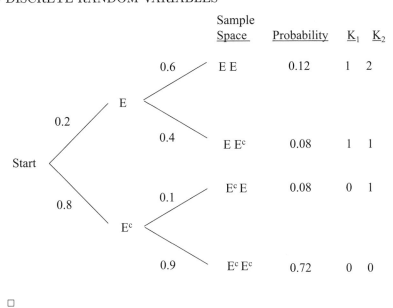

This example serves to motivate the need for joint treatment of two random variables. In the remainder of this section, we discuss various aspects of the analysis of discrete random variables while continuing to use this example for illustration.

5.1.1 The joint PMF

For problems involving two discrete random variables, the probability that the random variables take on any given pair of values is specified by the *joint* probability mass function (PMF), defined by[1]

$$f_{K_1 K_2}[k_1, k_2] \stackrel{\text{def}}{=} \Pr[K_1 = k_1, K_2 = k_2] \tag{5.1}$$

The joint PMF for the random variables described in the Example 5.1 is illustrated in Fig 5.1. The joint PMF is represented as a three-dimensional plot with values of (k_1, k_2) represented in the horizontal plane and values of their probabilities represented in the vertical direction. (Zero values of probability are not explicitly shown in Fig. 5.1.) In general the joint PMF could have non-zero values for $-\infty < k_1 < \infty$ and $-\infty < k_2 < \infty$. Since the set of all the possible values of k_1 and k_2 represent a mapping of all outcomes in the sample space, however, it follows that

$$\sum_{k_1=-\infty}^{\infty} \sum_{k_2=-\infty}^{\infty} f_{K_1 K_2}[k_1, k_2] = 1 \tag{5.2}$$

Needless to say, the individual values of $f_{K_1 K_2}[k_1, k_2]$, which represent probabilities, must all lie between 0 and 1.

Since K_1 and K_2 are both random variables, each can be described by its own PMF

---

[1] A more precise notation involving the sample space could be given (see (3.1) and the equation preceeding in Chapter 3). The notation of the right of (5.1) is common, however, and less cumbersome.

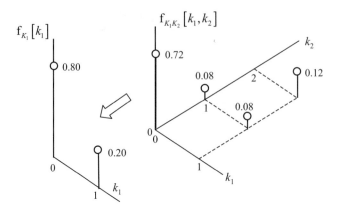

Figure 5.1 Joint PMF for the random variables K_1 and K_2 in Example 5.1 and marginal PMF for K_1.

$f_{K_i}[k_i]$. These so-called *marginal* PMFs are obtained from the equations

$$f_{K_1}[k_1] = \sum_{k_2=-\infty}^{\infty} f_{K_1 K_2}[k_1, k_2] \quad \text{(a)}$$

$$(5.3)$$

$$f_{K_2}[k_2] = \sum_{k_1=-\infty}^{\infty} f_{K_1 K_2}[k_1, k_2] \quad \text{(b)}$$

It can be seen that the marginal PMFs are obtained by summing over the other variable. As such, they can be considered to be the *projections* of the joint density on the margins of the area over which they are defined. Fig. 5.1 shows the marginal PMF for K_1, which is formed by projecting (summing) over values of K_2. The marginal PMF for the variable K_2 is formed in an analogous way and is shown in Fig. 5.1.1.

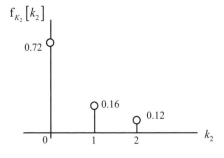

Figure 5.2 Marginal PMF for random variable K_2.

Equations 5.3 follow directly from the *principle of total probability* [(2.11) of Chapter 2]. To see this, let C be the event "$K_1 = k_1$" and A_i be the events "$K_2 = i$." Using these events in (2.11) produces (5.3)(a). Equation 5.3(b) can be obtained in a similar manner.

5.1.2 Independent random variables

Two discrete random variables are defined to be *independent* if their PMFs satisfy the condition

$$f_{K_1 K_2}[k_1, k_2] = f_{K_1}[k_1] \cdot f_{K_2}[k_2] \qquad \text{(for independent random variables)} \qquad (5.4)$$

The definition follows directly from the corresponding definition (2.12) for events. Notice that (5.4) is a *special condition* for random variables and does not apply in general! In particular, if two random variables are *not* independent, there is *no way* that the joint PMF can be inferred from the marginals. In that case the marginals are insufficient to describe any joint properties between K_1 and K_2.

To examine this point further, let us check out independence for our example. The two marginal densities and their product are represented in the matrix below on the left

	0.72	.0.16	0.12				
0.80	0.576	0.128	0.096		0.720	0.080	0
0.20	0.144	0.032	0.024		0	0.080	0.120

The values along the left and top edges of the matrix represent values of $f_{K_1}[k_1]$ and $f_{K_2}[k_2]$ computed earlier, while the other values in the matrix represent the product of these values $f_{K_1}[k_1] \cdot f_{K_2}[k_2]$. The matrix on the right represents the values of the joint PMF from Fig. 5.1. Clearly the values in the two matrices are not the same; hence the random variables fail to satisfy (5.4) and are *not* independent. This conclusion satisfies intuition because from the tree diagram in Example 5.1, the value of K_2 definitely depends upon the value of K_1.

5.1.3 Conditional PMFs for discrete random variables

When random variables are not independent, they are conditionally *dependent* upon one another. This leads to the need to define a conditional PMF for random variables. For discrete random variables, the conditional PMF is defined as

$$f_{K_1|K_2}[k_1|k_2] \stackrel{\text{def}}{=} \Pr[K_1 = k_1 | K_2 = k_2] \qquad (5.5)$$

(Recall that the vertical bar '|' is read as "given.") Since all of these PMFs (joint, marginal, conditional) represent probabilities directly, we can use the definition of conditional probability for events (2.13) to write

$$f_{K_1|K_2}[k_1|k_2] = \frac{f_{K_1 K_2}[k_1, k_2]}{f_{K_2}[k_2]} \qquad (5.6)$$

Let us illustrate this computation in terms of our example.

The observed values of K_1 and K_2 are clearly related as we have seen (i.e., they are not independent). Suppose you had somehow missed observing the value of K_1 but had observed the value of K_2. Then you might be interested in knowing the probabilities for K_1 *given* the value of K_2. That is, you would want to know the conditional PMF (5.6).

The values of the joint PMF are shown in Fig. 5.1 and the marginal PMF for K_2 is sketched in Fig 5.1.1. The conditional PMF is represented by a *set* of plots, shown in Fig. 5.3, corresponding to each possible value of k_2. The plots of Fig. 5.3 represent normalized *slices* of the joint PMF in Fig. 5.1. That is, for each value of k_2 with non-zero probability, $f_{K_1|K_2}[k_1|k_2]$ is computed from (5.6) by taking a slice of the joint PMF in the k_1 direction by dividing the values by $f_{K_2}[k_2]$ at that location. For example, to compute $f_{K_1|K_2}[k_1|0]$ shown in Fig. 5.3 (a), you would fix $k_2 = 0$ and divide the joint

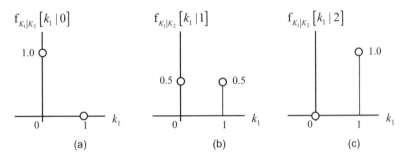

Figure 5.3 Conditional PMF $f_{K_1|K_2}[k_1|k_2]$. (a) $k_2 = 0$. (b) $k_2 = 1$. (c) $k_2 = 2$.

PMF along the line $k_2 = 0$ in Fig. 5.1 by the value $f_{K_2}[0] = 0.72$. A similar procedure is followed to compute the plots shown in Fig. 5.3 (b) and (c).

From Fig. 5.3 we see that the conditional PMF is represented by a function of a single variable (k_1) and *parameterized* by the other variable (k_2). It is indeed a probability mass function in the first variable in that

$$\sum_{k_1=-\infty}^{\infty} f_{K_1|K_2}[k_1|k_2] = 1 \tag{5.7}$$

for any value of k_2 (see Prob. 5.4). The summation

$$\sum_{k_2=-\infty}^{\infty} f_{K_1|K_2}[k_1|k_2]$$

however, over the conditioning variable, has absolutely no meaning.

5.1.4 Bayes' rule for discrete random variables

A final result to be mentioned here is a form of Bayes' rule for discrete random variables. Writing (5.6) in the form

$$f_{K_1|K_2}[k_1|k_2]f_{K_2}[k_2] = f_{K_1K_2}[k_1,k_2] = f_{K_2|K_1}[k_2|k_1]\,f_{K_1}[k_1]$$

and then rearranging yields

$$f_{K_1|K_2}[k_1|k_2] = \frac{f_{K_2|K_1}[k_2|k_1]\,f_{K_1}[k_1]}{f_{K_2}[k_2]} \tag{5.8}$$

Then since the denominator term can be computed as

$$f_{K_2}[k_2] = \sum_{k_1=-\infty}^{\infty} f_{K_1K_2}[k_1,k_2] = \sum_{k_1=-\infty}^{\infty} f_{K_2|K_1}[k_2|k_1]\,f_{K_1}[k_1]$$

substituting this result in (5.8) produces the formula

$$f_{K_1|K_2}[k_1|k_2] = \frac{f_{K_2|K_1}[k_2|k_1]\,f_{K_1}[k_1]}{\sum_{k_1=-\infty}^{\infty} f_{K_2|K_1}[k_2|k_1]\,f_{K_1}[k_1]} \tag{5.9}$$

This form of Bayes' rule is useful when it is desired to "work backward" and infer the probability of some fundamental (but usually unobservable) random variable from another related random variable that can be measured or observed directly. Equations 5.8 and 5.9 are not really new results if you consider that the various PMFs represent probabilities of events. In this case they are direct consequences of (2.18) and (2.19) of Chapter 2.

Let us close the discussion of discrete random variables with an example that further illustrates some of the ideas presented in this subsection.

Example 5.2: The Navy's new underwater digital communication system is not perfect. In any sufficiently long period of operation, the number of communication errors can be modeled as a Poisson random variable K_1 with PMF

$$f_{K_1}[k_1] = e^{-\alpha}\frac{\alpha^{k_1}}{k_1!} \qquad 0 \le k_1 < \infty$$

where the parameter α represents the average number of errors occuring in the time period.

The contractor responsible for developing this system, having realized that the communication system is not perfect, has implemented an error detection system at the receiver. Unfortunately, the error detection system is not perfect either. Given that K_1 errors occur, only K_2 of those errors are detected. K_2 is a random variable with a binomial distribution conditioned on K_1:

$$f_{K_2|K_1}[k_2|k_1] = \binom{k_1}{k_2}p^{k_2}(1-p)^{(k_1-k_2)} \qquad 0 \le k_2 \le k_1$$

where p is the probability of detecting a single error. To make matters worse, the contractor, Poseidon Systems Incorporated, who had never heard of the Poisson distribution, thought that their system was being dubbed by the Navy as a "Poseidon random variable," and threatened to sue.

The admiral in charge of development has called for a performance analysis of the overall system and has required the contractor to compute $f_{K_2}[k_2]$, the marginal PMF for the number of errors detected, and the conditional PMF $f_{K_1|K_2}[k_1|k_2]$ for the number of errors occuring given the number of errors corrected. The admiral is a graduate of the Naval Postgraduate School with a master's degree in operations research, and claims that both of these distributions are Poisson. Is he correct?

To find the marginal PMF, let us first write the joint PMF as

$$
\begin{aligned}
f_{K_1K_2}[k_1,k_2] &= f_{K_2|K_1}[k_2|k_1]\,f_{K_1}[k_1] \\
&= \binom{k_1}{k_2}p^{k_2}(1-p)^{(k_1-k_2)}\cdot e^{-\alpha}\frac{\alpha^{k_1}}{k_1!} \qquad 0 \le k_2 \le k_1 < \infty
\end{aligned}
$$

This joint PMF is non-zero only in the region of the k_1,k_2 plane illustrated below.

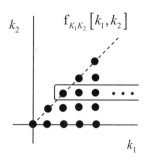

Then to find the marginal from the joint PMF, we expand the binomial coefficient in the expression for $f_{K_1 K_2}[k_1, k_2]$ and sum over k_1. (The region of summation is also shown in the figure.) The algebraic steps are as follows:

$$f_{K_2}[k_2] = \cdot e^{-\alpha} p^{k_2} \sum_{k_1 = k_2}^{\infty} \frac{\alpha^{k_1}}{k_1!} \frac{k_1!}{(k_1 - k_2)! \, k_2!} (1 - p)^{(k_1 - k_2)}$$

$$= \frac{e^{-\alpha} p^{k_2} \alpha^{k_2}}{k_2!} \sum_{k_1 = k_2}^{\infty} \frac{[\alpha(1 - p)]^{(k_1 - k_2)}}{(k_1 - k_2)!} = \frac{e^{-\alpha}(\alpha p)^{k_2}}{k_2!} \sum_{k=0}^{\infty} \frac{[\alpha(1-p)]^k}{k!}$$

$$= \frac{e^{-\alpha}(\alpha p)^{k_2}}{k_2!} e^{\alpha(1-p)} = \frac{e^{-\alpha p}(\alpha p)^{k_2}}{k_2!}$$

Thus the marginal PMF for the number of errors detected is in fact also a Poisson PMF, but with parameter αp.

The conditional PMF $f_{K_1 | K_2}$ can now be found from (5.8) or (5.9). From (5.8), we have

$$f_{K_1 | K_2}[k_1 | k_2] = \frac{f_{K_2 | K_1}[k_2 | k_1] \, f_{K_1}[k_1]}{f_{K_2}[k_2]}$$

$$= \frac{\binom{k_1}{k_2} p^{k_2} (1 - p)^{(k_1 - k_2)} \cdot e^{-\alpha} \frac{\alpha^{k_1}}{k_1!}}{\frac{e^{-\alpha p}(\alpha p)^{k_2}}{k_2!}} = e^{-\alpha(1-p)} \alpha^{(k_1 - k_2)} (1 - p)^{(k_1 - k_2)}$$

$$= e^{-\alpha(1-p)} [\alpha(1 - p)]^{(k_1 - k_2)} \qquad 0 \le k_2 \le k_1 < \infty$$

Therefore this conditional PMF is Poisson as well, with parameter $\alpha(1 - p)$; but the distribution is shifted to the right. This shift is understandable since the number of errors detected cannot be greater than the number of errors occuring in the data, i.e., the probability that $K_1 < K_2$ must be 0.

$\square$

5.2 Two Continuous Random Variables

The probabilistic characterization of two continuous random variables involves the joint probability density function (PDF). As in the case of a single random variable, we will approach the PDF via its relationship to the (joint) cumulative distribution function.

5.2.1 Joint distributions

The joint cumulative distribution function (CDF) for two random variables X_1 and X_2 is defined as

$$F_{X_1 X_2}(x_1, x_2) \stackrel{\text{def}}{=} \Pr[X_1 \le x_1, X_2 \le x_2] \qquad (5.10)$$

A typical CDF for two continuous random variables is plotted in Fig. 5.4(a) The function $F_{X_1 X_2}(x_1, x_2)$ goes to 0 as either variable x_1 or x_2 approaches minus infinity, satisfying our intuition that the probability of any variable being less than $-\infty$ should be 0. The CDF is then monotonically increasing in both variables and equals or approaches 1 as both variables approach infinity, again satisfying intuition. Although

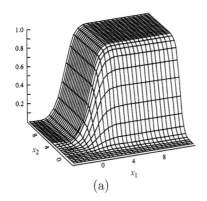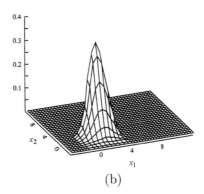

(a)   (b)

Figure 5.4 Typical joint CDF and PDF for continuous random variables. (a) Cumulative distribution function. (b) Probability density function.

the CDF shown in Fig. 5.4(a) for two *continuous* random variables is continuous, the definition (5.10) can apply to discrete or mixed random variables. In this case the CDF although monotonic, may have discontinuities. At such discontinuities, the function is *continuous from the right* in each variable.

The joint PDF is defined as the following second derivative of the CDF:

$$f_{X_1 X_2}(x_1, x_2) = \frac{\partial^2 F_{X_1 X_2}(x_1, x_2)}{\partial x_1 \partial x_2} \tag{5.11}$$

The PDF corresponding to the joint CDF of Fig. 5.4(a) is shown in Fig. 5.4(b) and resembles a small pile of sand placed on the x_1, x_2 plane. You can see from this picture, however, that the corresponding CDF, which is related to the PDF by integrating:

$$F_{X_1 X_2}(x_1, x_2) = \int_{-\infty}^{x_1} \int_{-\infty}^{x_2} f_{X_1 X_2}(u_1, u_2) du_2 du_1 \tag{5.12}$$

has the properties discussed above.

If impulses, or even line singularities[2] are allowed in the density, then discrete or mixed random variables can also be treated in terms of the PDF. The procedure is similar to what has been discussed in Chapter 3 for single random variables.

The interpretation of the joint PDF for continuous random variables in terms of probability is also extremely important. If a_1 and a_2 are two particular values of x_1 and x_2, and Δ_1 and Δ_2 are two small increments, then

$$\boxed{\Pr[a_1 < X_1 \le a_1 + \Delta_1, \ a_2 < X_2 \le a_2 + \Delta_2] \approx f_{X_1 X_2}(a_1, a_2) \Delta_1 \Delta_2} \tag{5.13}$$

This formula is illustrated in Fig. 5.5. The probability that random variables X_1 and X_2 are in the small rectangular region in the vicinity of a_1 and a_2 is equal to the density evaluated at (a_1, a_2) multiplied by the area of the small region. This interpretation is not a new result, but actually follows from the other equations presented so far. Given this result, it is easy to see that if $\mathcal{R}$ is any region of the x_1, x_2 plane, then the probability that the point (X_1, X_2) is in that region can be obtained by integration,

---

[2] By a line singularity we refer to a function that is singular in one dimension only; for example: $F(x_1, x_2) = \delta(x_1 - a)$. On the other hand, an *impulse* in the two-dimensional plane at coordinates (a_1, a_2) would have the form $F(x_1, x_2) = \delta(x_1 - a_1)\delta(x_2 - a_2)$.

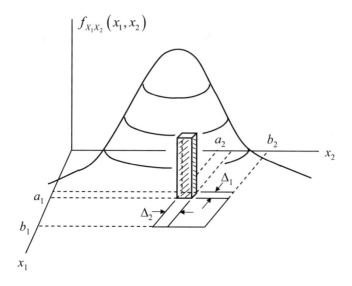

Figure 5.5 Interpretation of joint PDF as probability.

that is,

$$\Pr[(X_1, X_2) \in \mathcal{R}] = \int \int_{\mathcal{R}} f_{X_1 X_2}(x_1, x_2) dx_2 dx_1 \qquad (5.14)$$

For example, if (a_1, a_2) and (b_1, b_2) define the corners of a rectangular region (see Fig. 5.5), then

$$\Pr[a_1 < X_1 \leq b_1, \ a_2 < X_2 \leq b_2] = \int_{a_1}^{b_1} \int_{a_2}^{b_2} f_{X_1 X_2}(x_1, x_2) dx_2 dx_1$$

Two fundamental properties of the joint PDF are listed in Table 5.1. These are *defining* properties, in the sense that any function satisfying these conditions can be interpreted as a PDF. First, the joint density can never be negative. Otherwise, you

non-negative	$f_{X_1 X_2}(x_1, x_2) \geq 0$
unit volume	$\displaystyle\int_{-\infty}^{\infty} \int_{-\infty}^{\infty} f_{X_1 X_2}(x_1, x_2) dx_2 dx_1 = 1$

Table 5.1 Fundamental properties of the joint PDF.

could integrate the density over the region where it is negative and obtain negative probability! Second, the probability obtained by integrating the density over the entire x_1, x_2 plane is equal to 1. From (5.12), integrating the density from $-\infty$ to ∞ in both variables is equivalent to computing $F_{X_1 X_2}(\infty, \infty)$, which is also 1.

5.2.2 *Marginal PDFs: Projections of the joint density*

Since both X_1 and X_2 are random variables, each of these can be described by its own PDF $f_{X_i}(x_i)$. These so-called *marginal* densities for the random variables can be computed from the joint density as follows:

$$f_{X_1}(x_1) = \int_{-\infty}^{\infty} f_{X_1 X_2}(x_1, x_2)\, dx_2 \quad \text{(a)}$$

$$f_{X_2}(x_2) = \int_{-\infty}^{\infty} f_{X_1 X_2}(x_1, x_2)\, dx_1 \quad \text{(b)}$$

(5.15)

The marginals are computed by integrating over, or in a geometric sense, *projecting* along the other variable (see Fig. 5.6). To see why these last relations are valid, consider

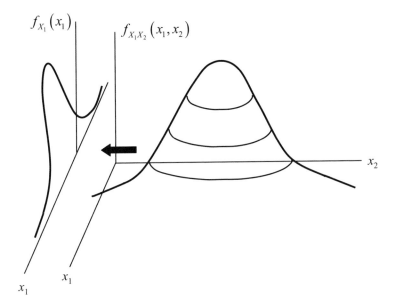

Figure 5.6 Interpretation of marginal PDF as a projection.

the event "$x_1 < X_1 \le x_1 + \Delta_1$" where Δ_1 is a small increment in x_1. The probability of this event is the same as the probability that (X_1, X_2) is in a very narrow strip of width Δ_1 in the x_1 direction and extending from $-\infty$ to ∞ in the x_2 direction. Thus using (5.14), we can write

$$
\begin{aligned}
\Pr[x_1 < X_1 \le x_1 + \Delta_1] &= \Pr[x_1 < X_1 \le x_1 + \Delta_1, \ -\infty < X_2 < \infty] \\
&= \int_{x_1}^{x_1 + \Delta_1} \int_{-\infty}^{\infty} f_{X_1 X_2}(u_1, u_2)\, du_2\, du_1 \\
&\approx \left[\int_{-\infty}^{\infty} f_{X_1 X_2}(x_1, u_2)\, du_2 \right] \Delta_1
\end{aligned}
$$

But we also know that

$$\Pr[x_1 < X_1 \le x_1 + \Delta_1] \approx f_{X_1}(x_1)\, \Delta_1$$

and this probability is unique for sufficiently small values of Δ_1. Therefore the two expressions must be equal and the PDF for X_1 is as given in (5.15)(a). A similar argument applies to computing the other marginal density function.

The example below illustrates the use of the joint PDF for computing probabilities for two random variables, and computation of the marginal PDFs.

Example 5.3: Two random variables are uniformly distributed over the region of the x_1, x_2 plane shown below. Find the constant C and the probability of the event "$X_1 > \frac{1}{2}$." Also find the two marginal PDFs.

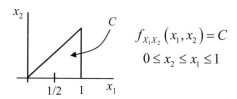

$$f_{X_1 X_2}(x_1, x_2) = C$$
$$0 \le x_2 \le x_1 \le 1$$

The constant C is found by integrating the density over the region where it is nonzero and setting that equal to 1.

$$\int_{-\infty}^{\infty}\int_{-\infty}^{\infty} f_{X_1 X_2}(x_1, x_2)\,dx_2\,dx_1 \;=\; \int_0^1 \int_0^{x_1} C\,dx_2\,dx_1$$

$$=\; C\int_0^1 x_1\,dx_1 = C\left.\frac{x_1^2}{2}\right|_0^1 = \frac{C}{2} = 1$$

Hence $C = 2$.

The probability of $X_1 > \frac{1}{2}$ is obtained by integrating the joint density over the appropriate region.

$$\Pr[X_1 > \tfrac{1}{2}] = \int_{1/2}^1 \int_0^{x_1} 2\,dx_2\,dx_1 = \int_{1/2}^1 2x_1\,dx_1 = \left.x_1^2\right|_{1/2}^1 = \tfrac{3}{4}$$

Finally, the marginals are obtained by integrating the joint PDF over each of the other variables. To obtain $f_{X_1}(x_1)$ the joint density function is integrated as follows:

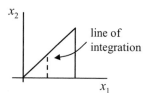

line of
integration

$$f_{X_1}(x_1) = \int_0^{x_1} 2\,dx_2 = 2x_1$$

The density function is sketched below.

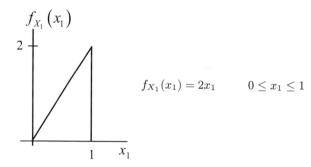

$$f_{X_1}(x_1) = 2x_1 \qquad 0 \le x_1 \le 1$$

To obtain $f_{X_2}(x_2)$ the integration is as shown below:

$$f_{X_2}(x_2) = \int_{x_2}^{1} 2\,dx_1 = 2(1-x_2)$$

The density function is sketched below.

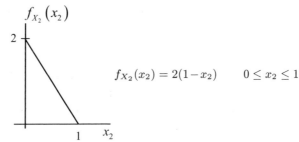

$$f_{X_2}(x_2) = 2(1-x_2) \qquad 0 \le x_2 \le 1$$

Notice that for both of these marginal densities the limits of definition $0 \le x_1 \le 1$ or $0 \le x_2 \le 1$ are very important.

□

5.2.3 Conditional PDFs: Slices of the joint density

Two random variables X_1 and X_2 are defined to be *independent* if their joint PDF is the product of the two marginals:

$$\boxed{f_{X_1 X_2}(x_1, x_2) = f_{X_1}(x_1) \cdot f_{X_2}(x_2) \qquad \text{(for independent random variables)}}$$

$$(5.16)$$

The definition follows from the analogous definition (2.12) for events. To show this, let us define the events:[3]

$$\begin{aligned} A_1 &: & x_1 < X_1 \le x_1 + \Delta_1 \\ A_2 &: & x_2 < X_2 \le x_2 + \Delta_2 \end{aligned}$$

If A_1 and A_2 are *independent* then

$$\Pr[A_1 A_2] = \Pr[A_1]\Pr[A_2] \approx (f_{X_1}(x_1)\Delta_1)(f_{X_2}(x_2)\Delta_2) = f_{X_1}(x_1)f_{X_2}(x_2)\,\Delta_1\,\Delta_2$$

But from (5.13) the probability of $A_1 A_2$ is also given by

$$f_{X_1 X_2}(x_1, x_2)\,\Delta_1\,\Delta_2$$

These last two equations motivate the definition (5.16).

It is important to note that (5.16) is a definition and *test* for independence. It is *not* a condition that holds for all random variables. If two random variables are

---

[3] The 'events' A_1 and A_2 are most correctly thought of as events in the sample space such that the indicated conditions on the X_i obtain.

not independent, then there is no way to compute a unique joint density from the marginals. Papoulis and Pillai [1, p. 218] give an example of a two different joint PDFs that have the same marginals.

Although the joint PDF is not always a product of the marginals, the joint PDF for any two random variables can always be expressed as the product of a marginal and a *conditional* density function:

$$f_{X_1 X_2}(x_1, x_2) = f_{X_1|X_2}(x_1|x_2) \, f_{X_2}(x_2) \tag{5.17}$$

The conditional density function $f_{X_1|X_2}$ is thus *defined* by[4]

$$f_{X_1|X_2}(x_1|x_2) \stackrel{\text{def}}{=} \frac{f_{X_1 X_2}(x_1, x_2)}{f_{X_2}(x_2)} \tag{5.18}$$

Recall that the conditional PMF for discrete random variables can be interpreted as a (normalized) discrete *slice* through the joint PMF. Likewise, for any given value of x_2 the conditional PDF, $f_{X_1|X_2}(x_1|x_2)$, can be interpreted as a slice through the joint PDF normalized by the value $f_{X_2}(x_2)$ (see Fig. 5.7). The conditional density can thus

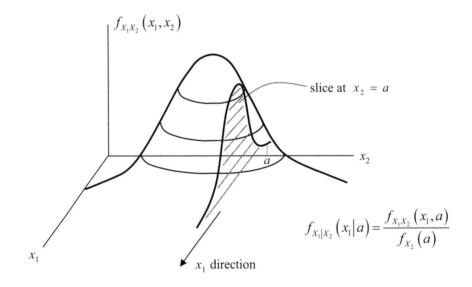

Figure 5.7 Interpretation of conditional PDF as a slice through the joint density.

be interpreted in the following way: the quantity

$$f_{X_1|X_2}(x_1|x_2) \, \Delta_1$$

is the probability that X_1 is in a small region $x_1 < X_1 \le x_1 + \Delta_1$ *given* that X_2 is in a correspondingly small region $x_2 < X_2 \le x_2 + \Delta_2$. Let us see why this interpretation is valid.

For convenience, define the events A_1 and A_2 as above, but let us *not* assume that these events are independent. Then according to the definition (2.13) for conditional

---

[4] Notice that when X_1 and X_2 are independent, (5.16) applies so that (5.18) reduces to $f_{X_1|X_2}(x_1|x_2) = f_{X_1}(x_1)$. In other words, when the random variables are independent, the conditional density is the same as the unconditional density.

probability,

$$\Pr[A_1|A_2] = \frac{\Pr[A_1 A_2]}{\Pr[A_2]} \approx \frac{f_{X_1 X_2}(x_1, x_2)\, \Delta_1 \, \Delta_2}{f_{X_2}(x_2)\, \Delta_2}$$

The ratio of PDFs in this equation is the conditional density defined in (5.18). Thus, after cancelling the common term Δ_2 we have the result

$$\boxed{\Pr[x_1 < X_1 \le x_1 + \Delta_1 \mid x_2 < X_2 \le x_2 + \Delta_2] \approx f_{X_1|X_2}(x_1|x_2)\, \Delta_1} \qquad (5.19)$$

as claimed.

If it seems strange to you that Δ_2 does not appear in the above equation, just remember that the conditional density $f_{X_1|X_2}$ is a PDF for the random variable X_1 but not for the random variable X_2 (again, see Fig. 5.7). In this PDF, X_2 acts like a *parameter* that may control size and shape of the function, but has no other role. In particular, the conditional PDF for X_1 satisfies the condition (see Prob. 5.4)

$$\int_{-\infty}^{\infty} f_{X_1|X_2}(x_1|x_2)dx_1 = 1 \qquad (5.20)$$

but the integral over the other (conditioning) variable

$$\int_{-\infty}^{\infty} f_{X_1|X_2}(x_1|x_2)dx_2$$

has no meaning.

Example 5.4 below explores the concept of independence for random variables and conditional densities.

Example 5.4: Let us continue with the random variables described in Example 5.3. In particular, we will check for independence of the random variables and determine the two conditional PDFs.

The product of the marginal densities computed in the previous example is

$$f_{X_1}(x_1) \cdot f_{X_2}(x_2) = 2x_1 \cdot 2(1 - x_2)$$

This is clearly not equal to the joint density $f_{X_1 X_2}(x_1, x_2)$, which is a constant $(C = 2)$ over the region of interest. Therefore the random variables are *not* independent.

The conditional density for X_1 can be computed from the definition (5.18).

$$f_{X_1|X_2}(x_1|x_2) = \frac{f_{X_1 X_2}(x_1, x_2)}{f_{X_2}(x_2)} = \frac{2}{2(1 - x_2)} = \frac{1}{1 - x_2} \qquad x_2 \le x_1 \le 1 \quad (1)$$

The conditional PDF is sketched below. It is found to be a *uniform* PDF with

limits that depend upon the conditioning variable x_2. Notice that the region of definition $x_2 \le x_1 \le 1$ in equation (1) above is extremely important. (These limits are derived from the original limits on the joint PDF: $0 \le x_2 \le x_1 \le 1$ (see the sketch of

$f_{X_1 X_2}$ in Example 5.3). Without these limits, the density function would not appear as shown above and would not integrate to 1.

The conditional density for X_2 is computed in a similar manner:

$$f_{X_2|X_1}(x_2|x_1) = \frac{f_{X_1 X_2}(x_1, x_2)}{f_{X_1}(x_1)} = \frac{2}{2x_1} = \frac{1}{x_1} \qquad 0 \le x_2 \le x_1$$

The result is sketched below.

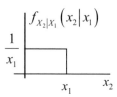

□

5.2.4 Bayes' rule for continuous random variables

A form of Bayes' rule can be derived for random variables that is very useful for signal detection and classification problems in electrical engineering. First, the joint PDF can be written using the conditional density for X_2 as

$$f_{X_1 X_2}(x_1, x_2) = f_{X_2|X_1}(x_2|x_1) \, f_{X_1}(x_1) \tag{5.21}$$

Substituting this result in (5.18) yields

$$f_{X_1|X_2}(x_1|x_2) = \frac{f_{X_2|X_1}(x_2|x_1) \, f_{X_1}(x_1)}{f_{X_2}(x_2)} \tag{5.22}$$

The denominator of this equation can be computed from (5.21) and the expression for marginal density in Table 5.1 as

$$f_{X_2}(x_2) = \int_{-\infty}^{\infty} f_{X_1 X_2}(x_1, x_2) dx_1 = \int_{-\infty}^{\infty} f_{X_2|X_1}(x_2|x_1) \, f_{X_1}(x_1) dx_1$$

Thus, substituting this result in (5.22) yields

$$f_{X_1|X_2}(x_1|x_2) = \frac{f_{X_2|X_1}(x_2|x_1) \, f_{X_1}(x_1)}{\displaystyle\int_{-\infty}^{\infty} f_{X_2|X_1}(x_2|x_1) \, f_{X_1}(x_1) dx_1} \tag{5.23}$$

This is Bayes' rule for continuous random variables.

The use of this formula for density functions can become a little involved. The following example illustrates the procedure.

Example 5.5:   The power transmitted by station KNPS is rather erratic. On any particular day the power transmitted is a continuous random variable X_1 uniformly distributed between 2 and 10 kilowatts. A listener in the area typically experiences fading. Thus the signal strength at the receiver is a random variable X_2 which depends on the power transmitted. To keep the problem simple, let us assume that $0 < X_2 \le X_1$, although in reality there should be a scale factor, and assume that the density for X_2 conditioned on X_1 is given by

$$f_{X_2|X_1}(x_2|x_1) = \frac{2}{x_1^2} x_2, \qquad 0 < x_2 \le x_1$$

The two density functions are shown below.

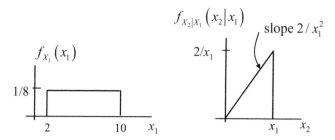

It is desired to find the conditional density $f_{X_1|X_2}(x_1|x_2)$ for the power transmitted, given the power received.

This is an application of Bayes' rule for densities, where the result can be computed directly from (5.23). The procedure is clearer, however, if we break the problem down into the individual steps leading to (5.23).

First we compute the joint density function $f_{X_1X_2}(x_1, x_2)$ as in (5.21)

$$f_{X_1X_2}(x_1, x_2) = f_{X_2|X_1}(x_2|x_1)\, f_{X_1}(x_1)$$

$$= \frac{2}{x_1^2} x_2 \cdot \frac{1}{8} = \frac{x_2}{4x_1^2}, \qquad 0 < x_2 \le x_1;\ 2 < x_1 \le 10$$

Notice that the inequalities defining where the joint density is nonzero are extremely important. This region is shown shaded in the figure below.

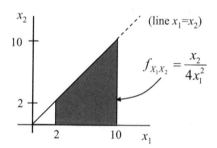

The marginal density function for x_2 can be computed by integrating the joint density over x_1. Notice from the above figure that there are two cases. For $0 < x_2 \le 2$ we have

$$f_{X_2}(x_2) = \int_2^{10} \frac{x_2}{4x_1^2}\, dx_1 = x_2 \left[-\frac{1}{4x_1}\right]_2^{10} = \frac{x_2}{10}, \qquad 0 < x_2 \le 2$$

while for $2 < x_2 \le 10$ we have

$$f_{X_2}(x_2) = \int_{x_2}^{10} \frac{x_2}{4x_1^2}\, dx_1 = x_2 \left[-\frac{1}{4x_1}\right]_{x_2}^{10} = \frac{10 - x_2}{40}, \qquad 2 < x_2 \le 10$$

The marginal density is plotted below.

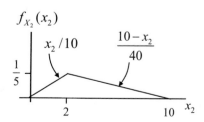

Finally, we can compute $f_{X_1|X_2}$ as in (5.22). (This is equivalent to (5.23).) Again, there are two cases. For $0 < x_2 \le 2$ we have

$$f_{X_1|X_2}(x_1|x_2) = \frac{x_2/4x_1^2}{x_2/10} = \frac{5}{2} \cdot \frac{1}{x_1^2} \quad 0 < x_2 \le 2;\ 2 < x_1 \le 10$$

while for $2 < x_2 \le 10$ we have

$$f_{X_1|X_2}(x_1|x_2) = \frac{x_2/4x_1^2}{(10-x_2)/40} = \frac{10x_2}{10-x_2} \cdot \frac{1}{x_1^2} \quad 2 < x_2 \le 10;\ 2 < x_1 \le 10$$

The conditional density $f_{X_1|X_2}(x_1|x_2)$ is plotted below. The conditional density can be used to estimate the actual power transmitted. This use is demonstated in Example 5.7, which appears in the next subsection.

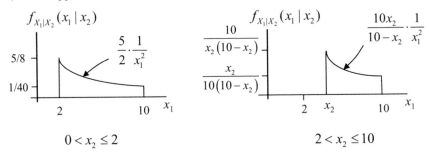

$\square$

5.3 Expectation and Correlation

Where two random variables are involved, expectation involves the *joint* PDF or PMF of the random variables. Without going back to general principles as we did in the previous chapter, we simply state that for any function $g(X_1, X_2)$ the expectation is computed from the formula

$$E\{g(X_1, X_2)\} = \int_{-\infty}^{\infty} \int_{-\infty}^{\infty} g(x_1, x_2) f_{X_1 X_2}(x_1, x_2) dx_2 dx_1 \qquad (5.24)$$

if the X_i are continuous, and

$$E\{g(X_1, X_2)\} = \sum_{\forall x_1} \sum_{\forall x_2} g(x_1, x_2) f_{X_1 X_2}[x_1, x_2] \qquad (5.25)$$

if the X_i are discrete. Note that this definition of expectation is consistent with that defined for a single random variable if $g(X_1, X_2)$ is just a function of one of the variables but not the other. For example, if $g(X_1, X_2) = X_1$, then $E\{X_1\}$ is the mean of X_1. Assuming X_1 is a continuous random variable, this is computed according to (5.24) as

$$E\{X_1\} = \int_{-\infty}^{\infty} \int_{-\infty}^{\infty} x_1 f_{X_1 X_2}(x_1, x_2) dx_2 dx_1 = \int_{-\infty}^{\infty} x_1 \left(\int_{-\infty}^{\infty} f_{X_1 X_2}(x_1, x_2) dx_2 \right) dx_1$$

Since the term in brackets is just the marginal density $f_{X_1}(x_1)$, we are left with

$$E\{X_1\} = \int_{-\infty}^{\infty} x_1 f_{X_1}(x_1) dx_1$$

which matches the expression (4.4) given for the mean in Chapter 4. From here on, we continue to treat expectation $E\{\cdot\}$ as a linear operator (see discussion in Section 4.1.4), but keep in mind the one- or two-dimensional integrals or sums that the expectation operator represents.

5.3.1 Correlation and covariance

With two random variables, some additional types of moments are important. The quantity

$$r = \mathcal{E}\{X_1 X_2\} \tag{5.26}$$

is called the *correlation* of X_1 and X_2. The product $X_1 X_2$ is a second order term and thus the correlation is a type of second order moment that measures the similarity between X_1 and X_2, on the average. If the correlation is high, then in a large collection of independent experiments, X_1 and X_2 will have similar values.

Correlation can be thought of as an (approximate) linear relationship between random variables. Figure 5.8 shows a scatter plot of the height data for fathers and sons

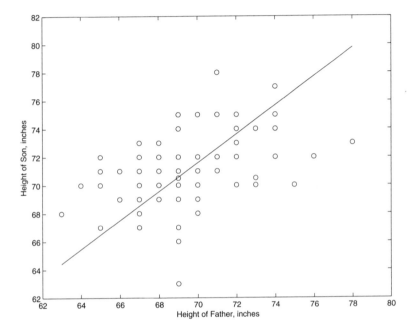

Figure 5.8 Scatter plot of father and son height data and regression line, exhibiting correlation.

collected by Sir Francis Galton in 1885. The straight line fitted to this data, which has a positive slope, indicates that a (positive) correlation exists between the two random variables (father height and son height). The straight line allows one to estimate the value of one random variable (son's height) from knowlege of the other (father's height). The subject of linear *regression analysis*, which deals with fitting such straight lines to data, was actually invented by Galton in his study of this data in the nineteenth century [2].

An important type of second moment related to the correlation is the *covariance c*, defined by

$$c = \mathrm{Cov}\,[X_1, X_2] \stackrel{\text{def}}{=} \mathcal{E}\{(X_1 - m_1)(X_2 - m_2)\} \tag{5.27}$$

where m_1 and m_2 are the means of X_1 and X_2. This is a type of *central* moment

(moment about the means) which is frequently more useful than the correlation. The correlation between two random variables may be large just due to the fact that the two random variables have large mean values. We may be more interested, however, in comparing the similarity of small *variations* about the mean. In this case the covariance is more relevant. For example, in an electronic circuit, amplifiers are often supplied with a large DC voltage, or bias, to put them into their desired region of operation. The information signal to be amplified rides on top of this DC bias and is rather small compared to the bias voltage, which just adds to the mean. In comparing two such waveforms within an amplifier, you would be more interested in the covariance, which focuses on just the information signal, rather than the correlation, which includes the bias voltage in the comparison.

The relation between correlation and covariance can be neatly expressed as

$$\mathrm{Cov}\,[X_1, X_2] = \mathcal{E}\{X_1 X_2\} - m_1 m_2 \qquad (5.28)$$

(The proof of this result is almost identical to the proof of (4.15) in Chapter 4.) This formula represents a convenient way to compute covariance. Moreover, by writing this equation in the form

$$\mathcal{E}\{X_1 X_2\} = \mathrm{Cov}\,[X_1, X_2] + m_1 m_2$$

it is clear how the mean value of the random variables influences correlation.

Another very useful quantity in the analysis of two random variables is the *correlation coefficient* ρ. This quantity is the result of normalizing the covariance c by the standard deviations σ_1 and σ_2 of the two random variables:

$$\rho \overset{\text{def}}{=} \frac{\mathrm{Cov}\,[X_1, X_2]}{\sigma_1 \sigma_2} = \frac{c}{\sigma_1 \sigma_2} \qquad (5.29)$$

It has the unique property that

$$-1 \le \rho \le 1 \qquad (5.30)$$

(See Problem 5.17 for proof of this fact.) The correlation coefficient is useful because it provides a *fixed scale* on which two random variables can be compared. Correlation and covariance on the other hand, can have any values from $-\infty$ to ∞. Thus the numerical value of r or c can only be judged in a relative sense as it relates to the particular random variables.

The three types of joint second moments we have been discussing are summarized in Table 5.2 for ease of comparison. Each of these has uses in the appropriate type of analysis.

quantity	definition	symbol
Correlation	$\mathcal{E}\{X_1 X_2\}$	r
Covariance, $\mathrm{Cov}\,[X_1, X_2]$	$\mathcal{E}\{(X_1 - m_1)(X_2 - m_2)\}$	c
Correlation Coefficient	$\dfrac{\mathrm{Cov}\,[X_1, X_2]}{\sigma_1 \sigma_2}$	ρ

Table 5.2 Three related joint second moments.

Some terminology related to correlation is appropriate to discuss, because it can be somewhat confusing. Two terms are listed in Table 5.3 that pertain to random variables. Confusion sometimes arises because the random variables are said to be

term	meaning
uncorrelated	$\rho = c = 0$
orthogonal	$r = 0$

Table 5.3 Terms related to correlation.

uncorrelated when the *covariance* (c) is 0. When the *correlation* (r) is 0, the random variables are said to be *orthogonal*. Fortunately all of these distinctions disappear, however, when at least one of the random variables has 0 mean.

The term "orthogonal" is derived from a vector space representation of random variables [3] and is beyond the scope of this text. The use of the term "uncorrelated" when the covariance is 0 can be understood by setting (5.28) to 0 and writing that equation as

$$\mathcal{E}\{X_1 X_2\} = m_1 m_2 = \mathcal{E}\{X_1\}\mathcal{E}\{X_2\} \quad \text{(for } uncorrelated \text{ random variables)} \quad (5.31)$$

In this case the expectation of the product is seen to distribute over the two random variables. Equation (5.31) can in fact be taken as the *definition* of the term "uncorrelated."

It is important to note that *independent random variables are automatically uncorrelated*. To see this, just observe that the PDF for independent random variables factors as a product of the marginals; therefore we can write

$$\mathcal{E}\{X_1 X_2\} = \int_{-\infty}^{\infty} \int_{-\infty}^{\infty} x_1 x_2 f_{X_1}(x_1) f_{X_2}(x_2) dx_2 dx_1$$
$$= \left(\int_{-\infty}^{\infty} x_1 f_{X_1}(x_1) dx_1 \right) \left(\int_{-\infty}^{\infty} x_2 f_{X_2}(x_2) dx_2 \right) = \mathcal{E}\{X_1\}\mathcal{E}\{X_2\}$$

Uncorrelated random variables are not necessarily independent, however. A special situation occurs in the case of jointly Gaussian random variables (see Section 5.4). For the Gaussian case, uncorrelated random variables are also independent.

The following example demonstrates the calculation of the entire set of first- and second-moment parameters for two random variables.

Example 5.6: Two random variables are described by the joint density function shown below.

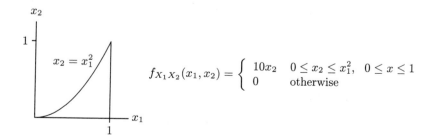

$$f_{X_1 X_2}(x_1, x_2) = \begin{cases} 10x_2 & 0 \leq x_2 \leq x_1^2, \ 0 \leq x \leq 1 \\ 0 & \text{otherwise} \end{cases}$$

The mean of random variable X_1 is given by

$$m_1 = \mathcal{E}\{X_1\} = \int_0^1 \int_0^{x_1^2} x_1 \cdot 10x_2 \, dx_2 \, dx_1 = 10 \int_0^1 x_1 \int_0^{x_1^2} x_2 \, dx_2 \, dx_1 = \frac{5}{6}$$

Likewise the mean of X_2 is given by

$$m_2 = \int_0^1 \int_0^{x_1^2} x_2 \, 10x_2 \, dx_2 \, dx_1 = 10 \int_0^1 \int_0^{x_1^2} x_2^2 \, dx_2 \, dx_1 = \frac{10}{21}$$

The variances of the two random variables are computed as

$$\mathcal{E}\{X_1^2\} = \int_0^1 \int_0^{x_1^2} x_1^2 \, 10x_2 \, dx_2 \, dx_1 = \frac{5}{7}$$

$$\sigma_1^2 = \mathcal{E}\{X_1^2\} - m_1^2 = \frac{5}{7} - \left(\frac{5}{6}\right)^2 = \frac{5}{252}$$

and

$$\mathcal{E}\{X_2^2\} = \int_0^1 \int_0^{x_1^2} x_2^2 \, 10x_2 \, dx_2 \, dx_1 = \frac{5}{18}$$

$$\sigma_2^2 = \mathcal{E}\{X_2^2\} - m_2^2 = \frac{5}{18} - \left(\frac{10}{21}\right)^2 = \frac{5}{98}$$

The correlation r is given by

$$r = \mathcal{E}\{X_1 X_2\} = \int_0^1 \int_0^{x_1^2} x_1 x_2 \, 10x_2 \, dx_2 \, dx_1 = \frac{5}{12}$$

The covariance is then computed using (5.28) as

$$c = \mathcal{E}\{X_1 X_2\} - m_x m_y = \frac{5}{12} - \left(\frac{5}{6}\right)\left(\frac{10}{21}\right) = \frac{5}{252}$$

Finally, the correlation coefficient ρ is computed as the normalized covariance.

$$\rho = \frac{c}{\sigma_1 \sigma_2} = \frac{5/252}{\sqrt{5/252}\sqrt{5/98}} = \sqrt{\frac{7}{18}}$$

□

5.3.2 Conditional expectation

Sometimes it is useful to perform expectation in stages. This leads to the concept of conditional expectation, where the expectation operation is performed with one variable remaining fixed. Let us be careful about notation here, however. Suppose the random variable X_2 is observed to have some particular value x_2. Then we could define the expectation of X_1 *given* x_2 to be the mean of the conditional density:

$$\mathcal{E}\{X_1|x_2\} = \int_{-\infty}^{\infty} x_1 f_{X_1|X_2}(x_1|x_2) dx_1$$

More frequently, the conditioning variable is also considered to be a random variable, however. We would then write

$$\mathcal{E}\{X_1|X_2\} = \int_{-\infty}^{\infty} x_1 f_{X_1|X_2}(x_1|X_2) dx_1 \tag{5.32}$$

The conditional expectation as defined by (5.32) is a random variable because of its dependence on the *random variable* X_2.

More generally, for any function $g(X_1, X_2)$, we can define

$$\mathcal{E}\{g(X_1, X_2)|X_2\} = \int_{-\infty}^{\infty} g(x_1, x_2) f_{X_1|X_2}(x_1|X_2) dx_1 \qquad (5.33)$$

Since the quantity on the left is a random variable, we can further define its expectation as

$$\mathcal{E}\{\mathcal{E}\{g(X_1, X_2)|X_2\}\} = \int_{-\infty}^{\infty} \mathcal{E}\{g(X_1, X_2)|X_2\} f_{X_2}(x_2) dx_2$$

Substituting (5.33) in the last equation and recognizing (5.17) then yields the result

$$\boxed{\mathcal{E}\{\mathcal{E}\{g(X_1, X_2)|X_2\}\} = \mathcal{E}\{g(X_1, X_2)\}} \qquad (5.34)$$

where the right hand side is the usual expectation (5.24) using the joint density function.[5]

As an illustration of conditional expectation, consider the following brief example.

Example 5.7: In many problems in engineering, it is desired to infer the value of one random variable (which cannot be measured directly) from another related random variable, which *can* be measured. This is the problem of statistical *estimation*. As an example of this, consider the situation of station KNPS and the listener described in Example 5.5. It is shown there that the probability density function for the power transmitted (X_1) given the power received (X_2) is

$$f_{X_1|X_2}(x_1|x_2) = \begin{cases} \dfrac{5}{2}\dfrac{1}{x_1^2} & 0 < x_2 \le 2;\ 2 < x_1 \le 10 \\[2mm] \dfrac{10x_2}{10 - x_2}\dfrac{1}{x_1^2} & 2 < x_2 \le 10;\ 2 < x_1 \le 10 \\[2mm] 0 & \text{otherwise} \end{cases}$$

Suppose we want to estimate the power transmitted from the power received. One well-known way to do this is to find the mean of the conditional density function. Denote this estimated value of X_1 by $\hat{X}_1$. Then

$$\hat{X}_1 = \mathcal{E}\{X_1|X_2\} = \int_{-\infty}^{\infty} x_1 f_{X_1|X_2}(x_1|x_2) dx_1$$

Substituting the expressions for the conditional density and integrating produces

$$\hat{X}_1 = \int_2^{10} x_1 \frac{5}{2}\frac{1}{x_1^2} dx_1 = \frac{5}{2}\int_2^{10}\frac{1}{x_1} dx_1 = \tfrac{5}{2}\ln 5 \approx 4.0236 \qquad 0 < X_2 \le 2$$

and

$$\hat{X}_1 = \int_{X_2}^{10} x_1 \frac{10X_2}{10 - X_2} \cdot \frac{1}{x_1^2} dx_1 = \frac{10X_2}{10 - X_2}\int_{X_2}^{10}\frac{1}{x_1} dx_1$$

$$= \frac{10}{(10/X_2) - 1}\ln(10/X_2) \qquad 2 < X_2 \le 10$$

---

[5] If the double expectation seems confusing, just notice that (5.34) can be written as

$$\int_{-\infty}^{\infty}\left(\int_{-\infty}^{\infty} g(x_1, x_2) f_{X_1|X_2}(x_1|x_2) dx_1\right) f_{X_2}(x_2) dx_2 = \int_{-\infty}^{\infty}\int_{-\infty}^{\infty} g(x_1, x_2) f_{X_1 X_2}(x_1, x_2) dx_1 dx_2$$

which is its true meaning.

A plot of this estimated value as a function of x_2 is given below.

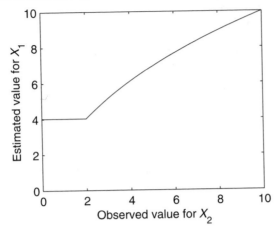

□

5.4 Gaussian Random Variables

One of the most common joint density functions is the bivariate density function for two jointly Gaussian random variables. The joint PDF for two jointly Gaussian random variables is given by the formidable expression[6]

$$f_{X_1 X_2}(x_1, x_2) = \frac{1}{2\pi\sigma_1\sigma_2\sqrt{1-\rho^2}} \times \tag{5.35}$$

$$\exp -\frac{1}{2(1-\rho^2)} \left\{ \frac{(x_1-m_1)^2}{\sigma_1^2} - 2\rho\frac{(x_1-m_1)(x_2-m_2)}{\sigma_1\sigma_2} + \frac{(x_2-m_2)^2}{\sigma_2^2} \right\}$$

and depicted in Fig. 5.9. The bivariate Gaussian density is frequently represented in

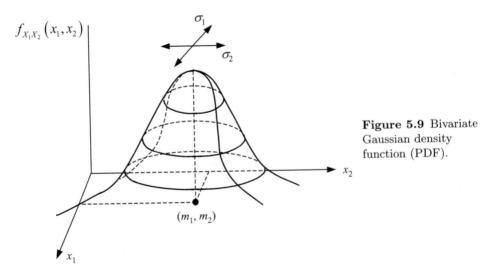

Figure 5.9 Bivariate Gaussian density function (PDF).

---

[6] A notationally more concise formula using vectors and matrices is given in Section 5.7 on random vectors.

terms of *contours* in the x_1, x_2 plane defined by

$$f_{X_1 X_2}(x_1, x_2) = \text{constant}$$

(for various chosen constants), which turn out to be ellipses (see Prob. 5.40). Some typical contours are shown in Fig. 5.10.

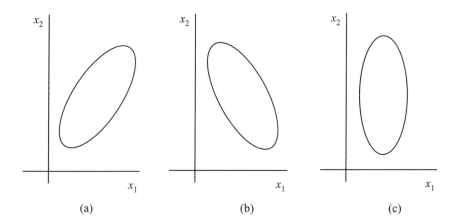

Figure 5.10 Contours of the bivariate Gaussian density function. (a) $\rho > 0$ (b) $\rho < 0$ (a) $\rho = 0$

There are five parameters that characterize the bivariate Gaussian PDF. The parameters m_i and σ_i^2 are the means and variances of the respective Gaussian random variables. That is, the marginal densities have the form

$$f_{X_1}(x_1) = \frac{1}{\sqrt{2\pi\sigma_1^2}} e^{-\frac{(x_1 - m_1)^2}{2\sigma_1^2}} \quad ; \quad f_{X_2}(x_2) = \frac{1}{\sqrt{2\pi\sigma_2^2}} e^{-\frac{(x_2 - m_2)^2}{2\sigma_2^2}} \tag{5.36}$$

which was introduced in Section 3.4. The remaining parameter ρ, is the *correlation coefficient*, defined in Section 5.3.1 of this chapter. Notice in Fig. 5.10 that when $\rho > 0$ the major axis of the ellipse representing the contours has a positive slope and when $\rho < 0$ the major axis of the ellipse has a negative slope. When $\rho = 0$, i.e., when the random variables are *uncorrelated*, the ellipse has its axes parallel to the x_1 and x_2 coordinate axes.

It is easy to show that for *Gaussian* random variables (only), "uncorrelated" implies "independent." To see this, notice that when $\rho = 0$ the product of the marginals in (5.36) is equal to the joint density (5.35). Thus, when $\rho = 0$, the random variables satisfy the condition (5.16) for independence. The fact that uncorrelated *Gaussian* random variables are also independent is a special property, unique to Gaussian random variables, that was mentioned earlier.

Another important fact about jointly Gaussian random variables, is that the *conditional* density functions are Gaussian. Thus both projections and slices of the joint PDF have a Gaussian form. The marginal densities (5.36) can be obtained from (5.35) by direct integration; then the conditional densities are obtained by algebraically dividing (5.35) by (5.36) and simplifying. These topics are explored in Prob. 5.23.

The conditional density for X_1 is given by

$$f_{X_1|X_2}(x_1|x_2) = \frac{1}{\sqrt{2\pi\sigma^2}} e^{-\frac{(x_1 - \mu(x_2))^2}{2\sigma^2}} \tag{5.37}$$

where the parameters σ and $\mu(x_2)$ are given by

$$\sigma^2 = \sigma_1^2(1-\rho^2) \tag{5.38}$$

$$\mu(x_2) = m_1 + \rho\frac{\sigma_1}{\sigma_2}(x_2 - m_2) \tag{5.39}$$

The general form of the conditional density is shown in Fig. 5.11 below.

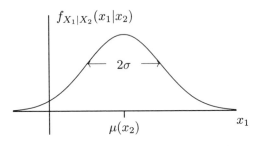

Figure 5.11 Conditional PDF for Gaussian random variables.

Now examine equation (5.38). Since the magnitude of ρ is less than 1 (as it must be if σ^2 is to remain positive) (5.38) shows that the variance σ^2 of the conditional density is less than the original variance σ_1^2. The variance σ^2 does not depend on the conditioning variable x_2 however, so the conditional density retains its same shape and size for all values of x_2. The mean μ of the conditional density is a function of x_2 however. Thus the position of the density along the x_1 axis depends (linearly) on the conditioning variable.

5.5 Multiple Random Variables

Sets of multiple random variables often occur in electrical and computer engineering problems. For example, a random signal may be sampled at N points in time and digitized. This section provides an introduction to methods dealing with multiple random variables. (A general more formal description is given in Section 5.7). In addition, several topics associated with sums of random variables are discussed in detail. Sums of random variables occur frequently in applications, and the results developed here turn out to be very useful.

5.5.1 PDFs for multiple random variables

The probabilistic description for multiple random variables is similar (in principle) to that for just two random variables. The joint PDF, PMF, or cumulative distribution function become a function of the n random variables $X_1, X_2, \ldots, X_n$. Marginal density functions are formed by integrating the joint density over the other variables, and conditional densities are defined as the ratio of the joint density to the marginals. Further, the concept of expectation is extended to multiple random variables as

$$\mathcal{E}\{g(x_1, x_2, \ldots, x_n)\} = \int_{-\infty}^{\infty} \cdots \int_{-\infty}^{\infty} f_{X_1 X_2 \cdots X_n}(x_1, x_2, \ldots, x_n)dx_1 \ldots dx_n \tag{5.40}$$

This is consistent with the earlier definitions of expectation and is seen to be a linear operation. Various moments can then be defined, but this is left as a topic for Section 5.7.

Multiple random variables can be cumbersome to deal with in practice because of the dimensionality. Nevertheless, three special cases commonly arise that simplify the procedure and make the analysis more tractable. These special cases are:

1. When the random variables are independent.

2. When the random variables form a Markov process.

3. When the random variables are jointly Gaussian.

Let us consider each of these cases briefly.

When the random variables are independent, the PDF is a product of the marginal PDFs for the individual random variables. That is,[7]

$$f_{X_1 X_2 \cdots X_n}(x_1, x_2, \ldots, x_n) = f_{X_1}(x_1) \cdot f_{X_2}(x_2) \cdots f_{X_n}(x_n) = \prod_{i=1}^{n} f_{X_i}(x_i)$$

(for independent random variables)   (5.41)

While this case seems quite restrictive, is is useful in a surprisingly large number of problems. With the assumption of independence, a whole variety of different PDFs (such as those described in earlier chapters) can be applied to problems involving multiple random variables.

The Markov model is useful when the random variables naturally form a sequence, as in samples of a signal or a time series. The model can be explained as follows. By using the definition of conditional probability, it is always possible to write (say, for $n = 4$)

$$f_{X_1 X_2 X_3 X_4}(x_1, x_2, x_3, x_4) = f_{X_4 | X_1 X_2 X_3}(x_4 | x_1, x_2, x_3) f_{X_1 X_2 X_3}(x_1, x_2, x_3)$$

Continuing this recursively produces

$$\begin{aligned} f_{X_1 X_2 X_3 X_4}(x_1, x_2, x_3, x_4) &= f_{X_4 | X_1 X_2 X_3}(x_4 | x_1, x_2, x_3) \\ &\times f_{X_3 | X_1 X_2}(x_3 | x_1, x_2) f_{X_2 | X_1}(x_2 | x_1) f_{X_1}(x_1) \end{aligned}$$

The Markov condition states that the conditional density for X_i depends only on the previous random variable in the sequence. That is $f_{X_i | X_1 X_2 \cdots X_{i-1}} = f_{X_i | X_{i-1}}$. Thus for Markov random variables, the joint density can be written as

$$f_{X_1 X_2 \cdots X_n}(x_1, x_2, \ldots, x_n) = f_{X_1}(x_1) \prod_{i=2}^{n} f_{X_i | X_{i-1}}(x_i | x_{i-1}) \qquad (5.42)$$

(for Markov random variables)

This can be a great simplification.

The Markov model is most appropriate when the random variables are observed sequentially, but the conditions of the problem cannot justify independence. Markov chains are discussed at greater length in Chapter 8.

The last case listed is the Gaussian case. You have aleady seen an expression for the joint Gaussian density for two random variables in Section 5.4. That joint density can be extended to the case of n random variables using vector and matrix notation. The fact that general expressions exist for the PDF of Gaussian random variables is fortunate because signals and noise with Gaussian statistics are ubiquitous in electrical engineering applications. The jointly Gaussian PDF for multiple random variables is discussed in detail under the topic of *Random Vectors* which appears in Section 5.7.

---

[7] An analogous result can be written for discrete random variables in terms of the PMF.

5.5.2 Sums of random variables

Certain properties of sums of random variables are important because sums of random variables occur frequently in applications.

Two random variables

Let us begin by considering the random variable

$$Y = X_1 + X_2$$

Since expectation is a linear operation, it follows that

$$\mathcal{E}\{Y\} = \mathcal{E}\{X_1\} + \mathcal{E}\{X_2\}$$

or

$$m_y = m_1 + m_2 \tag{5.43}$$

Thus the mean of the sum is the *sum of the means*.

Now consider the variance, which by definition is

$$\text{Var}\,[Y] = \mathcal{E}\left\{(Y - m_y)^2\right\} = \mathcal{E}\left\{X_1 + X_2 - (m_1 - m_2))^2\right\}$$

Regrouping terms and expanding yields

$$\begin{aligned}
\text{Var}\,[Y] &= \mathcal{E}\left\{((X_1 - m_1) + (X_2 - m_2))^2\right\} \\
&= \mathcal{E}\left\{(X_1 - m_1)^2\right\} + 2\mathcal{E}\left\{(X_1 - m_1)(X_2 - m_2)\right\} + \mathcal{E}\left\{(X_1 - m_1)^2\right\}
\end{aligned}$$

Thus we have the result

$$\boxed{\text{Var}\,[Y] = \text{Var}\,[X_1] + \text{Var}\,[X_2] + \text{Cov}\,[X_1, X_2]} \tag{5.44}$$

An especially important case of this formula is when the random variables are independent (or just uncorrelated). In this case $\text{Cov}\,[X_1, X_2] = 0$ and *the variance of the sum is the sum of the variances.*

Now let us consider the PDF for the random variable Y. A simple way to approach this computation is via the CDF. In particular, $F_Y(y)$ can be computed as

$$\begin{aligned}
F_Y(y) &= \Pr[Y \le y] = \Pr[X_1 + X_2 \le y] \\
&= \int_{-\infty}^{\infty} \int_{-\infty}^{y - x_1} f_{X_1 X_2}(x_1, x_2)\,dx_2 dx_1
\end{aligned}$$

(The region of integration for this event is sketched in Fig. 5.12.) Since $f_Y(y)$ is the

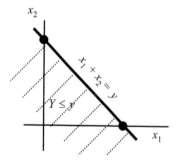

Figure 5.12 The event $X_1 + X_2 \le y$.

derivative of $F_Y(y)$, the PDF can be obtained by differentiating inside of the integral (assuming this is allowed):

$$f_Y(y) = \frac{dF_Y(y)}{dy} = \int_{-\infty}^{\infty} \left[\frac{d}{dy} \int_{-\infty}^{y-x_1} f_{X_1 X_2}(x_1, x_2) dx_2 \right] dx_1$$

or

$$f_Y(y) = \int_{-\infty}^{\infty} f_{X_1 X_2}(x_1, y - x_1) dx_1 \qquad (5.45)$$

An analogous result holds for discrete random variables in terms of the joint PMF with the integral replaced by a summation.

A very important special case of (5.45) occurs when X_1 and X_2 are independent. In this case the joint density factors into a product of marginals and leaves

$$f_Y(y) = \int_{-\infty}^{\infty} f_{X_1}(x) f_{X_2}(y - x) dx \qquad (5.46)$$

(We have dropped the subscript '1' on x since it is just a dummy variable of integration.) You may recognize the operation on the right hand side as *convolution*. We write this special result as

$$f_Y = f_{X_1} \circledast f_{X_2} \qquad \text{(for independent random variables)} \qquad (5.47)$$

where the special symbol in this equation is the convolution operator. The result also holds for independent discrete random variables using the PMF and the discrete form of convolution.

Let us illustrate the concepts discussed so far in this section with an example.

Example 5.8: Suppose that X_1 and X_2 are two independent uniform random variables with the density functions shown below.

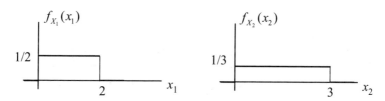

Since X_1 and X_2 are independent, (5.47) applies; the convolution is given by the integral (5.46). The arguments of the integral are shown below for a fixed value of y in three different cases.

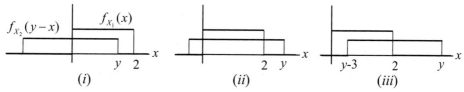

In each case we need to integrate only over the "overlap" region, i.e., the region where both functions are nonzero. The specific integrations are:

(i) for $0 \le y \le 2$ $\qquad\qquad f_Y(y) = \int_0^y \frac{1}{2} \cdot \frac{1}{3} \, dx = \frac{1}{6}y$

(ii) for $2 < y \leq 3$ $f_Y(y) = \int_0^2 \frac{1}{2} \cdot \frac{1}{3}\, dx = \frac{1}{3}$

(iii) for $3 < y \leq 5$ $f_Y(y) = \int_{y-3}^2 \frac{1}{2} \cdot \frac{1}{3}\, dx = \frac{1}{6}(5 - y)$

For all other values of y the integrands have no overlap, so f_Y is 0. The complete density function for Y is sketched below.

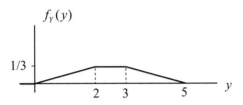

$\square$

Two or more random variables

When Y is the sum of n random variables, i.e.,

$$Y = X_1 + X_2 + \cdots + X_n \tag{5.48}$$

the results of the previous discussion generalize. In particular, because expectation is a linear operator, it follows that

$$\mathcal{E}\{Y\} = \sum_{i=1}^n \mathcal{E}\{X_i\}$$

or

$$m_Y = \sum_{i=1}^n m_i \tag{5.49}$$

The expression for the variance follows from a procedure similar to that which was used to derive (5.44). The corresponding result for n random variables is

$$\boxed{\operatorname{Var}[Y] = \sum_{i=1}^n \operatorname{Var}[X_i] + 2\sum_{i=1}^n \sum_{j=1}^{i-1} \operatorname{Cov}[X_i, X_j]} \tag{5.50}$$

Again, probably the most important case is for the case of *independent* random variables, where (5.50) reduces to

$$\sigma_y^2 = \sum_{i=1}^n \sigma_i^2 \tag{5.51}$$

The joint density function for a sum of n *independent* random variables is the n-fold convolution of the densities. Probably the easiest way to see this is by using the moment generating function (MGF). The MGF for Y is defined as

$$M_Y(s) = \mathcal{E}\{e^Y\} = \mathcal{E}\left\{e^{(X_1+X_2+\cdots+X_n)}\right\}$$

When the X_i are independent, this expectation of a product of random variables becomes a product of the expectations; that is,

$$M_Y(s) = \mathcal{E}\{e^Y\} = \mathcal{E}\{e^{X_1}\} \cdot \mathcal{E}\{e^{X_2}\} \cdots \mathcal{E}\{e^{X_n}\} = \prod_{i=1}^n M_{X_i}(s) \tag{5.52}$$

The product of transforms is the convolution of the original functions, so that[8]

$$f_Y = f_{X_1} \circledast f_{X_2} \circledast \cdots \circledast f_{X_n} \qquad \text{(for independent random variables)} \qquad (5.53)$$

The n-fold convolution can be conveniently carried out in the transform domain, however, by using (5.52) and inverting.

IID random variables

A final case that deserves some special mention is when the random variables X_i are *independent* and *identically distributed* (IID). In other words, all of the random variables are described by identical PDFs. In this case the mean and variance of the sum of these random variables is given by

$$
\begin{align}
m_Y &= n m_X & (5.54)\\
\sigma_Y^2 &= n \sigma_X^2 & (5.55)
\end{align}
$$

where m_X and σ_X^2 are the mean and variance of any one of the X_i. The standard deviation of Y then becomes

$$\sigma_Y = \sqrt{n}\,\sigma_X$$

The MGF of Y from (5.52) becomes

$$M_Y(s) = (M_X(s))^n$$

which can be inverted to obtain the PDF. A famous result applies here, however, which is known as the central limit theorem (see Chapter 6). Loosely stated, this theorem claims in effect, that for any original densities $f_X(x_i)$ and any reasonably large values of n, the density function of Y approaches that of a Gaussian random variable!

5.6 Sums of Some Common Types of Random Variables

As we have seen, the density function for a sum of independent random variables is a convolution of densities. This leads to some interesting relations, which are discussed here.

5.6.1 Bernoulli random variables

Suppose Y in (5.48) is a sum of IID Bernoulli random variables. That is, each X_i is discrete and takes on only values of 1 or 0 with probability p and $q = 1 - p$ respectively. The probability generating function (PGF) for the Bernoulli random variable reduces to

$$G_X(z) = \sum_{k=-\infty}^{\infty} f_X[k]z^k = q + pz$$

The PGF of Y is then given by

$$G_Y(z) = (G_X(z))^n = (q + pz)^n$$

This PGF is the nth power of a binomial, which can be expanded as

$$G_Y(z) = \sum_{k=0}^{n} \binom{n}{k} q^{(n-k)}(pz)^k = \sum_{k=0}^{n} \left[\binom{n}{k} p^k q^{(n-k)} \right] z^k$$

---

[8] A similar convolution for discrete random variables can be derived using the PGF.

The term in square brackets is thus $f_Y[k]$, which is the binomial PMF (see (3.5)).

It is really not surprising that Y turns out to be a binomial random variable. Since the X_i are either 1 or 0, Y simply counts the number of 1's in the sum. You have already seen when we introduced the binomial PMF that this distribution can be used to specify the probability of k 1's in a sequence of n binary digits.

5.6.2 Geometric random variables

Let K represent the sum of n IID geometric random variables of Type I (Equation(3.6)). Then the PMF for K can be found (using the PGF) to be

$$f_K[k] = \binom{k-1}{n-1} p^n (1-p)^{(k-n)}; \quad k \geq n \tag{5.56}$$

This is known as the *Pascal* PMF.

If the geometric PMF describes the number of bits to the *first* error in a stream of binary data, then the Pascal PMF describes the number of bits to the nth error in the stream. More generally, when observing discrete events, $f_K[k]$ is the probability that the nth event of interest occurs at the kth event. Some authors refer to this distribution or a variation of it as the *negative binomial*. However, while there is no dispute about the definition of the Pascal PMF, there does not seem to be precise agreement about the definition of the negative binomial [1, 4, 5, 6, 7].

Example 5.9: Digital data is transmitted one byte at a time over a computer interface. Errors in each byte occur and are detected with probability p. If six errors occur within a sequence of ten bytes, a retransmission is requested as soon as the sixth error is detected. What is the probability that a retransmission is requested?

In this example $n = 6$ and the probability of a retransmission is given by

$$P_r = \sum_{k=6}^{10} f_K[k] = \sum_{k=6}^{10} \binom{k-1}{5} p^6 (1-p)^{(k-6)}$$

Suppose that on the average, errors occur at the rate of 1 error in 100 bytes. Then by taking $p = 1/100$, the above expression evaluates to $P_r = 2.029 \times 10^{-10}$. This corresponds to one retransmission in about 5 megabytes of data.

$\square$

5.6.3 Exponential random variables

When Y is the sum of n IID exponential random variables, the PDF of Y is given by

$$f_Y(y) = \frac{\lambda^n y^{n-1}}{(n-1)!} e^{-\lambda y}; \quad y \geq 0 \tag{5.57}$$

This is known as the *Erlang* density function. To interpret this PDF, recall the use and interpretation of the exponential density function. In waiting for events that could occur at any time on a continuous time axis with average arrival rate λ, the exponential density characterizes the waiting time to the next such event. A sum of n exponential random variables X_i can therefore be interpreted as the total waiting time to the nth event. The Erlang density characterizes this total waiting time.

The Erlang PDF is most easily derived by starting with the MGF for the exponential, namely

$$M_X(s) = \frac{\lambda}{\lambda - s}$$

raising it to the nth power, and inverting. Further use of this density is made in Chapter 8 in the section on the Poisson process.

The integral of the density (5.57) is also a useful formula. In particular, the cumulative density function is given by

$$F_Y(y) = \int_0^y \frac{\lambda^n u^{n-1}}{(n-1)!} e^{-\lambda u} du = 1 - \sum_{k=0}^{n-1} \frac{(\lambda y)^k}{k!} e^{-\lambda y}; \qquad y \geq 0 \qquad (5.58)$$

The Erlang is a special case of the *Gamma* PDF which is given by

$$f_X(x) = \frac{\lambda^a x^{a-1}}{\Gamma(a)} e^{-\lambda x}; \qquad x \geq 0 \qquad (5.59)$$

where the parameter a is any positive real number and the Gamma function $\Gamma(a)$ is defined by

$$\Gamma(a) = \int_0^\infty u^{a-1} e^{-u} du \qquad (5.60)$$

When a is a positive integer n, this integral reduces to $\Gamma(n) = (n-1)!$, so the Gamma PDF reduces to the Erlang.

Example 5.10: Consider a supply of four light bulbs. The lifetime of each bulb is an exponential random variable with a mean of $1/\lambda = 2$ months. One light bulb is placed in service; when a bulb burns out it is immediately discarded and replaced with another. Let us find the probability that these four light bulbs last more than one year.

Let X_i represent the lifetime of the ith bulb. Then the total lifetime Y is given by

$$Y = X_1 + X_2 + X_3 + X_4$$

which is an Erlang random variable.

The probability that the total lifetime (in months) is more than 12 is given by

$$\begin{aligned} \Pr[Y > 12] &= 1 - \Pr[Y \leq 12] \\ &= 1 - F_Y(12) = \sum_{k=0}^{3} \frac{(12/2)^k}{k!} e^{-12/2} = 0.1512 \end{aligned}$$

where the last part of the equation comes from using (5.58) with $n = 4$, $\lambda = 1/2$, and $y = 12$.

The astute reader may notice the sum on the right side looks like a Poisson PMF, and indeed it is. You can think of "burnouts" as arrival of events. Then the probability of less than 4 burnouts in a period of 12 months is given by $\sum_{k=0}^{3} f_K[k]$ where f_K is the Poisson PMF given by (3.8) of Chapter 3.

□

5.6.4 Gaussian random variables

In the case where Y is the sum of independent Gaussian random variables (not necessarily identically distributed), Y is also a Gaussian random variable. *This is an important fact to remember.* The proof of this result using the MGF is explored in Prob. 5.30 at the end of this chapter. If the random variables X_i have means m_i and variances σ_i^2 then Y has a Gaussian PDF with mean and variance

$$m_Y = \sum_{i=1}^{n} m_i \qquad \sigma_Y^2 = \sum_{i=1}^{n} \sigma_i^2$$

5.6.5 Squared Gaussian random variables

A final slightly different sum of random variables is

$$Y = X_1^2 + X_2^2 + \cdots + X_n^2 \tag{5.61}$$

where the X_i are *zero-mean unit-variance Gaussian random variables* (i.e., $m_i = 0$, $\sigma_i^2 = 1$). In this case it can be shown that the PDF of Y is

$$\boxed{f_Y(y) = \frac{y^{(n-2)/2} e^{-y/2}}{2^{n/2}\Gamma(n/2)}} \tag{5.62}$$

where the Gamma function is defined by (5.60). This PDF is known as a *Chi-square* density function with "n degrees of freedom." This density is best known for its use in the "Chi-squared test," a statistical procedure used to test for Gaussian random variables. The Chi-squared density is also a special case of the Gamma density function given above.

5.7 Random Vectors

When dealing with a set of two or more random variables, it is sometimes convenient to group those random variables into a vector and use vector and matrix notation. A vector whose components are random variables is called a *random vector*. In this section a random vector is denoted by a bold upper case symbol such as[9]

$$\boldsymbol{X} = [X_1, X_2, \ldots, X_n]^T$$

where the X_i are random variables, and the realization of the random vector (i.e., the value that it takes on) is denoted by a bold italic lower case symbol

$$\boldsymbol{x} = [x_1, x_2, \ldots, x_n]^T$$

The notation is thus consistent with that used in the earlier parts of the text. To keep the section brief, we discuss random vectors in the context of continuous random variables, although a similar development could be done for purely discrete random variables if needed.

5.7.1 Cumulative distribution and density functions

Let us define the notation $\boldsymbol{X} \le \boldsymbol{x}$ to mean

$$\boldsymbol{X} \le \boldsymbol{x}: \quad X_1 \le x_1, X_2 \le x_2, \ldots, X_n \le x_n \tag{5.63}$$

---

[9] Throughout this text, vectors are represented by column vectors and the superscript T denotes vector or matrix transpose.

The cumulative distribution function (CDF) for the random vector is then defined as

$$F_{\boldsymbol{X}}(\boldsymbol{x}) = F_{X_1 X_2 \ldots X_n}(x_1, x_2, \ldots, x_n) \overset{\text{def}}{=} \Pr[\boldsymbol{X} \le \boldsymbol{x}] \tag{5.64}$$

As in the one- and two-dimensional cases, this function is monotonic and continuous from the right in each of the n arguments. It equals or approaches 0 as *any* of the arguments X_i approach minus infinity and equals or approaches one when *all* of the X_i approach infinity.

The PDF for the random vector (which is in reality a *joint* PDF among all of its components) is given by

$$f_{\boldsymbol{X}}(\boldsymbol{x}) = \frac{\partial}{\partial x_1} \frac{\partial}{\partial x_2} \cdots \frac{\partial}{\partial x_n} F_{\boldsymbol{X}}(\boldsymbol{x}) \tag{5.65}$$

The PDF for a random vector has the probabilistic interpretation

$$\Pr\left[\boldsymbol{x} < \boldsymbol{X} \le \boldsymbol{x} + \boldsymbol{\Delta}\right] \approx f_{\boldsymbol{X}}(\boldsymbol{x}) \Delta_1 \Delta_2 \ldots \Delta_n \tag{5.66}$$

where $\boldsymbol{x} < \boldsymbol{X} \le \boldsymbol{x} + \boldsymbol{\Delta}$ denotes the event

$$\boldsymbol{x} < \boldsymbol{X} \le \boldsymbol{x} + \boldsymbol{\Delta}: \quad x_1 < X_1 \le x_1 + \Delta_1, \ldots, x_n < X_n \le x_n + \Delta_n \tag{5.67}$$

and where the Δ_i are small increments in the components of the random vector.

The probability that $\boldsymbol{X}$ is any region of the n-dimensional space is given by the multidimensional integral

$$\Pr[\boldsymbol{X} \in \mathcal{R}] = \int_{\mathcal{R}} f_{\boldsymbol{X}}(\boldsymbol{x}) d\boldsymbol{x} \overset{\text{def}}{=} \int\!\!\int \cdots \int_{\mathcal{R}} f_{\boldsymbol{X}}(\boldsymbol{x}) dx_n \ldots dx_2\, dx_1 \tag{5.68}$$

In particular, if $\mathcal{R}$ is the entire n-dimensional space then

$$\int_{-\infty}^{\infty} f_{\boldsymbol{X}}(\boldsymbol{x})\, d\boldsymbol{x} = \int_{-\infty}^{\infty} \cdots \int_{-\infty}^{\infty} \int_{-\infty}^{\infty} f_{\boldsymbol{X}}(\boldsymbol{x})\, dx_n \ldots dx_2\, dx_1 = 1 \tag{5.69}$$

5.7.2 *Expectation and moments*

If $\mathbf{G}$ is any (scalar, vector, matrix) quantity that depends on $\boldsymbol{X}$, then the expectation of $\mathbf{G}(\boldsymbol{X})$ can be defined as

$$\mathcal{E}\{\mathbf{G}(\boldsymbol{X})\} = \int_{-\infty}^{\infty} \mathbf{G}(\boldsymbol{x})\, f_{\boldsymbol{X}}(\boldsymbol{x})\, d\boldsymbol{x} \tag{5.70}$$

If $\mathbf{G}$ is a vector or matrix, (5.70) is interpreted as the application of the expectation operation to every component of the vector or element of the matrix. Thus, the result of taking the expectation of a vector or matrix is another vector or matrix of the same size.

The quantities usually of most interest for a random vector are the first and second moments. The *mean vector* is defined by taking $\mathbf{G}(\boldsymbol{X}) = \boldsymbol{X}$:

$$\mathbf{m}_{\boldsymbol{X}} = \mathcal{E}\{\boldsymbol{X}\} = \int_{-\infty}^{\infty} \boldsymbol{x} f_{\boldsymbol{X}}(\boldsymbol{x}) d\boldsymbol{x} \tag{5.71}$$

The mean vector is a vector of constants

$$\mathbf{m}_{\boldsymbol{X}} = \begin{bmatrix} m_1 \\ m_2 \\ \vdots \\ m_n \end{bmatrix} \quad \text{with} \quad m_i = \mathcal{E}\{X_i\} \tag{5.72}$$

The *correlation matrix* is defined by

$$\mathbf{R_X} = \mathcal{E}\{\mathbf{X}\mathbf{X}^T\} \tag{5.73}$$

Note that the term $\mathbf{X}\mathbf{X}^T$ is an *outer* product rather than an inner product of vectors; thus the result is a matrix, not a scalar. This matrix has the form

$$\mathbf{R_X} = \begin{bmatrix} \mathcal{E}\{X_1^2\} & \mathcal{E}\{X_1 X_2\} & \cdots & \mathcal{E}\{X_1 X_n\} \\ \mathcal{E}\{X_2 X_1\} & \mathcal{E}\{X_2^2\} & \cdots & \mathcal{E}\{X_2 X_n\} \\ \vdots & \vdots & & \vdots \\ \mathcal{E}\{X_n X_1\} & \mathcal{E}\{X_n X_2\} & \cdots & \mathcal{E}\{X_n^2\} \end{bmatrix}$$

The off-diagonal terms represent the correlations $r_{ij} = \mathcal{E}\{X_i X_j\}$ between all pairs of vector components, and the diagonal terms are the second moments of the components. The mean vector and correlation matrix thus provide a complete description of the random vector using moments up to second order.

The *covariance matrix* is the matrix of second *central* moments of the random vector and is defined by

$$\mathbf{C_X} = \mathcal{E}\{(\mathbf{X} - \mathbf{m_X})(\mathbf{X} - \mathbf{m_X})^T\} \tag{5.74}$$

This matrix has the form

$$\mathbf{C_X} = \begin{bmatrix} \sigma_1^2 & c_{12} & \cdots & c_{1n} \\ c_{21} & \sigma_2^2 & \cdots & c_{1n} \\ \vdots & \vdots & & \vdots \\ c_{n1} & c_{n2} & \cdots & \sigma_n^2 \end{bmatrix} \qquad \text{where} \qquad \begin{aligned} \sigma_i^2 &= \text{Var}\,[X_i] \\ c_{ij} &= \text{Cov}\,[X_i, X_j] \end{aligned}$$

The covariance and correlation matrices are related as

$$\mathbf{C_X} = \mathbf{R_X} - \mathbf{m_X}\mathbf{m_X}^T \tag{5.75}$$

(The proof of this is similar to the proof of (4.15).) For the case of a two-dimensional random vector this equation has the form

$$\begin{bmatrix} \sigma_1^2 & c_{12} \\ c_{21} & \sigma_2^2 \end{bmatrix} = \begin{bmatrix} \mathcal{E}\{X_1^2\} & \mathcal{E}\{X_1 X_2\} \\ \mathcal{E}\{X_2 X_1\} & \mathcal{E}\{X_2^2\} \end{bmatrix} - \begin{bmatrix} m_1 \\ m_2 \end{bmatrix} \begin{bmatrix} m_1 & m_2 \end{bmatrix}$$

This is just the matrix embodiment of the relations (4.15) and (5.28), so it is not anything new.

As in the case of a single random variable, it is ofter easier to compute the covariance matrix using (5.75) than from the definition (5.74). The following example illustrates this computation.

Example 5.11: The two jointly-distributed random variables described in Example 5.6 are taken to be components of a random vector

$$\mathbf{X} = [X_1, X_2]^T$$

The mean vector and correlation matrix for this random vector are given by

$$\mathbf{m_X} = \begin{bmatrix} \mathcal{E}\{X_1\} \\ \mathcal{E}\{X_2\} \end{bmatrix} = \begin{bmatrix} 5/6 \\ 10/21 \end{bmatrix}$$

and

$$\mathbf{R_X} = \begin{bmatrix} \mathcal{E}\{X_1^2\} & \mathcal{E}\{X_1 X_2\} \\ \mathcal{E}\{X_2 X_1\} & \mathcal{E}\{X_2^2\} \end{bmatrix} = \begin{bmatrix} 5/7 & 5/12 \\ 5/12 & 5/18 \end{bmatrix}$$

where the numerical values of the moments are taken from Example 5.6. The covariance matrix can then be computed from (5.75) as

$$\mathbf{C_X} = \begin{bmatrix} 5/7 & 5/12 \\ 5/12 & 5/18 \end{bmatrix} - \begin{bmatrix} 5/6 \\ 10/21 \end{bmatrix} \begin{bmatrix} 5/6 & 10/21 \end{bmatrix} = \begin{bmatrix} 5/252 & 5/252 \\ 5/252 & 5/98 \end{bmatrix}$$

The elements σ_1^2, c, and σ_2^2 correspond to the numerical values computed in the earlier Example (5.6).

$\square$

5.7.3 Multivariate Gaussian density function

The multidimensional or *multivariate* Gaussian density function for an n-dimensional random vector X is defined by

$$f_X(x) = \frac{1}{(2\pi)^{\frac{n}{2}} |C_X|^{\frac{1}{2}}} \exp -\tfrac{1}{2}(x - m_X)^T C_X^{-1}(x - m_X) \qquad (5.76)$$

In the 2-dimensional case m_X and C_X are given by

$$m_X = \begin{bmatrix} m_1 \\ m_2 \end{bmatrix} \qquad C_X = \begin{bmatrix} \sigma_1^2 & c_{12} \\ c_{21} & \sigma_2^2 \end{bmatrix} = \begin{bmatrix} \sigma_1^2 & \rho\sigma_1\sigma_2 \\ \rho\sigma_1\sigma_2 & \sigma_2^2 \end{bmatrix}$$

where ρ is the correlation coefficient defined by (5.29). In this case the explicit inverse of the matrix can be found as

$$C_X^{-1} = \frac{1}{1 - \rho^2} \begin{bmatrix} \dfrac{1}{\sigma_1^2} & -\dfrac{\rho}{\sigma_1\sigma_2} \\ -\dfrac{\rho}{\sigma_1\sigma_2} & \dfrac{1}{\sigma_2^2} \end{bmatrix}$$

while the determinant is given by

$$|C_X| = \sigma_1^2\sigma_2^2(1 - \rho^2)$$

Substituting these last two equations in (5.76) yields the bivariate density function for two jointly Gaussian random variables given as equation (5.35).

In Section 5.4 it is shown that the marginal and conditional densities derived from the bivariate density are all Gaussian. This property also extends to the multivariate Gaussian density with $n > 2$. In particular, any set of components of the random vector has a jointly Gaussian PDF, and any set of components conditioned on any other set is also jointly Gaussian. The multivariate Gaussian density is especially important for problems in signal processing and communications. A more extensive treatment including the case of complex random variables can be found in [3].

5.7.4 Transformations of random vectors

Transformations of moments

Translations and linear transformations have easily formulable effects on the first and second order moments of random vectors because the order of the moments is preserved. The results of nonlinear transformations are not easy to formulate because first and second order moments of the new random variable generally require higher order moments of the original random variable. Our discussion here is thus restricted to translations and *linear* transformations.

When a random vector is translated by adding a constant, this has the effect of adding the constant to the mean. Consider

$$Y = X + b \qquad (5.77)$$

where $\mathbf{b}$ is a constant vector. Then taking the expectation yields

$$\mathcal{E}\{Y\} = \mathcal{E}\{X\} + \mathcal{E}\{\mathbf{b}\}$$

or

$$\boxed{\mathbf{m}_Y = \mathbf{m}_X + \mathbf{b}} \qquad (5.78)$$

It will be seen shortly, that while a translation has an effect on the correlation matrix, it has *no effect on the covariance matrix* because the covariance matrix is defined by subtracting the mean.

Consider now a linear transformation of the form

$$Y = AX \qquad (5.79)$$

where $\mathbf{A}$ is an $m \times n$ matrix.[10] The mean of Y is easily found as

$$\mathcal{E}\{Y\} = \mathcal{E}\{AX\} = A\mathcal{E}\{X\}$$

or

$$\boxed{\mathbf{m}_Y = \mathbf{A}\mathbf{m}_X} \qquad (5.80)$$

The correlation matrix of Y can be found from the definition

$$\mathbf{R}_Y = \mathcal{E}\{YY^T\} = \mathcal{E}\{(AX)(AX)^T\} = A\mathcal{E}\{XX^T\}A^T$$

or

$$\boxed{\mathbf{R}_Y = \mathbf{A}\mathbf{R}_X\mathbf{A}^T} \qquad (5.81)$$

It can easily be shown that the covariance matrix satisfies an identical relation (see Prob. 5.37). Note that the matrix A in this transformation does not need to be invertible or does not even need to be a square matrix. In other words, the linear transformation may result in a compression or expansion of dimensionality. The results are valid in any case.

The complete set of first and second order moments under a combination of linear transformation and translation is listed for convenience in Table 5.4. Notice that the

transformation	$Y = AX + b$
mean vector	$\mathbf{m}_Y = A\mathbf{m}_X + b$
correlation matrix	$\mathbf{R}_Y = A\mathbf{R}_X A^T + A\mathbf{m}_X b^T + b\mathbf{m}_X^T A^T + bb^T$
covariance matrix	$\mathbf{C}_Y = A\mathbf{C}_X A^T$

Table 5.4 Transformation of moments.

formula for the correlation matrix is quite complicated when $\mathbf{b} \neq \mathbf{0}$ and $\mathbf{m}_X \neq \mathbf{0}$ while the expression for the covariance matrix is much simpler. This is because for the covariance you eliminate the mean value before taking the expectation (see (5.74)).

---

[10] We assume here that the dimension m of Y is not necessarily the same as the dimension n of X.

Tranformation of the density function

Let us now consider a more general transformation of a random vector of the form[11]

$$Y = g(X) \tag{5.82}$$

The function $g(\cdot)$ may be nonlinear, but it is assumed to be invertible, i.e., X can be written as

$$X = g^{-1}(Y) \tag{5.83}$$

The relation between the densities f_X and f_Y is derived following essentially the same procedure used as when dealing with single random variables (see Chapter 3). The probability that X is in a small rectangular region of the $\mathcal{X}$ space is given by (5.66). Let us denote the volume of this small region by

$$\Delta V_X = \Delta_1 \Delta_2 \dots \Delta_n$$

and let ΔV_Y represent the volume of the corresponding region in the $\mathcal{Y}$ space (see Fig. 5.13). Note that this region need not be a rectangular region. Since the probability

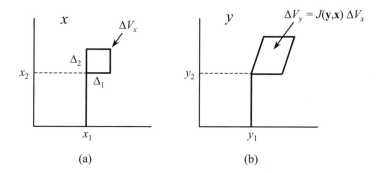

(a)   (b)

Figure 5.13 Mapping of small region of $\mathcal{X}$ space to $\mathcal{Y}$ space. (a) Small region of $\mathcal{X}$ space. (b) Corresponding region in $\mathcal{Y}$ space.

that X is in the small region of volume ΔV_X is the same as the probability that Y is in the small image region of volume ΔV_Y, the two probabilities can be equated:

$$f_X(x)\Delta V_X = f_Y(y)\Delta V_Y \tag{5.84}$$

The volume of the small region in the $\mathcal{Y}$ space is given by (see, e.g., [8])

$$\Delta V_Y = J(y, x)\Delta V_X \tag{5.85}$$

---

[11] The notation in this equation for a vector-valued function is shorthand for

$$\begin{bmatrix} Y_1 \\ Y_2 \\ \vdots \\ Y_n \end{bmatrix} = \begin{bmatrix} g_1(X) \\ g_2(X) \\ \vdots \\ g_n(X) \end{bmatrix} = \begin{bmatrix} g_1(X_1, X_2, \dots, X_n) \\ g_2(X_1, X_2, \dots, X_n) \\ \vdots \\ g_n(X_1, X_2, \dots, X_n) \end{bmatrix}.$$

The g_i are the component functions of the vector valued function g, and it is assumed that the dimension of Y is the same as that of X.

where $J(y, x)$ is the Jacobian of the transformation:

$$J(y, x) \overset{\text{def}}{=} \text{abs} \begin{vmatrix} \frac{\partial g_1(x)}{\partial x_1} & \frac{\partial g_2(x)}{\partial x_1} & \cdots & \frac{\partial g_n(x)}{\partial x_1} \\ \frac{\partial g_1(x)}{\partial x_2} & \frac{\partial g_2(x)}{\partial x_2} & \cdots & \frac{\partial g_n(x)}{\partial x_2} \\ \vdots & \vdots & & \vdots \\ \frac{\partial g_1(x)}{\partial x_n} & \frac{\partial g_2(x)}{\partial x_n} & \cdots & \frac{\partial g_n(x)}{\partial x_n} \end{vmatrix} \tag{5.86}$$

and 'abs' denotes the absolute value of the determinant. Substituting (5.85) in (5.84) yields

$$f_X(x)\Delta V_X = f_Y(y)J(y, x)\Delta V_X$$

Then cancelling the term ΔV_X and applying (5.83) produces the desired relation

$$\boxed{f_Y(y) = \frac{1}{J(y, x)} f_X(x)\bigg|_{x = g^{-1}(y)}} \tag{5.87}$$

In the special case when g is a nonsingular transformation, in the form

$$Y = AX + b \tag{5.88}$$

(as in Table 5.4), the Jacobian is equal to the absolute value of the determinant of A. Therefore the transformed density function is given by

$$f_Y(y) = \frac{1}{\text{abs}|A|} f_X(A^{-1}(y - b)) \tag{5.89}$$

where A^{-1} is the matrix inverse.

Example 5.12: Consider an invertible transformation of the form (5.88) where X has the Gaussian PDF given by (5.76). Applying the formula (5.89) yields

$$\begin{aligned} f_Y(y) &= \frac{1}{\text{abs}|A|} \frac{1}{(2\pi)^{\frac{n}{2}}|C_X|^{\frac{1}{2}}} \exp -\tfrac{1}{2}(A^{-1}(y - b) - m_X)^T C_X^{-1}(A^{-1}(y - b) - m_X) \\ &= \frac{1}{(2\pi)^{\frac{n}{2}}[|A|^2|C_X|]^{\frac{1}{2}}} \exp -\tfrac{1}{2}(y - b - Am_X)^T (A^{-1})^T C_X^{-1} A^{-1}(y - b - Am_X) \end{aligned}$$

This appears to have the form of a multivariate Gaussian PDF. Let us investigate further.

From Table 5.4, the mean and covariance of Y are given by

$$m_Y = Am_X + b \quad \text{and} \quad C_Y = AC_X A^T \tag{2}$$

The inverse covariance matrix is thus

$$C_Y^{-1} = (AC_X A^T)^{-1} = (A^T)^{-1} C_X^{-1} A^{-1} = (A^{-1})^T C_X^{-1} A^{-1}$$

while the determinant is

$$|C_Y| = |AC_X A^T| = |A||C_X||A^T| = |A|^2|C_X|$$

By comparing these results with the terms in (1) you can verify that $f_Y(y)$ in fact has the Gaussian form

$$f_Y(y) = \frac{1}{(2\pi)^{\frac{n}{2}}|C_Y|^{\frac{1}{2}}} \exp -\tfrac{1}{2}(y - m_Y)^T C_Y^{-1}(y - m_Y)$$

with parameters m_Y and C_Y given in (2).

$\square$

This example proves that (at least when the linear transformation is invertible) a linear transformation and/or translation of a Gaussian random vector is another Gaussian random vector. The result happens to be true even when the linear transformation is not invertible, but the proof is a bit more complicated. Another example illustrates a *nonlinear* transformation of a Gaussian random vector.

Example 5.13: In a communications problem the in-band noise causes an amplitude and phase modulation of the carrier. The noise signal is of the form

$$N(t) = A(t) \cos(2\pi f_c t + \Theta(t))$$

where f_c is the carrier frequency and $N(t)$ and $\Theta(t)$ represent the amplitude and phase (respectively) of the noise, which for any time 't' are random variables. If the trigonometric formula for the cosine of a sum of terms is used, the noise signal can be written in rectangular or "quadrature" form:

$$N(t) = X_1(t) \cos(2\pi f_c t) + X_2(t) \sin(2\pi f_c t)$$

where

$$X_1(t) = A(t) \cos \Theta(t) \quad \text{and} \quad X_2(t) = A(t) \sin \Theta(t)$$

Let us consider sampling this noise signal at a given time 't' and thus drop the dependence of all of the variables on time. It is known that the random variables X_1 and X_2 representing the quadrature components of the noise are typically independent zero-mean Gaussian random variables each with variance σ^2. Thus the joint density for X_1 and X_2 is given by

$$f_{X_1 X_2}(x_1, x_2) = \frac{1}{\sqrt{2\pi\sigma^2}} e^{-x_1^2/2\sigma^2} \cdot \frac{1}{\sqrt{2\pi\sigma^2}} e^{-x_2^2/2\sigma^2} = \frac{1}{2\pi\sigma^2} e^{-(x_1^2+x_2^2)/2\sigma^2} \quad (1)$$

It is desired to find the density functions for the random variables A and Θ and to determine if these random variables are independent.

We can proceed by defining the random vectors

$$\mathbf{X} = \begin{bmatrix} X_1 \\ X_2 \end{bmatrix} \qquad \mathbf{Y} = \begin{bmatrix} A \\ \Theta \end{bmatrix}$$

and using the formula (5.87) for the transformation of the joint density. The transformation $\mathbf{Y} = \mathbf{g}(\mathbf{X})$ is given by

$$\begin{bmatrix} A \\ \Theta \end{bmatrix} = \begin{bmatrix} \sqrt{X_1^2 + X_2^2} \\ \tan^{-1}(X_2/X_1) \end{bmatrix} = \begin{bmatrix} g_1(\mathbf{X}) \\ g_2(\mathbf{X}) \end{bmatrix}$$

The Jacobian of the transformation is given by

$$J(\mathbf{y}, \mathbf{x}) = \text{abs} \begin{vmatrix} \frac{\partial g_1(\mathbf{x})}{\partial x_1} & \frac{\partial g_2(\mathbf{x})}{\partial x_1} \\ \frac{\partial g_1(\mathbf{x})}{\partial x_2} & \frac{\partial g_2(\mathbf{x})}{\partial x_2} \end{vmatrix} = \text{abs} \begin{vmatrix} \frac{1}{2}\frac{2x_1}{\sqrt{x_1^2+x_2^2}} & \frac{-x_2/x_1^2}{1+x_2^2/x_1^2} \\ \frac{1}{2}\frac{2x_2}{\sqrt{x_1^2+x_2^2}} & \frac{1/x_1}{1+x_2^2/x_1^2} \end{vmatrix}$$

$$= \text{abs} \begin{vmatrix} \cos\theta & -(\sin\theta)/a \\ \sin\theta & (\cos\theta)/a \end{vmatrix} = \frac{1}{a}(\cos^2\theta + \sin^2\theta) = \frac{1}{a}$$

From (5.87) (and using (1) above) we thus have

$$f_{A\Theta}(a, \theta) = a \frac{1}{2\pi\sigma^2} e^{-a^2/2\sigma^2}$$

This is the joint density function.

The marginal density of A is then given by

$$f_A(a) = \int_{-\pi}^{\pi} f_{A\Theta}(a,\theta)d\theta = \frac{a}{2\pi\sigma^2}e^{-a^2/2\sigma^2}\int_{-\pi}^{\pi} d\theta = \frac{a}{\sigma^2}e^{-a^2/2\sigma^2}; \qquad a \geq 0$$

This PDF is known as the Rayleigh probability density with parameter $\alpha = 1/\sigma$.

The marginal density for Θ is then

$$f_\Theta(\theta) = \int_{-\pi}^{\pi} f_{A\Theta}(a,\theta)da = \frac{1}{2\pi}\int_0^\infty \frac{a}{\sigma^2}e^{-a^2/2\sigma^2}da = \frac{1}{2\pi} \qquad -\pi < \theta \leq \pi$$

Therefore the phase is *uniformly* distributed. You can observe that $f_{A\Theta}(a,\theta) = f_A(a) \cdot f_\Theta(\theta)$, so the random variables A and Θ are *independent*.

This example derives the well-known result that a signal which at any time 't' has identically-distributed zero-mean Gaussian components in a rectangular coordinate system, has independent magnitude and phase. The magnitude has a Rayleigh distribution and the phase has a uniform distribution.

□

5.8 An Application to Signal Detection

Section 3.8.1 of Chapter 3 introduced the problem of optimal detection for a signal in additive noise. The discussion there was limited to the case where the "signal" is a single random variable. A more realistic situation can now be considered that serves as the model for detection in communications systems as well as in radar and sonar.

The signal to be detected is assumed to have a known form and has been sampled and represented in discrete time. A typical such signal is shown in Fig. 5.14. It will be

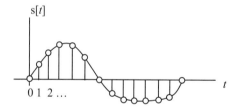

Figure 5.14 Depiction of signal to be detected in noise.

assumed that time is normalized, so t takes on the values $0, 1, 2, \ldots$ and that n samples of the signal are taken and formed into a vector

$$\mathbf{s} = \begin{bmatrix} s[0] & s[1] & \cdots & s[n-1] \end{bmatrix}^T$$

Random Gaussian noise is present, so the receiver observes a random vector $\mathbf{X}$ and must choose one of two hypotheses:

$$H_0: \quad \mathbf{X} = \mathbf{N} \qquad \text{no signal is present (just noise)}$$

$$H_1: \quad \mathbf{X} = \mathbf{s} + \mathbf{N} \quad \text{signal is present (in additive noise)}$$

where $\mathbf{N}$ is a vector of the added noise samples with density function

$$f_{\mathbf{N}}(\boldsymbol{\eta}) = \frac{1}{(2\pi)^{\frac{n}{2}}|\mathbf{C}_{\mathbf{N}}|^{\frac{1}{2}}} \exp -\tfrac{1}{2}\boldsymbol{\eta}^T\mathbf{C}_{\mathbf{N}}^{-1}\boldsymbol{\eta}$$

(Note that the noise has 0 mean.)

Under the hypothesis H_0, the PDF for the observation vector is the same as the PDF for the noise; therefore, the density function for $\boldsymbol{X}$ given H_0 is

$$f_{\boldsymbol{X}|H_0}(\boldsymbol{x}|H_0) = \frac{1}{(2\pi)^{\frac{n}{2}}|\boldsymbol{C_N}|^{\frac{1}{2}}} \exp -\tfrac{1}{2}\boldsymbol{x}^T\boldsymbol{C_N}^{-1}\boldsymbol{x} \tag{5.90}$$

where $\boldsymbol{C_N}$ is the covariance matrix for the noise. Under hypothesis H_1, the observation vector consists of noise which has undergone a translation due to presence of the signal. Therefore the conditional density function has a shift in the mean from 0 to $\mathbf{s}$:

$$f_{\boldsymbol{X}|H_1}(\boldsymbol{x}|H_1) = \frac{1}{(2\pi)^{\frac{n}{2}}|\boldsymbol{C_N}|^{\frac{1}{2}}} \exp -\tfrac{1}{2}(\boldsymbol{x}-\mathbf{s})^T\boldsymbol{C_N}^{-1}(\boldsymbol{x}-\mathbf{s}) \tag{5.91}$$

As discussed in Section 3.8.1, the posterior probabilities can be written as

$$\Pr[H_i|\boldsymbol{x}] = \frac{f_{\boldsymbol{X}|H_i}(\boldsymbol{x}|H_i)\Pr[H_i]}{f_{\boldsymbol{X}}(\boldsymbol{x})} ; \qquad i = 0, 1 \tag{5.92}$$

where $\Pr[H_i]$ are the prior probabilities and $f_{\boldsymbol{X}}(\boldsymbol{x})$ is the unconditional density of the observations. Recall then, the procedure that minimizes the probability of error is to choose H_0 or H_1 according to

$$\Pr[H_1|\boldsymbol{x}] \underset{H_0}{\overset{H_1}{\gtrless}} \Pr[H_0|\boldsymbol{x}]$$

Upon substitution of (5.92) into this equation, cancelling the common term $f_{\boldsymbol{X}}(\boldsymbol{x})$, and rearranging, we arrive at the *likelihood ratio test*

$$\boxed{\frac{f_{\boldsymbol{X}|H_1}(\boldsymbol{x}|H_1)}{f_{\boldsymbol{X}|H_0}(\boldsymbol{x}|H_0)} \underset{H_0}{\overset{H_1}{\gtrless}} \frac{P_0}{P_1}} \tag{5.93}$$

which represents the optimal decision rule.

As in the case of a single random variable for the observation, the decision rule (5.93) can be considerably simplified algebraically. First, by substituting (5.90) and (5.91) in (5.93), cancelling common terms and taking logarithms, we can write

$$\ln\frac{f_{\boldsymbol{X}|H_1}(\boldsymbol{x}|H_1)}{f_{\boldsymbol{X}|H_0}(\boldsymbol{x}|H_0)} = -\tfrac{1}{2}(\boldsymbol{x}-\mathbf{s})^T\boldsymbol{C_N}^{-1}(\boldsymbol{x}-\mathbf{s}) + \tfrac{1}{2}\boldsymbol{x}^T\boldsymbol{C_N}^{-1}\boldsymbol{x} \underset{H_0}{\overset{H_1}{\gtrless}} \ln\frac{P_0}{P_1}$$

Then by expanding the first term and simplifying, the decision rule can be put in the form

$$\mathbf{w}^T\boldsymbol{x} \underset{H_0}{\overset{H_1}{\gtrless}} \tau \tag{5.94}$$

where

$$\mathbf{w} = \boldsymbol{C_N}^{-1}\mathbf{s}$$
$$\tau = \tfrac{1}{2}\mathbf{s}^T\boldsymbol{C_N}^{-1}\mathbf{s} + \ln\frac{P_0}{P_1} \tag{5.95}$$

It can be seen that the optimal decision rule forms an inner product of the observation vector with some weight vector $\mathbf{w}$ and compares the result to a scalar threshold τ. If the threshold is exceeded, then the receiver declares that a signal is present.

An important special case occurs when the samples of the noise are uncorrelated.[12]

---

[12] This is the so-called "white noise" case.

The noise covariance matrix then has the form

$$
\mathbf{C_N} = \begin{bmatrix} \sigma_o^2 & 0 & \cdots & 0 \\ 0 & \sigma_o^2 & \cdots & 0 \\ \vdots & \vdots & & \vdots \\ 0 & 0 & \cdots & \sigma_o^2 \end{bmatrix} = \sigma_o^2 \mathbf{I}
$$

where σ_o^2 is the variance of the noise. In this case the parameters (5.95) in the optimal receiver reduce to

$$
\mathbf{w} = \frac{1}{\sigma_o^2}\mathbf{s}
$$

$$
\tau = \frac{|\mathbf{s}|^2}{2\sigma_o^2} + \ln\frac{P_0}{P_1}
$$

so the weight vector is proportional to the signal vector. If the prior probabilities of the two hypotheses are equal, then the logarithmic term drops out, and the decision rule can be written as

$$
\mathbf{s}^T \mathbf{x} \underset{H_0}{\overset{H_1}{\gtrless}} \tfrac{1}{2}|\mathbf{s}|^2 \tag{5.96}
$$

Thus the observation vector is compared to the signal vector (by forming an inner product) and the result is compared to a threshold, which represents half of the total signal energy. Note that under these conditions the optimal decision rule is independent of the noise level σ_o^2. In this form the receiver is sometimes referred to as a "correlation receiver" because the inner product on the left is analogous to performing a correlation between the known signal and the received observation.

An especially convenient realization for the receiver is to build a filter whose impulse response is the signal turned around (reversed) in time. This realization is known as a "matched filter" and is shown in Fig. 5.15. The convolution of the reversed signal

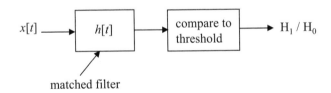

matched filter

Figure 5.15 Matched filter realization of the optimal detector.

and the input performed by the matched filter implements the inner product in (5.96) which is then compared to the threshold.

The matched filter implementation has one further advantage. The expected time of arrival of the signal does not need to be known in order for the inner product to be computed. The filter processess the observation sequence continuously. When a signal arrives and causes the output to exceed the threshold, a detection is announced. Further the estimated signal time of arrival is given by the time at which the matched filter has the *largest* output. This principle is used in all radar and sonar systems.

5.9 *Summary*

Many problems in electrical and computer engineering involve sets of two or more random variables. A probabilistic description of these multiple random variables requires consideration of the joint PMF, the joint PDF, or joint CDF. If the random variables happen to be independent, then the joint distribution is the product of the marginals. Although the marginal PMFs or PDFs can always be derived from the joint PDF by summing or integrating, the joint PMF or PDF cannot always be derived from just the marginals. A joint PDF can be derived from marginals and conditional PDFs, however, e.g., $f_{X_1 X_2} = f_{X_1|X_2} f_{X_2}$. Conditional PDFs such as $f_{X_1|X_2}$ represent a density function for X_1 with X_2 acting as a parameter. Similar statements apply to the PMF of discrete random variables.

The extension of expectation to two or more random variables is straightforward and involves summation or integration with the joint density function. An important new concept however is that of *correlation* between two random variables. Correlation, covariance, and the correlation coefficent ρ are all related quantities that describe probabilistic "similarity" between a pair of random variables. The PDF of Gaussian random variables is completely described by mean, variance, and covariance parameters.

Sums of multiple random variables also occur frequently in practical applications. Important results exist for the mean, variance, and PDF of sums of random variables. Four new standard distributions that apply to the sums of random variables are the binomial PMF, the Pascal PMF, the Erlang PDF, and the Chi-squared PDF.

Sets of related random variables, such as samples of a random signal, are frequently best treated as random vectors. With this formulation, well-known results from matrix algebra can be applied to describe the random variables and help in their analysis. An introduction to methods of random vectors is provided in this chapter. These methods allow us to address the application of signal detection, introduced earlier in Chapter 3, in a more realistic context where multiple samples of the received signal are available.

References

[1] Athanasios Papoulis and S. Unnikrishna Pillai. *Probability, Random Variables, and Stochastic Processes.* McGraw-Hill, New York, fourth edition, 2002.

[2] R. S. Cowan. *Sir Francis Galton and the Study of Heredity in the Nineteenth Century.* Garland, New York, 1985.

[3] Charles W. Therrien. *Discrete Random Signals and Statistical Signal Processing.* Prentice Hall, Inc., Upper Saddle River, New Jersey, 1992.

[4] William Feller. *An Introduction to Probability Theory and Its Applications - Volume I.* John Wiley & Sons, New York, second edition, 1957.

[5] Harold J. Larson and Bruno O. Shubert. *Probabilistic Models in Engineering Sciences - Volume I.* John Wiley & Sons, New York, 1979.

[6] Alvin W. Drake. *Fundamentals of Applied Probability Theory.* McGraw-Hill, New York, 1967.

[7] Alberto Leon-Garcia. *Probability and Random Processes for Electrical Engineering.* Addison-Wesley, Reading, Massachusetts, second edition, 1994.

[8] Wilfred Kaplan. *Advanced Calculus.* Addison-Wesley, Reading, Massachusetts, fifth edition, 2003.

[9] Gilbert Strang. *Introduction to Applied Mathematics.* Wellesley Cambridge Press, Wellesley, MA, 1986.

Problems

Two discrete random variables

5.1   Given a joint discrete PDF

$$f_{I_1 I_2}[i_1, i_2] = \begin{cases} Cp(1-p)^{i_1} & 0 \le i_1 \le 9, \ 0 \le i_2 \le 9 \\ 0 & \text{otherwise,} \end{cases}$$

Let $p = 1/8$.

(a) Find C to make this a valid PMF.

(b) Find the marginal density functions of I_1 and I_2.

(c) Determine the CDF.

5.2   A joint PMF is defined by

$$f_{K_1 K_2}[k_1, k_2] = \begin{cases} C & 0 \le k_1 \le 5 \quad 0 \le k_2 \le k_1 \\ 0 & \text{otherwise} \end{cases}$$

(a) Find the constant C.

(b) Compute and *sketch* the marginal PMFs $f_{K_1}[k_1]$ and $f_{K_2}[k_2]$. Are the random variables K_1 and K_2 independent?

(c) Compute the conditional PMFs $f_{K_1 | K_2}[k_1 | k_2]$ and $f_{K_2 | K_1}[k_2 | k_1]$.

5.3   A thesis student measures some data K, which depends on setting a parameter α in the experiment. Unfortunately the student forgets to record the setting of the parameter α; but he knows that it was set to either 0.4 or 0.6 and that these values are equally likely. The data is described by the PMF

$$f_{K | \alpha}[k | \alpha] = \begin{cases} \alpha(1-\alpha)^{k-1} & k \ge 1 \\ 0 & \text{otherwise.} \end{cases}$$

From the advisor's point of view, what is the PDF for data, $f_K[k]$?

5.4   By using using the definitions (5.3) and (5.6) for marginal and conditional densities, show that (5.7) is true.

Two continuous random variables

5.5   The joint PDF $f_{X_1, X_2}(x_1, x_2)$ of two random variables X_1 and X_2 is shown below.

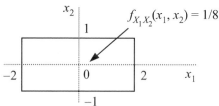

(a) Find the marginal PDFs.

(b) Determine the following probabilities:

(i) $\Pr[0 \le X_1 \le 1, -0.5 \le X_2 \le 2]$

(ii) $\Pr[X_1 < X_2]$

(iii) $\Pr[X_1^2 + X_2^2 \le 1]$

5.6 The joint PDF for two random variables $f_{X_1 X_2}(x_1, x_2)$ is equal to a constant C for X_1 and X_2 in the region

$$0 \le X_1 \le 1 \quad \text{and} \quad 0 \le X_2 \le 2(1 - X_1)$$

and $f_{X_1 X_2}(x_1, x_2)$ is zero outside of this region.

(a) Sketch the region in the X_1, X_2 plane where the PDF is nonzero.

(b) What is the constant C?

(c) Compute the *marginal* density functions $f_{X_1}(x_1)$ and $f_{X_2}(x_2)$.

(d) Compute the *conditional* density functions $f_{X_1 | X_2}(x_1 | x_2)$ and $f_{X_2 | X_1}(x_2 | x_1)$.

(e) Are the two random variables statistically independent?

5.7 (a) Compute the joint CDF $F_{X_1 X_2}(x_1, x_2)$ corresponding to the following joint PDF

$$f_{X_1 X_2}(x_1, x_2) = \begin{cases} e^{-(x_1 + x_2)} & 0 \le x_1 < \infty \quad 0 \le x_2 < \infty \\ 0 & \text{otherwise} \end{cases}$$

(b) The marginal cumulative distributions for the random variables are defined by $F_{X_i}(x_i) = \Pr[X_i \le x_i]$. Find the marginal CDFs $F_{X_1}(x_1)$ and $F_{X_2}(x_2)$ for this joint density.

(c) Find the marginal PDFs $f_{X_1}(x_1)$ and $f_{X_2}(x_2)$ by differentiating the marginal CDFs.

(d) Are the random variables X_1 and X_2 independent?

5.8 Consider the PDF given in Problem 5.5.

(a) Determine the CDF.

(b) From the CDF, compute $\Pr[X_1 > 1/2, X_2 < 1/2]$.

5.9 The joint PDF of two random variables is given by

$$f_{X_1 X_2}(x_1, x_2) = \begin{cases} C(3 - x_1 x_2) & 0 \le x_1 \le 1; \ 0 \le x_2 \le 3 \\ 0 & \text{otherwise} \end{cases}$$

(a) Find C to make it a valid PDF.

(b) Find the marginal density functions.

(c) Determine the CDF $F_{X_1 X_2}(x_1, x_2)$.

5.10 Refer to the joint PDF $f_{X_1 X_2}(x_1, x_2)$ shown in Prob. 5.5.

(a) Compute the conditional PDF $f_{X_2 | X_1}(x_2 | x_1)$.

(b) Are X_1 and X_2 independent?

5.11 Given the joint PDF of two random variables X_1 and X_2:

$$f_{X_1 X_2}(x_1, x_2) = \begin{cases} x_1 e^{-x_1(C + x_2)} & x_1 > 0, x_2 > 0 \\ 0 & \text{otherwise,} \end{cases}$$

determine the following:

(a) The constant C.

(b) The marginal PDFs $f_{X_1}(x_1)$ and $f_{X_2}(x_2)$.

(c) If X_1 and X_2 are independent.

(d) The conditional PDF $f_{X_1 | X_2}(x_1 | x_2)$.

(e) The conditional PDF $f_{X_2|X_1}(x_2|x_1)$.

5.12 Given a joint PDF

$$f_{X_1X_2}(x_1, x_2) = \begin{cases} x_1 x_2 + C & x_1 + x_2 \leq 1, \ x_1 \geq 0, \ x_2 \geq 0 \\ 0 & \text{otherwise.} \end{cases}$$

(a) Find C.
(b) Compute the marginal PDFs: $f_{X_1}(x_1)$ and $f_{X_2}(x_2)$.
(c) Are the random variables independent?
(d) Determine the conditional PDF $f_{X_1|X_2}(x_1|x_2)$.

5.13 The probability that the first message arrives at a certain network node at an absolute time X given that the node comes online at an absolute time Y is given by the conditional density

$$f_{X|Y}(x|y) = \begin{cases} e^{-(x-y)} & 0 \leq y \leq x < \infty \\ 0 & x < y \end{cases}$$

(a) Suppose that it is known that the node comes online at time $Y = 0.5$, what is the (approximate) probability that the first message arrives between $X = 2$ and $X = 2.01$? Hint: 0.01 is a "small increment."
(b) If the probability that the node comes online at time Y is uniformly distributed between 0 and 1, what is the joint PDF $f_{XY}(x, y)$? Specify the region where the density is nonzero and sketch this region in the xy plane.
(c) What is the unconditional density $f_X(x)$? Be sure to specify the region where the density is nonzero.

5.14 The joint PDF for random variables X_1 and X_2 is given by

$$f_{X_1,X_2}(x_1, x_2) = \begin{cases} 1/4 & \text{if } |x_1| + 2|x_2| \leq 2 \\ 0 & \text{otherwise} \end{cases}$$

(a) Sketch the region of the x_1, x_2 plane where the joint density is nonzero.
(b) Find and sketch the marginal density $f_{X_1}(x_1)$.
(c) Find and sketch the conditional density $f_{X_1|X_2}(x_1|x_2)$.

Expectation and correlation

5.15 The joint PDF for two random variables $f_{X_1X_2}(x_1, x_2)$ is given by

$$f_{X_1X_2}(x_1, x_2) = \begin{cases} 1 & 0 \leq x_1 \leq 1 \quad \text{and} \quad 0 \leq x_2 \leq 2(1 - x_1) \\ 0 & \text{otherwise} \end{cases}$$

Sketch the region where the density is nonzero and compute the following quantities:

(a) $m_{X_1} = E\{X_1\}$
(b) $m_{X_2} = E\{X_2\}$
(c) $E\{X_1^2\}$
(d) $E\{X_2^2\}$
(e) $E\{X_1X_2\}$

(f) $\sigma_{X_1}^2 = \text{Var}[X_1]$
(g) $\sigma_{X_2}^2 = \text{Var}[X_2]$
(h) $c_{X_1X_2} = \text{Cov}[X_1, X_2]$
(i) $E\{X_1|X_2\}$
(j) $E\{X_2|X_1\}$

Hint: You may want to use some of the results from Prob. 5.6 to help in some of these computations.

5.16 Random variables X and Y are independent and described by the density functions shown below:

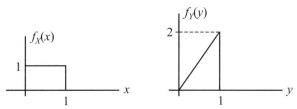

(a) Give an algebraic expression for the joint density $f_{XY}(x,y)$ (including limits) and sketch the region where it is nonzero below.

(b) What is $\Pr[X \leq Y]$?

(c) What is the variance of Y?

(d) What is the correlation coefficient ρ_{XY}?

5.17 The Schwartz inequality [9] for functions of two variables can be written as

$$\left(\int_{-\infty}^{\infty} \int_{-\infty}^{\infty} h(x_1, x_2) g(x_1, x_2) dx_1 dx_2 \right)^2 \leq$$

$$\left(\int_{-\infty}^{\infty} \int_{-\infty}^{\infty} h^2(x_1, x_2) dx_1 dx_2 \right) \left(\int_{-\infty}^{\infty} \int_{-\infty}^{\infty} g^2(x_1, x_2) dx_1 dx_2 \right)$$

where h and g are any two square integrable functions. The inequality holds with *equality* when h and g are linearly related, i.e., when $h(x,y) = ag(x,y)$ where a is a constant.

Apply the Schwartz inequality to the definition

$$c = \text{Cov}\,[X_1, X_2] = \int_{-\infty}^{\infty} \int_{-\infty}^{\infty} (x_1 - m_1)(x_2 - m_2) f_{X_1 X_2}(x_1, x_2) dx_1 dx_2$$

for the covariance of two random variables to show that

$$c^2 \leq \sigma_1^2 \sigma_2^2$$

Use this to prove (5.30). Under what condition is ρ *equal* to ± 1?

5.18 Two random variables X_1 and X_2 are described by the joint density function shown below.

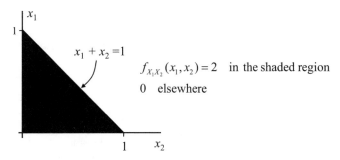

(a) Find and sketch $f_{X_2}(x_2)$.

(b) Find and sketch $f_{X_1|X_2}(x_1|x_2)$.

(d) Determine $\mathcal{E}\{X_1|X_2\}$.

Gaussian random variables

5.19 Two zero-mean unit-variance random variables X_1 and X_2 are described by the joint Gaussian PDF

$$f_{X_1X_2}(x_1, x_2) = \frac{1}{\pi\sqrt{3}}e^{-(x_1^2 - x_1x_2 + x_2^2)2/3}$$

(a) Show by integration that the marginal density for X_1 is Gaussian and is equal to

$$f_{X_1}(x_1) = \frac{1}{\sqrt{2\pi}}e^{-x_1^2/2}$$

Hint: Use the technique of "completing the square" to write

$$x_1^2 - x_1x_2 + x_2^2 = \tfrac{3}{4}x_1^2 + \tfrac{1}{4}x_1^2 - x_1x_2 + x_2^2 = \tfrac{3}{4}x_1^2 + (x_2 - \tfrac{1}{2}x_1)^2$$

Then form the definite integral over x_2 and make the substitution of variables $u = x_2 - \tfrac{1}{2}x_1$. Use the fact that the one-dimensional Gaussian form integrates to 1 in order to evaluate the integral.

(b) What is the conditional density $f_{X_2|X_1}(x_2|x_1)$? Observe that it has the form of a Gaussian density for X_2 but that the mean is a function of x_1. What *are* the mean and variance of this conditional density?

5.20 Two zero-mean unit-variance random variables X_1 and X_2 are described by the joint Gaussian PDF

$$f_{X_1X_2}(x_1, x_2) = \frac{1}{2\pi\sqrt{1-\rho^2}}e^{-(x_1^2 - 2\rho x_1 x_2 + x_2^2)/2(1-\rho^2)}$$

Two new random variables Y_1 and Y_2 are defined through the transformation

$$\begin{bmatrix} Y_1 \\ Y_2 \end{bmatrix} = \begin{bmatrix} 1 & 0 \\ a & 1 \end{bmatrix}\begin{bmatrix} X_1 \\ X_2 \end{bmatrix}$$

(a) Find the constant the Jacobian $J(\mathbf{y})$ of the transformation.
(b) Solve for X_1 and X_2 in terms of Y_1 and Y_2 and find $f_{Y_1Y_2}(y_1, y_2)$.
(c) Let $\rho = 1/2$. For what value of a are the random variables Y_1 and Y_2 independent? What then are the variances of Y_1 and Y_2?

5.21 Consider the random variables described in Prob. 5.20. For a general value of ρ with $(-1 \le \rho \le +1)$ find a in terms of ρ so that Y_1 and Y_2 are independent. What are the variances of Y_1 and Y_2 in terms of ρ?

5.22 Determine the joint PDF of jointly Gaussian random variables for the following cases.

(a) $m_1 = m_2 = 0$, $\sigma_1 = \sigma_2 = 2$, and $\rho = 0.8$.
(b) $m_1 = m_2 = 0$, $\sigma_1 = \sigma_2 = s$, and $\rho = 0$.
(c) $m_1 = m_2 = 1$, $\sigma_1 = \sigma_2 = 3$, and $\rho = 0$.

5.23 By integrating (5.35) over one variable, show that the marginal densities for jointly Gaussian random variables are given by (5.36). Also show that the conditional densities are Gaussian and have the form (5.37).

Multiple random variables

5.24 Let K be the discrete random variable representing the number rolled on a pair or dice. Compute the mean and variance of K using the fact that K is the sum of two random variables. Hint: The easy way to do this problem is to use the results of Prob. 4.8 (Chapter 4).

5.25 Stations A and B are connected by two parallel message channels which operate *independently*. A message is sent from A to B *over both channels at the same time*. Random variables X and Y, which are independent, represent the delays over these two channels. Each of these delays is an exponential random variable with mean of 1 hour.

A message is considered "received" as soon as it arrives on *any one* channel; the message is considered "verified" as soon as it has arrived over *both* channels.

(In order to answer the following questions, you may find it useful to draw the continuous sample space involving the two random variables X and Y and depict the events in the sample space.)

(a) What is the probability that a message is *received* within 15 minutes (1/4 hour) after it is sent?

(b) What is the probability that a message is *received and verified* within 15 minutes after it is sent?

(c) What is the probability that a message is verified within 15 minutes after it is received?

(d) Given that a message is received within 15 minutes after it is sent, what is the probability that it is verified within 15 minutes after it was sent?

[This problem is a modified version of a problem from Alvin W. Drake, *Fundamentals of Applied Probability Theory*, McGraw-Hill, New York, 1967. Reproduced by permission.]

5.26 Consider the situation in Prob. 5.25, but assume the delays X and Y are each *uniformly* distributed between 0 and 1 hour. Answer the following questions.

(a) What is the probability that a message is received within 15 minutes after it has been sent?

(b) What is the probability that a message is received but not verified within 15 minutes after it has been sent?

(c) Assume that a message is received on one channel and the chief in the communications room at point B goes on a 15 minute coffee break immediately following receipt of the message. What is the probability that the message will not be verified (i.e., that he will miss the second message)?

5.27 Random variables X, Y, and Z are described by the joint density function

$$f_{XYZ}(x, y, z) = \begin{cases} (x + 3y)z & \text{for } 0 \le x \le 1; \ 0 \le y \le 1; \ 0 \le z \le 1 \\ 0 & \text{otherwise} \end{cases}$$

Determine the following:

(a) The joint density for X and Y (without Z).

(b) The marginal density for Z.

(c) The conditional density for X and Y given Z.

(d) The probability that $X < 3$.

(e) The expected value of Z.

(f) Whether or not X and Y are independent.

5.28 K_1 and K_2 are two independent, identically-distributed (IID) Poisson random variables. Each PMF is given by

$$f_{K_i}[k] = \begin{cases} \dfrac{\alpha^k}{k!} e^{-\alpha} & k \geq 0 \\ 0 & \text{otherwise} \end{cases}$$

A new random variable J is equal to the sum $J = K_1 + K_2$.

(a) Show that the PGF for the random variable K_i is given by

$$G_{K_i}(z) = e^{\alpha(z-1)}$$

(b) What is the PGF $G_J(z)$ for the random variable J?

(c) What is the PMF $f_J[j]$? What form does it take?

(d) Find the mean and variance of J.

5.29 Two random variables X_1 and X_2 are independent and each is uniformly distributed in the interval $0 \leq X_i \leq 1$.

(a) What is the PDF for the random variable $Y = X_1 + X_2$?

(b) Find the MGF $M_Y(s)$ for Y. Hint: What is the MGF for X_i? What happens to the MGFs when you convolve the densities?

(c) If a message must wait in two successive queues for service and X_i is the time in seconds for service in the ith queue, then Y is the total service time. What is the probability that the total service time is longer than 1.5 seconds?

5.30 In this problem you will show that the sum of two or more independent Gaussian random variables is Gaussian. Let

$$Y = X_1 + X_2 + \cdots + X_n$$

where the X_i are *independent* Gaussian random variables with means m_i and variances σ_i^2. The density for Y is obtained from the n-fold convolution of the individual Gaussian density functions (see (5.53)).

(a) What is the MGF for the ith random variable X_i? Hint: See Prob. 4.22.

(b) Since convolution of densities corresponds to multiplication of the MGFs, find the MGF for Y by multiplying the MGFs from part (a).

(c) Notice that the MGF for Y has the form of a Gaussian MGF. Thus Y must be Gaussian. What are the mean and variance of Y?

5.31 The sum of two independent random variables X_1 and X_2 is given by:

$$X = X_1 + X_2$$

where X_1 is a uniform random variable in the range $[0, 1]$, and X_2 is a uniform random variable $[-1, 0]$.

(a) Determine the mean and the variance of X.

(b) Determine and sketch the PDF of X.

(c) Using the PDF for X obtained in (b), compute its mean and variance. On your sketch of the PDF of X, indicate the mean and the standard deviation.

Sums of common random variables

5.32 In a certain computer network, you observe that your message is fifth in line for service, i.e., there are four others ahead of you. The processing time for each message is described by an exponential random variable with expected value $1/\lambda = 2.5$ sec.

(a) What is the mean waiting time until your meassage has been processed?

(b) What is the *variance* of the random variable that represents the waiting time until completion of processing?

5.33 A sum is formed of n independent and identically-distributed (IID) random variables:
$$T = T_1 + T_2 + \cdots T_n$$
where T_i is exponentially distributed with parameter λ. Under these conditions, the sum T is an Erlang random variable.

(a) Using the fact that T is the sum of n independent exponential random variables, determine the mean and the variance of the Erlang random variable T.

(b) The MGF of an individual exponential random variable T_i is given by
$$M(s) = \frac{\lambda}{\lambda - s}.$$
What is the MGF of T?

(c) Use the result of (b) to compute the mean and the variance of T.

5.34 The average life of a set of batteries used in an electronic flash unit is 6 months. What is the probability that four sets of batteries will last longer than 2 years?

Random vectors

5.35 (a) What are the mean vector and the correlation matrix for the random vector
$$X = \begin{bmatrix} X_1 \\ X_2 \end{bmatrix}$$
where X_1 and X_2 are described in Prob. 5.15?

(b) What is the covariance matrix?

5.36 The mean vector and covariance matrix for a Gaussian random vector X are given by
$$m_X = \begin{bmatrix} 1 \\ 1 \end{bmatrix} \qquad R_X = \begin{bmatrix} 4 & -1 \\ -1 & 3 \end{bmatrix}$$

(a) Compute the covariance matrix C_X.

(b) What is the correlation coefficient $\rho_{X_1 X_2}$?

(c) Invert the covariance matrix and write an explicit expression for the Gaussian density function for X.

5.37 Using the definition (5.74) for the covariance matrix and assuming the linear transformation $Y = AX$, show that
$$C_Y = AC_X A^T$$
Repeat the proof, this time using the relation $C_Y = R_Y - m_Y m_Y^T$ together

with (5.80) and (5.81). Also show for a square matrix $\mathbf{A}$ that $|\mathbf{R}_Y| = |\mathbf{A}|^2|\mathbf{R}_X|$. (The symbol $|\cdot|$ denotes the determinant.)

5.38 The joint PDF of two random variables X_1 and X_2 is given by

$$f_{X_1X_2}(x_1, x_2) = \begin{cases} 1/12 & -1 \le x_1 \le 1, \ 3 \le x_2 \le 3 \\ 0 & \text{otherwise.} \end{cases}$$

(a) Find the first moment of the random variables in vector notation.

(b) Determine the correlation matrix.

(c) Determine the covariance matrix.

Application to signal detection

5.39 Suppose that in a signal detection problem, the signal vector consists of five samples

$$\mathbf{s} = \tfrac{1}{3}\begin{bmatrix} 1 & 2 & 3 & 2 & 1 \end{bmatrix}^T$$

The noise samples are uncorrelated with variance $\sigma_o^2 = 1$ and mean 0.

(a) Find the explicit form of the decision rule if the prior probability for the signal is $P_1 = 2/3$.

(b) How does the result change if $P_1 = 1/2$?

5.40 The signal vector in a detection problem is given by

$$\mathbf{s} = \begin{bmatrix} 2 & 2 \end{bmatrix}^T$$

The noise has 0 mean and a covariance matrix of the form

$$\mathbf{C}_N = \sigma^2 \begin{bmatrix} 1 & \rho \\ \rho & 1 \end{bmatrix}$$

where $\sigma^2 = 2$ and $\rho = 1/\sqrt{2}$.

(a) Find the explicit form of the decision rule as a function of the parameter $\eta = \ln(P_0/P_1)$.

(b) The contour for the density of the received vector when no signal is present is defined by

$$f_{\mathbf{X}|H_0}(\mathbf{x}|H_0) = f_{X_1X_2|H_0}(x_1, x_2|H_0) = \text{constant}$$

From (5.90) this is equivalent to requiring

$$\mathbf{x}^T \mathbf{C}_N^{-1} \mathbf{x} = \text{constant}$$

Carefully sketch or plot this contour in the x_1, x_2 plane. You can set the constant equal to 1. Hint: To put the equation for the contour in the simplest form, consider rotating the coordinate system by the transformation $y_1 = (x_1 + x_2)/\sqrt{2}$ and $y_2 = (x_2 - x_1)/\sqrt{2}$ (i.e., y_1 and y_2 become the new coordinate axes).

(c) Sketch or plot the contour for the density of $\mathbf{X}$ when the signal *is* present. Hint: The form of the contour is the same except for a shift in the mean vector.

(d) The decision boundary between regions in the plane where we decide H_1 and where we decide H_0 is defined by the equation

$$\mathbf{w}^T \boldsymbol{x} = \tau$$

where $\mathbf{w}$ and τ are given by (5.95). Plot the decision boundary in the x_1, x_2 plane for $\eta = 0$, $\eta = 1/2$, and $\eta = -1/2$. Observe that since the decision boundary interacts with the two Gaussian density functions, changing η provides a trade-off betwen the probability of false alarm and the probability of detection.

Computer Projects

Project 5.1

In this project you will estimate the parameters (mean and covariance) of two-dimensional Gaussian data and plot the contours of the density functions.

The files x01.dat, x02.dat, and x03.dat from the data package for this book each contain 256 realizations of two-dimensional Gaussian random vectors. The kth line of the file contains the kth realization of the random vector:

$$\mathbf{x}^{(k)} = \left[\begin{array}{cc} x_1^{(k)} & x_2^{(k)} \end{array} \right]^T$$

An *estimate* for the mean vector and covariance matrix can be made using the following formulas:

$$\widehat{\mathbf{m}}_X = \frac{1}{256} \sum_{k=1}^{256} \mathbf{x}^{(k)} \qquad \widehat{\mathbf{C}}_X = \frac{1}{256} \sum_{k=1}^{256} \mathbf{x}^{(k)} \mathbf{x}^{(k)T} - \widehat{\mathbf{m}}_X \widehat{\mathbf{m}}_X^T$$

For each of the three data sets, do the following:

1. Estimate the mean vector and covariance matrix using the above formulas. Identify the parameters m_1, m_2, σ_1, σ_2, c_{12}, and ρ and list their values in a table.

2. Make a scatter plot of the data. That is, plot the vectors as points in the x_1, x_2 plane without any connecting lines.

3. Plot the ellipse representing the contour of the Gaussian density function (see Fig. 5.10) on the same plot. Choose the constant appropriately, and observe that most of the data fall within the ellipse.

4. Tell what the shape and orientation of the ellipse indicates for each data set.

MATLAB programming notes

You may use the function 'plotcov' from the software package for this book to do the plotting. Try setting the parameter to 2 standard deviations.

Project 5.2

Computer Project 3.2 considered a detection problem where the "signal" transmitted consisted of a single value (A) observed in noise. In this project, you will consider a more realistic detection problem where the signal is an entire *sequence* of values. The signal is described by the discrete-time sequence

$$s[k] = A \cos(2\pi k/12) \quad k = 0, 1, \ldots, 11$$

where the amplitude A is to be specified.

Following the procedure in Section 5.8, this sequence is formed into a 12-dimensional vector $\mathbf{s}$ which is transmitted in Gaussian noise. The noise is assumed to have mean 0 and autocorrelation function

$$R_N[l] = 2 \cdot (0.85)^{|l|} \qquad -\infty < l < \infty$$

The covariance matrix $\mathbf{C_N}$ thus has elements $c_{ij} = R_N[|i - j|] = 2 \cdot (0.85)^{|i-j|}$. Using these facts,

1. Develop and print the explicit form of the covariance matrix and its *inverse*.
2. Find the optimal weight vector $\mathbf{w}$ and threshold τ in the optimal decision rule (5.94) assuming $P_0 = P_1 = \frac{1}{2}$. Note that the weight vector depends on the signal amplitude A.
3. Observe that if we define $t = \mathbf{w}^T \mathbf{X}$, then t is a Gaussian random variable. (Tell why.) Plot the PDF for t under each hypothesis, H_0 and H_1. Assume that $A = 2$.
4. Set the threshold τ so that $p_{FA} = 0.01$. What is the corresponding p_D assuming $A = 2$? What value of A is necessary so that $p_D \geq 0.95$?
5. For the value of A determined in the last step, plot the ROC for this problem by sweeping over suitable values of the threshold τ.

6 Inequalities, Limit Theorems, and Parameter Estimation

This chapter deals with a number of topics that on one hand seem abstract and theoretical but on the other hand are essential for the practical application of probability to real life problems. We begin with the topic of inequalities. The inequalities developed here are well known and are used to estimate or bound the probability for random variables when only partial information (e.g., mean and variance) is known for the distribution. As such they provide tools for calculating or bounding probability from limited real data and also for proving general results that make as few assumptions as possible about the properties of the random variables.

Next we move on to the topic of stochastic convergence and limit theorems. While an advanced treatment of convergence is well beyond the scope of this chapter, some appreciation of types of convergence for a sequence of random variables is of practical importance when dealing with the number n of measurments that may be taken in an engineering application. After discussing types of convergence, we develop two well known asymptotic results: the law of large numbers and the central limit theorem. The former provides theoretical assurance that in practice expectations can be replaced by averages of a sufficiently large number of random variables. The latter states the almost incredible result that any sum of n independent random variables for large enough n can be described by a *Gaussian* distribution.

The last half of the chapter deals with estimation of parameters. This problem arises when the conditions of a problem tell you that a certain type of probabilistic distribution is appropriate but you do not know the numerical values for the parameters. For example, you may know that a Poisson model is the correct model for describing "hits" on a web page but you do not know the arrival rate λ. Although parameter estimation usually falls in the realm of a course in statistics, the basic ideas presented here are useful in practical applications of probability. We first discuss properties of estimates, and illustrate these via two fundamental statistics known as the sample mean and sample variance. We then introduce the topic of an *optimal* estimate known as the maximum likelihood estimate and show that in some typical cases the maximum likelihood estimate satisfies our intuition about what should be done with the data to estimate the desired parameter. The chapter closes with description of an application to signal detection.

6.1 Inequalities

Suppose that the mean and variance for a given random variable X are known; but neither the PDF nor the CDF is known. For example, given sizeable measured data belonging to this random variable, estimated values can be computed for the mean and variance according to methods discussed later in this chapter even if the distribution is completely unknown.

Knowing only the mean and variance of a random variable, in general, we cannot determine the exact probabilities of particular events; but bounds on these probabilities can be obtained using the Markov and Chebyshev inequalities.[1]

6.1.1 Markov inequality

For a random variable X that takes only non-negative values, the Markov inequality is stated as

$$\Pr[X \geq a] \leq \frac{\mathcal{E}\{X\}}{a}, \tag{6.1}$$

where $a > 0$.

In order to prove the statement in the above equation, we start with

$$\mathcal{E}\{X\} = \int_0^\infty x\, f_X(x)\, dx = \int_0^a x\, f_X(x)\, dx + \int_a^\infty x\, f_X(x)\, dx$$

Since all integrals have finite *positive* values, by discarding the first integral on the right, we have the inequality

$$\mathcal{E}\{X\} \geq \int_a^\infty x\, f_X(x)\, dx$$

Then replacing x in the integrand by a smaller but constant value a yields

$$\mathcal{E}\{X\} \geq \int_a^\infty a\, f_X(x)\, dx = a \int_a^\infty f_X(x)\, dx = a \Pr[X \geq a].$$

which proves the result.

Let us illustrate the Markov bound with a short example.

Example 6.1:   Average occupancy of packets in the buffer of a switch in the Internet is known to be 40%. Without knowing the PDF of the buffer occupancy X, a random variable, what can you say about the probability that buffer occupancy equals or exceeds 60%?

Using the Markov inequality (6.1)

$$\Pr[X \geq 0.60] \leq \frac{\mathcal{E}\{X\}}{0.60} = \frac{0.40}{0.60} = 0.6667.$$

□

Observe that the result produced by the bound is reasonable, but not necessarily close to the correct probability since the bound covers a wide range of possible distributions. The following example compares the Markov bound to the true probability in a specific case.

Example 6.2:   A voltage measured across a certain resistor is a uniform random variable in the range [0,5] volts. The probability that the voltage measured across this resistor exceeds 4.0 volts is 0.2 (see figure following). Using the Markov inequality, the upper bound for this probability is

$$\Pr[X \geq 4.0] \leq \frac{\mathcal{E}\{X\}}{4.0} = \frac{2.5}{4.0} = 0.625,$$

---

[1] In special cases, such as when the PDF is Gaussian or exponential, the probabilities can be computed exactly. In other cases, especially when the distribution is unknown, the inequalities may be the best mathematical tools available.

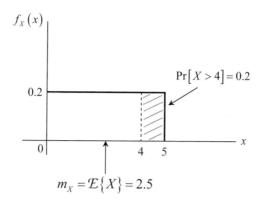

$$m_X = \mathcal{E}\{X\} = 2.5$$

which is nowhere close to the true probability! Nevertheless, the bound is correct, in that the true value of 0.2 is below the upper bound. Note that a bound takes the worst case scenario into account.

□

6.1.2 Chebyshev inequality

The Chebyshev inequality for a random variable X is given by

$$\Pr[|X - m_X| \geq a] \leq \frac{\sigma_X^2}{a^2} \tag{6.2}$$

where $a > 0$.

To prove the Chebyshev inequality, we note that $(X - m_X)^2$ is a random variable that takes on only non-negative values and apply the Markov inequality to this quantity:

$$\Pr[(X - m_X)^2 \geq a^2] \leq \frac{\mathcal{E}\{(X - m_X)^2\}}{a^2} = \frac{\sigma_X^2}{a^2}$$

By noting that $(X - m_X)^2 \geq a^2$ and $|X - m_X| \geq a$ are equivalent, the last equation is seen to be the same as (6.2).

Let us illustrate the Chebyshev bound using the random variable described in Example 6.2.

Example 6.3: The mean of the uniform variable described in Example 6.2 is $m_X = 2.5$. The variance can be computed (see Prob. 4.8) to be $\sigma_X^2 = 25/12$. Strictly speaking, we cannot compute a bound on $\Pr[X \geq 4.0]$ using the Chebyshev inequality. If someone tells us that the distribution is symmetric about the mean, however, we could notice that the probability of $X \geq 4.0$ must be one-half the probability of $|X - 2.50| \geq 1.5$. The upper bound for this latter probability is

$$\Pr[|X - 2.5| \geq 1.50] \leq \frac{\sigma_X^2}{(1.5)^2} = \frac{25}{12 \times 2.25} = 0.926$$

Thus the bound (assuming a symmetric distribution) would be $\Pr[X \geq 4.0] \leq 0.463$, which is closer to the true probability.

□

The Chebyshev inequality is generally more versatile than the Markov inequality because it does not make specific assumptions about the data. It presumes that we know the mean *and* variance, however, which helps tighten the bound. A version of the Chebyshev inequality pertaining to random variables on one side or the other of the mean is given below.

6.1.3 One-sided Chebyshev inequality

The one-sided Chebyshev inequality is an upper bound for probabilities of the form $\Pr[X - m_X \geq a]$, where $a > 0$, when only mean m_X and variance σ_X^2 of X are given. The one-sided Chebyshev inequality is written in one of two forms depending on whether X is above or below the mean:

$$\Pr[X - m_X \geq a] \leq \frac{\sigma_X^2}{\sigma_X^2 + a^2} \quad \text{for } X > m_X \quad (a)$$

$$\Pr[m_X - X \geq a] \leq \frac{\sigma_X^2}{\sigma_X^2 + a^2} \quad \text{for } X \leq m_X \quad (b) \qquad (6.3)$$

Example 6.4: Continuing with Example 6.2, let us determine the upper bound that the voltage is greater than 4.0 using the one-sided Chebyshev inequality. The probability of $X \geq 4.0$ is the same as the probability of $X - m_X \geq 1.5$ where $m_X = 2.5$ is the mean of the random variable. The upper bound using (6.3)(a) is thus

$$\Pr[X - 2.5 \geq 1.5] \leq \frac{\sigma_X^2}{\sigma_X^2 + (1.5)^2} = \frac{25}{25 + 27} = 0.481.$$

which is somewhat tighter than the Markov result, which was found to be

$$\Pr[X \geq 4.0] \leq 0.625.$$

This is not surprising since the one-sided Chebyshev inequality uses both the mean and variance of the distribution to form the bound, while the Markov inequality is based on the mean alone.

□

We may just notice in passing that since the random variable used in these last three examples has a distribution that is *symmetric about the mean* an even tighter bound can be achieved by dividing the standard Chebysev bound by two:

$$\Pr[X - 2.5 \geq 1.5] \leq 0.926/2 = 0.463$$

Again this is not surprising since the one-sided inequalities (6.3) use the same information (mean and variance) but make no assumption of symmetry.

6.1.4 Other inequalities

There are two other important inequalities, Hölder's inequality and Miknowski's inequality, that are frequently encountered in the literature. Proofs can be found in several more advanced texts (e.g., [1]).

Hölder's inequality.[2] Consider two random variables X_1 and X_2 with means m_1 and m_2 and variances σ_1^2 and σ_2^2, respectively. Given that $k, l > 1$ and $k^{-1} + l^{-1} = 1$, Hölder's inequality states that

$$\mathcal{E}\{X_1 X_2\} \leq \left(\mathcal{E}\{|X_1^k|\}\right)^{1/k} \left(\mathcal{E}\{|X_2^l|\}\right)^{1/l} \qquad (6.4)$$

For $k = l = 2$, Hölder's inequality becomes the Cauchy-Schwartz inequality, which provides a bound on the correlation term $\mathcal{E}\{X_1 X_2\}$ in terms of the second moments of X_1 and X_2:

$$\mathcal{E}\{X_1 X_2\} \leq \sqrt{\mathcal{E}\{X_1^2\}\mathcal{E}\{X_2^2\}},$$

By subtracting the mean from each random variable, the Cauchy-Schwartz inequality can also be extended to the covariance:

$$\mathcal{E}\{(X_1 - m_1)(X_2 - m_2)\} \leq \sqrt{\mathcal{E}\{(X_1 - m_1)^2\}\mathcal{E}\{(X_2 - m_2)^2\}},$$

or

$$\mathrm{Cov}\,[X_1, X_2] \leq \sigma_1 \sigma_2.$$

Rewriting this equation, we can obtain the following bound on the correlation coefficient:

$$\rho_{12} = \frac{\mathrm{Cov}\,[X_1, X_2]}{\sigma_1 \sigma_2} \leq 1$$

The correlation coefficient can be negative, however, in which case the last equation is a useless tautology. The complete bound, which is given by

$$\boxed{|\rho_{12}| \leq 1} \tag{6.5}$$

can be obtained by redefining one of the random variables as its negative and repeating the procedure.

Minkowski's inequality.[3] Given that $k \geq 1$, Minkowski's inequality states that

$$\left(\mathcal{E}\{|X_1 + X_2|^k\}\right)^{1/k} \leq \left(\mathcal{E}\{|X_1^k|\}\right)^{1/k} + \left(\mathcal{E}\{|X_2^k|\}\right)^{1/k}. \tag{6.6}$$

For $k = 2$, Minkowski's inequality relates the second moment of $X_1 + X_2$ to those of X_1 and X_2:

$$\sqrt{\mathcal{E}\{(X_1 + X_2)^2\}} \leq \sqrt{\mathcal{E}\{X_1^2\}} + \sqrt{\mathcal{E}\{X_2^2\}}.$$

Again, by subtraction of the mean we can extend the result to covariances

$$\sqrt{\mathcal{E}\{((X_1 - m_1) + (X_2 - m_2))^2\}} \leq \sqrt{\mathcal{E}\{(X_1 - m_1)^2\}} + \sqrt{\mathcal{E}\{(X_2 - m_2)^2\}},$$

which can be written as

$$\boxed{(\mathrm{Var}\,[X_1 + X_2])^{1/2} \leq \sigma_1 + \sigma_2} \tag{6.7}$$

By employing (5.44), this also can be used to show the bound (6.5) on the correlation coefficient.

6.2 Convergence and Limit Theorems

A sequence of (non-random) numbers x_n converges to a limit x if for any $\epsilon > 0$ there exists some sufficiently large integer n_c such that

$$|x_n - x| < \epsilon, \quad \text{for every } n > n_c$$

In this section, however, we are interested in the convergence of sequences of *random variables* and present definitions of different modes of convergence of a sequence of random variables.

Consider the experiment where a sequence of random variables is formed:

$$X_1, X_2, \ldots, X_n, \ldots$$

Since each outcome s of the random experiment results in a sequence of numbers $X_1(s)$, $X_2(s), \ldots, X_n(s), \ldots$, the sequence of random variables $X_1, X_2, \ldots, X_n, \ldots$ forms a family of such sequences of numbers. The convergence of a sequence of random variables, therefore, must consider criteria for convergence of not just one realization but the entire family of these sequences. Consequently, there are a number of modes of convergence of sequences of random variables, some strong and some weak. Our interest here is whether convergence of a random sequence obeys one of: convergence with probability 1, convergence in probability, convergence in the mean-square sense, and convergence in distribution. These types of convergence are defined below.

Convergence with probability 1. This type of convergence, also known as "convergence almost everywhere," states that the sequence of random variables X_n converges to the random variable X with probability 1 for all outcomes $s \in \mathcal{S}$:

$$\lim_{n \to \infty} \Pr[X_n(s) \to X(s)] = 1, \quad \text{for all } s \in \mathcal{S}. \tag{6.8}$$

Mean-square convergence. The sequence of random variables X_n converges to the random variable X in the mean-square sense if:

$$\lim_{n \to \infty} E[|X_n(s) - X(s)|^2] = 0, \quad \text{for all } s \in \mathcal{S}. \tag{6.9}$$

Convergence in probability. This type of convergence is also known as stochastic convergence or convergence in measure. The sequence of random variables X_n converges to the random variable X in probability if for any $\epsilon > 0$:

$$\lim_{n \to \infty} \Pr[|X_n(s) - X(s)| > \epsilon] = 0, \quad \text{for all } s \in \mathcal{S}. \tag{6.10}$$

From the Chebyshev inequality (6.2), we have

$$\Pr[|X_n - X| > \epsilon] \leq \frac{E[|X_n - X|^2]}{\epsilon^2},$$

which suggests that if a sequence of random variables converges in the mean square sense, then it also converges in probability.

Convergence in distribution. The sequence of random variables X_n with cumulative distribution functions $F_{X_n}(x)$ converges in distribution to the random variable X with cumulative distribution function $F_X(x)$ if:

$$\lim_{n \to \infty} F_{X_n}(x) = F_X(x) \tag{6.11}$$

for all x for which $F_X(x)$ is continuous. The central limit theorem discussed later in this chapter is an example of convergence in distribution. Convergence in distribution makes no reference to the sample space, $\mathcal{S}$, but is only a condition on the cumulative distribution functions of the random variables in the sequence.

A Venn diagram representing the different convergence modes is shown in Figure 6.1. Each oval indicates all sequences of random variables that converge according to the indicated mode; no sequence converges outside of the largest oval. The reader may note that convergence in distribution is the weakest mode of convergence. If a sequence converges with probability 1, then it also converges in distribution (but not vice versa).

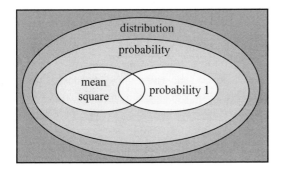

Figure 6.1 Comparison of different convergence modes.

6.2.1 Laws of large numbers

We now consider a important theoretical result known as the "law of large numbers." This result is important because it links the mean of a random variable to the average value of a number of realizations of the random variable, and thus justifies computing the mean experimentally as an average. Specifically, assume that we have outcomes or "observations" $X_1, X_2, \ldots, X_n$ resulting from n independent repetitions of the experiment that produces a certain random variable X. Thus the X_i are independent and identically-distributed (IID). The *sample mean* is defined as the average of these n random variables:

$$M_n = \frac{X_1 + X_2 + \ldots + X_n}{n} = \frac{1}{n} \sum_{i=1}^{n} X_i \tag{6.12}$$

Note that since each of the X_i is a random variable, the sample mean is also a random variable. The law of large numbers states that the sample mean converges to the mean m_X of the distribution as n approaches infinity. The law is usually stated in the two forms given below.

Weak law of large numbers

The weak law of large numbers provides a precise statement that the sample mean M_n is *close to* the distribution mean m_X in the following sense:

$$\lim_{n \to \infty} \Pr\left[|M_n - m_X| \geq \epsilon\right] = 0 \tag{6.13}$$

where ϵ is any small positive number. In other words, as n gets large, the sample mean converges *in probability* to the true mean m_X.

To prove this result, first consider the mean and variance of M_n. The mean is given by

$$\mathcal{E}\{M_n\} = \mathcal{E}\left\{\frac{1}{n} \sum_{i=1}^{n} X_i\right\} = \frac{1}{n} \sum_{i=1}^{n} \mathcal{E}\{X_i\} = m_X \tag{6.14}$$

(The last equality follows from the fact that all of the X_i have the same mean m_X.) It is seen from (6.14) that the mean of the sample mean is equal to the distribution mean m_x. The variance can also be easily computed by noting that the X_i are IID. Thus (see Table 4.2 and Equation 5.50) the variance is given by

$$\text{Var}[M_n] = \text{Var}\left[\frac{1}{n} \sum_{i=1}^{n} X_i\right] = \left(\frac{1}{n}\right)^2 \sum_{i=1}^{n} \text{Var}[X_i] = \frac{\sigma_X^2}{n} \tag{6.15}$$

where σ_X^2 is the variance of the distribution.

The proof of the weak law (6.13) follows by applying the Chebyshev inequality (6.2) to M_n:

$$\Pr\left[|M_n - m_X| \geq \epsilon\right] \leq \frac{\mathrm{Var}(M_n)}{\epsilon^2} = \frac{\sigma_X^2}{n\epsilon^2}.$$

Clearly, for any chosen value of ϵ, as $n \to \infty$, the right side approachs zero, which proves the result.

Strong law of large numbers

The strong law of large numbers states that as $n \to \infty$, the sample mean M_n converges to the distribution mean m_X with probability one:

$$\Pr\left[\lim_{n\to\infty} M_n = m_X\right] = 1. \tag{6.16}$$

The proof of the strong law of large numbers is beyond the scope of this discussion. Nevertheless, you may note that the weak law shows that the sample mean remains very close to the distribution mean but does not take into account the question of what actually happens to the sample mean as $n \to \infty$, which is considered by the strong law of large numbers. Figure 6.1 provides a comparison of the two modes of convergence.

6.2.2 Central limit theorem

The central limit theorem is an important theoretical result in probability theory. The essence of the central limit theorem is that the PDF of the sum of a large number of independent continuous random variables approaches the shape of a Gaussian density function. Here we present the statement of the central limit theorem without proof.

Consider a sequence of independent and identically-distributed random variables X_1, X_2 ... X_n, each with a mean of m_X and variance σ_X^2. The sum $Y_n = X_1 + X_2 + \ldots + X_n$ has a mean of nm_X and a variance of $n\sigma_X^2$. Define a normalized random variable

$$Z_n = \frac{Y_n - nm_X}{\sigma_X \sqrt{n}},$$

which has a mean of 0 and variance 1. Its cumulative distribution function is

$$F_{Z_n}(z) = \Pr[Z_n \leq z].$$

The CDF of the normalized random variable Z_n approaches that of a Gaussian random variable with zero mean and unit variance:

$$\lim_{n\to\infty} F_{Z_n}(z) = \Phi(z) = \frac{1}{\sqrt{2\pi}} \int_{-\infty}^{z} e^{-\frac{v^2}{2}} dv, \tag{6.17}$$

Since $\Phi(z)$ is the normalized Gaussian CDF discussed in Chapter 3, the sequence of normalized random variables Z_n *converges in distribution* to a Gaussian random variable.

The central limit theorem is true for a sum of a large number of any sequence of IID random variables (having finite mean and finite variance) of any distribution. The theorem, in fact, is a result of convolving a large number of positive-valued functions and is used in a wide variety of applications. Extensions to the basic theorem are possible where various conditions stated above are relaxed.

Example 6.5: Consider a sequence of 100 binary values (IID Bernoulli random variables) with distribution parameter $p = \Pr[1] = 0.6$. We shall determine (i) the probability of more than 65 1's in the sequence, and (ii) the probability that the number of 1's is between 50 and 70.

The mean and the variance of each binary random variable are are found to be 0.6 and 0.24, respectively. The number of 1's in this sequence is the random variable

$$Y_{100} = \sum_{i=1}^{100} X_i.$$

The mean of the sum is $nm_X = 60$ and the variance $n\sigma_X^2 = 24$. Since the normalized representation of Y_{100} is

$$Z_{100} = \frac{Y_{100} - 60}{\sqrt{24}},$$

we have

$$
\begin{aligned}
\Pr[Y_{100} > 65] &= \Pr\left[Z_{100} > \frac{65 - 60}{\sqrt{24}}\right] \\
&= Q\left(\frac{5}{\sqrt{24}}\right) = Q(1.0206) \approx 0.16,
\end{aligned}
$$

where the value for $Q(1.0206)$ is obtained from Table A.

To determine the probability that Y_{100} is between 50 and 70, we write

$$
\begin{aligned}
\Pr[50 \le Y_{100} \le 70] &= \Pr\left[\frac{50 - 60}{\sqrt{24}} \le Z_{100} \le \frac{70 - 60}{\sqrt{24}}\right] \\
&= \Pr[-2.0412 \le Z_{100} \le 2.0412] = 1 - 2Q(2.0412) \\
&\approx 1 - 2 \times 0.021 = 0.958.
\end{aligned}
$$

□

6.3 *Estimation of Parameters*

In order to apply probabilistic ideas in practice, it is necessary to have numerical values for the various parameters that describe the distribution. For example, for a Gaussian random variable, we need to know the mean and variance; for a binomial random variable we need to know the parameter p, and so on. In most cases, these parameters are not known *a priori* but must be *estimated* experimentally from observations of the random variable.

We have already introduced the *sample mean* as an estimate of the mean of a distribution. In the remainder of the chapter we develop estimation procedures for other parameters as well, and develop some basic theoretical results about the properties of these procedures.

6.3.1 Estimates and properties

Let us first establish some notation. We consider a parameter θ pertaining to the distribution for a particular random variable X. As in Section 6.2.1 we let $X_1, X_2, \ldots, X_n$ represent the outcomes resulting from n independent repetitions of the experiment that produces the random variable. We define a *statistic* or an *estimate* for the parameter θ by[2]

$$\Theta_n = \Theta(X_1, X_2, \ldots, X_n)$$

where Θ on the right represents a known function of random variables $X_1, X_2, \ldots, X_n$. Note that we use a capital letter for the estimate Θ_n since it depends on the random variables $X_1, X_2, \ldots, X_n$ and is thus a random variable itself. We also attach the subscript 'n' to the estimate because the dependence on the number of observations n is important. Many authors use a 'hat' (^) over the symbol to indicate an estimate, but in our limited treatment it is not necessary to do that.

Three important properties of estimates are worth considering:

1. An estimate Θ_n is *unbiased* if

$$\mathcal{E}\{\Theta_n\} = \theta$$

   where θ is the true value of the parameter. Otherwise the estimate is *biased* with bias $b(\theta) = \mathcal{E}\{\Theta_n\} - \theta$.

   An estimate is *asymptotically unbiased* if

$$\lim_{n \to \infty} \mathcal{E}\{\Theta_n\} = \theta$$

2. An estimate is said to be *efficient* with respect to another estimate if it has a lower variance. If an estimate Θ_n is simply said to be efficient (without any qualification), this means that Θ_n is efficient with respect to Θ_{n-1}.

3. An estimate Θ_n is *consistent* if

$$\lim_{n \to \infty} \Pr[|\Theta_n - \theta| < \epsilon] = 1$$

   for any arbitrarily small number ϵ. Thus the estimate is consistent if it converges in probability to the true value of the parameter θ.

Consistency is obviously a desirable property.[3] It implies that the estimate becomes closer and closer to the true parameter as the number of observations (n) is increased. We can make the following important statement:

If Θ_n is unbiased and efficient, then Θ_n is a consistent estimate.

The proof of this statement follows from the Chebyshev inequality (6.2). Since Θ_n is unbiased, the mean is equal to θ. Equation 6.2 in this case becomes

$$\Pr[|\Theta_n - \theta| \geq \epsilon] \leq \frac{\text{Var}[\Theta_n]}{\epsilon^2}$$

[2] A *statistic* is any function of the random variables $X_1, X_2, \ldots, X_n$. An *estimate* is a statistic that is intended to approximate some particular parameter θ.

[3] In general, an estimate is consistent if it converges in *some* sense to the true value of the parameter. Other possible (stronger) ways to define consistency are convergence with probability 1 or convergence in mean-square.

Thus if the variance of Θ_n decreases with n, the probability that $|\Theta_n - \theta| \geq \epsilon$ approaches 0 as $n \to \infty$. In other words the probability that $|\Theta_n - \theta| < \epsilon$ approaches 1. This last property is illustrated in Fig. 6.2. As n becomes large, the PDF for Θ_n

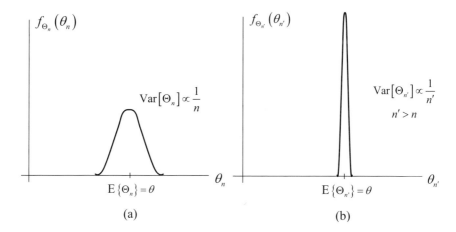

Figure 6.2 Illustration of consistency for an unbiased estimate whose variance decreases with n. (a) PDF of the estimate Θ_n. (b) PDF of the estimate $\Theta_{n'}$ with $n' > n$.

becomes more concentrated about the true parameter value θ. In the limit, the density takes on the character of an impulse located at θ.

The properties discussed in this section are illustrated below for some specific statistics.

6.3.2 Sample mean and variance

The sample mean M_n (already introduced) and the sample variance Σ_n^2 are statistics commonly used to estimate the distribution mean $m_X = \mathcal{E}\{X\}$ and variance $\sigma_X^2 = \mathcal{E}\{(X - m_X)^2\}$ of a random variable when those theoretical parameters are not known. Let us examine the properties of these particular statistics according to the definitions provided in the previous subsection.

Sample mean

The sample mean M_n is defined by (6.12). It is shown in Section 6.2.1 that the mean of M_n is equal to the distribution mean m_X (see (6.14)). Therefore the sample mean is *unbiased*.

The variance of the sample mean is also derived in Section 6.2.1 (see (6.15)) and is given by σ_X^2/n. Since the variance decreases as a function of n, the estimate is *efficient*. Finally, since the estimate is unbiased and efficient, the sample mean is a *consistent estimate* for the mean.

Sample variance

The sample variance is defined as

$$\Sigma_n^2 = \frac{(X_1 - M_n)^2 + \ldots + (X_n - M_n)^2}{n} = \frac{1}{n}\sum_{i=1}^{n}(X_i - M_n)^2 \qquad (6.18)$$

(Note that in estimating the variance, we use the sample mean M_n rather than the true mean m_X, since the latter is generally not known.) The mean of the sample variance is given by

$$\mathcal{E}\{\Sigma_n^2\} = \frac{1}{n}\sum_{i=1}^{n}\mathcal{E}\{(X_i - M_n)^2\}$$

This expectation is complicated by the fact that M_n, which appears on the right, is also a random variable. Let us first write

$$X_i - M_n = (X_i - m_X) - (M_n - m_X)$$

Substituting this in the previous equation yields

$$\mathcal{E}\{\Sigma_n^2\} = \frac{1}{n}\sum_{i=1}^{n}\mathcal{E}\{(X_i - m_X)^2\} + \frac{1}{n}\sum_{i=1}^{n}\mathcal{E}\{(M_n - m_X)^2\}$$
$$-\frac{2}{n}\mathcal{E}\left\{\sum_{i=1}^{n}(M_n - m_X)(X_i - m_X)\right\}$$

This equation can be simplified to

$$\mathcal{E}\{\Sigma_n^2\} = \sigma_X^2 + \mathcal{E}\{(M_n - m_X)^2\} - \frac{2}{n}\mathcal{E}\left\{(M_n - m_X)\sum_{i=1}^{n}(X_i - m_X)\right\}$$

where we have recognized that the first term is the distribution variance σ_X^2 and that the quantity $M_n - m_X$ does not depend on the summation variable i and can be removed from the sum.

Now observe that the last term can be written as:

$$\frac{2}{n}\mathcal{E}\left\{(M_n - m_X)\sum_{i=1}^{n}(X_i - m_X)\right\} = 2\mathcal{E}\left\{(M_n - m_X)\left(\frac{1}{n}\sum_{i=1}^{n}X_i - \frac{1}{n}\sum_{i=1}^{n}m_X\right)\right\}$$
$$= 2\mathcal{E}\{(M_n - m_X)(M_n - m_X)\} = 2\mathcal{E}\{(M_n - m_X)^2\}$$

Upon substituting this in the previous equation, we find

$$\mathcal{E}\{\Sigma_n^2\} = \sigma_X^2 - \mathcal{E}\{(M_n - m_X)^2\}$$

Finally, by recognizing that $\mathcal{E}\{(M_n - m_X)^2\} = \mathrm{Var}\,[M_n] = \sigma_X^2/n$ (see (6.15)), we arrive at the equation

$$\mathcal{E}\{\Sigma_n^2\} = \left(1 - \frac{1}{n}\right)\sigma_X^2,$$

This result indicates that the sample variance, although *not* unbiased, is nevertheless *asymptotically* unbiased since $\mathcal{E}\{\Sigma_n^2\} \to \sigma_X^2$ as $n \to \infty$.

An alternate form of the sample variance is sometimes used. This estimate is given by

$$(\Sigma_n^2)' = \frac{1}{n-1}\sum_{i=1}^{n}(X_i - M_n)^2 \tag{6.19}$$

This differs from the previous definition (6.18) by a factor of $n/(n-1)$ and so is *unbiased*. Both forms of sample variance can often be found in statistical analysis programs and on hand-held calculators.

The variance of the sample variance can also be computed and shown to approach 0 as n increases. Thus the sample variance is an efficient estimate and can be shown to be consistent. We omit the details here since they are beyond the scope of the discussion.

6.4 Maximum Likelihood Estimation

Having now been introduced to at least a couple of different examples of estimates, you may wonder how the formulas for these estimates are derived. One answer to this question is through the procedure known as *maximum likelihood* estimation.

Consider the following hypothetical situation. You have a random variable X with a density function similar to one of those depicted in Fig. 6.3, but you do not know

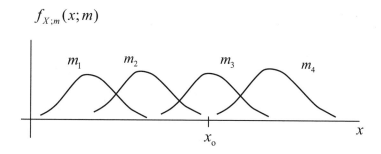

$$f_{X;m}(x;m)$$

Figure 6.3 Choices in estimation of the mean.

the mean. Further, you have an observation of the random variable X; let's call that observation x_o. On the basis of that single observation, you want to estimate the mean of the density. Referring to Fig. 6.3, one could argue that the value selected for the mean should be m_3, because it makes the given observation (x_o) the *most likely* observation. After all, why would we choose m_1 or m_4? These choices would imply that our given observation x_o has low probability.

This reasoning can be extended to what we shall call the *principle of maximum likelihood*. This principle states that in choosing an estimate for a parameter, we should choose the value (of the parameter) that makes our observed data the *most likely result*. Let us formulate that principle mathematically.

First, in most practical estimation problems there is more than one observation of the data. We denote the various data samples or observations by $X_1, X_2, \ldots, X_n$, as we have in the previous sections. For notational convenience, we also represent this set of observations by a random vector $\boldsymbol{X}$. If θ is the parameter that decribes the density function for X, then we denote the joint density function for all the observations by $f_{\boldsymbol{X};\theta}(\boldsymbol{x};\theta)$. The *maximum likelihood (ML) estimate* for the parameter θ is defined by

$$\Theta_{ml}(\boldsymbol{X}) = \underset{\theta}{\operatorname{argmax}} \; f_{\boldsymbol{X};\theta}(\boldsymbol{X};\theta) \tag{6.20}$$

In other words, it is the value of θ that maximizes $f_{\boldsymbol{X};\theta}$, given the observations. In line with the discussion in Section 6.3, we use capital letters for both Θ_{ml} and $\boldsymbol{X}$, since they are random variables. We will illustrate how this definition can be used to derive explicit formulas for estimates of particular parameters; however, let us first make some remarks about the procedure in general:

1. In maximum likelihood estimation the quantity $f_{\boldsymbol{X};\theta}(\boldsymbol{x};\theta)$ is considered to be a function of the parameter θ for fixed (known) values of the observations $\boldsymbol{X}$. When viewed *as a function of* θ, $f_{\boldsymbol{X};\theta}$ is called the *"likelihood function"* rather than the PDF.

2. The maximum likelihood estimate, resulting from the solution of (6.20), is the estimate with smallest variance (called a "minimum-variance" estimate). Thus the maximum likelihood estimate is an optimal estimate in that it is the *most efficient* estimate for the parameter θ.

3. Since the logarithm is a monotonic transformation, maximizing $\ln f_{\boldsymbol{X};\theta}$ is equivalent to maximizing $f_{\boldsymbol{X};\theta}$. Because many distributions involve powers of the variables, it is frequently most convenient to compute the ML estimate from the following:

$$\Theta_{ml}(\boldsymbol{x}) = \underset{\theta}{\mathrm{argmax}} \quad \ln f_{\boldsymbol{X};\theta}(\boldsymbol{x};\theta) \tag{6.21}$$

The quantity $\ln f_{\boldsymbol{X};\theta}$ is known as the *log likelihood functiion*.

With the foregoing results in mind, let us explore the procedure of computing a maximum likelihood estimate through a set of examples.

Example 6.6:   The time X between requests for service on a file server is a random variable with the exponential PDF

$$f_X(x) = \begin{cases} \lambda e^{-\lambda x} & x \geq 0 \\ 0 & \text{otherwise} \end{cases}$$

A total of $n+1$ requests are observed with independent interarrival times $X_1, X_2, \ldots X_n$. It is desired to find a maximum likelihood estimate for the arrival rate parameter λ.

Since the interarrival times are independent, the likelihood function for this problem is

$$f_{\boldsymbol{X};\lambda}(\boldsymbol{X};\lambda) = \prod_{i=1}^{n} \lambda e^{-\lambda X_i} = \lambda^n e^{-\lambda \sum_{i=1}^{n} X_i}$$

Since the form involves some exponential terms, it is easier to deal with the log likelihood function, which is given by

$$\ln f_{\boldsymbol{X};\lambda}(\boldsymbol{X};\lambda) = \ln \lambda^n e^{-\lambda \sum_{i=1}^{n} X_i} = n \ln \lambda - \lambda \sum_{i=1}^{n} x_i$$

To find the maximum likelihood estimate for λ, we can take the derivative of the log likelihood function and set it to 0. This yields

$$\frac{d}{d\lambda}\left[n \ln \lambda - \lambda \sum_{i=1}^{n} X_i \right] = \frac{n}{\lambda} - \sum_{i=1}^{n} X_i = 0$$

Solving this for λ produces the maximum likelihood estimate. Since the estimate is actually a random variable, we use a capital letter and write it as

$$\Lambda_{ml} = \frac{1}{\frac{1}{n}\sum_{i=1}^{n} X_i}$$

The estimate is seen to be the reciprocal of the average interarrival time.

□

When the distribution pertains to a discrete random variable, the procedure is similar. The likelihood function is derived from the PMF, however, rather than the PDF.

Example 6.7: The number of errors k occurring in a string of n binary bits is a binomial random variable with PMF

$$f_K[k] = \binom{n}{k} p^k (1-p)^{(n-k)} \qquad 0 \le k \le n$$

where the parameter p is the probability of a single bit error. Given an observation of k errors, what is the maximum likelihood estimate for the bit error parameter p?

The likelihood function for this problem is given by

$$f_{K;p}[K;p] = \binom{n}{K} p^K (1-p)^{(n-K)}$$

While we could use the log likelihood function here (see Prob. 6.18), it is equally easy in this case to deal with the likelihood function directly.

Taking the derivative with respect to the parameter and setting it to 0 yields

$$\begin{aligned}
\frac{df_{K;p}}{dp} &= \binom{n}{K}\left[Kp^{K-1}(1-p)^{n-K} - p^K(n-K)(1-p)^{n-K-1} \right] \\
&= \binom{n}{K} p^{K-1}(1-p)^{n-K-1}\left[K(1-p) - p(n-K) \right] = 0
\end{aligned}$$

Solving for p, we find the maximum likelihood estimate:

$$P_{ml} = \frac{K}{n}$$

Now, recalling that the mean of the binomial distribution is $\mathcal{E}\{K\} = np$, we can write

$$\mathcal{E}\{P_{ml}\} = \frac{\mathcal{E}\{K\}}{n} = p$$

thus demonstrating that this estimate is *unbiased*.

□

The following example is a little different from the previous two, and is intended to emphasize the distinction between the PDF and the likelihood function.

Example 6.8: A random variable X has the uniform density function

$$f_X(x) = \begin{cases} 1/a & 0 \le x \le a \\ 0 & \text{otherwise} \end{cases}$$

Given n independent observations $X_1, X_2, \ldots X_n$, it is desired to find a maximum likelihood estimate of the parameter a.

This problem is tricky, because you have to think carefully about the meaning of the likelihood function. One cannot proceed blindly to take derivatives as in the previous two examples, because the maximum of the likelihood function actually occurs at one of the boundaries of the region, where the derivative is *not* 0.

To proceed with this example, let us consider the simplest case, namely $n = 1$. The likelihood function is obtained and plotted by writing the above equation for the density, as a function of the variable a while the observation X is held fixed:

$$f_{X;a}(X;a) = \begin{cases} 1/a & a \geq X \\ 0 & \text{otherwise} \end{cases}$$

From the picture, we see that the maximum of the likelihood function occurs at the point $a = X$; hence the ML estimate for $n = 1$ is given by $A_{ml} = X$.

Next consider $n = 2$. Since the observations are independent, the joint PDF is given by

$$f_{X_1 X_2}(x_1, x_2) = \begin{cases} 1/a^2 & 0 \leq x_1 \leq a, 0 \leq x_2 \leq a \\ 0 & \text{otherwise} \end{cases}$$

Writing and plotting this as a function of a, for fixed values of the observations, we obtain the likelihood function:

$$f_{\boldsymbol{X};a}(\boldsymbol{X};a) = \begin{cases} 1/a^2 & a \geq X_1, a \geq X_2 \\ 0 & \text{otherwise} \end{cases}$$

Notice that since *both* of the inequalities $a \geq X_1$ and $a \geq X_2$ must be satisfied for the likelihood function to be nonzero, the maximum of the function occurs at $A_{ml} = \max\{X_1, X_2\}$.

Continuing to reason in this way, we find that for an arbitrary number n of samples, the likelihood function is equal to $1/a^n$ for $a \geq \max\{X_1, X_2, \ldots, X_n\}$ and 0 otherwise. The peak of this function and thus the maximum likelihood estimate occurs at

$$A_{ml} = \max\{X_1, X_2, \ldots, X_n\}$$

A little thought and reflection on this example should convince you that this is the intuitively correct answer.

□

A number of other cases arise where a maximum likelihood estimate is possible. For example, the sample mean, introduced in Section 6.2.1, is the ML estimate for the mean of the Gaussian distribution (and some other distributions as well). The proof is straightforward, so it is left to the problems at the end of this chapter.

6.5 *Application to Signal Estimation*

We have seen in earlier chapters that many problems in communication and related areas involve probability in the form of hypothesis testing. This is the basis of most signal detection and classification schemes. A comparably large number of applications however, involve *estimation* of signals and other parameters. That is the topic of the present section.

This section discusses the problem of estimating a signal in additive noise. The situation is similar to that of the signal detection application of Section 3.8.1. A signal with constant value s is transmitted to a receiver. There is noise in the channel, however, and the *received* signal is not constant, but is random due to the added noise (see Fig. 6.4). It is desired to estimate the value of s from the received data; we can

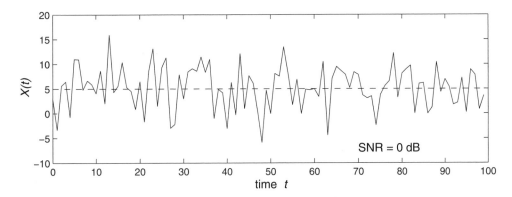

Figure 6.4 Received signal with additive noise. Transmitted signal is shown by dashed line ($s = 5$).

take as many samples of the received signal as necessary.

Assume that n samples of the received signal are taken. The ith sample is a random variable given by

$$X_i = s + N_i, \quad i = 1, 2, \ldots, n \tag{6.22}$$

where N_i denotes the ith sample of the noise. Further, assume that all of the random variables N_i are IID with 0 mean and variance σ^2. (In this context, σ^2 is the average noise *power*.) Then the observations X_i are IID random variables with mean s and variance σ^2. The signal-to-noise ratio in decibels can be defined as

$$\text{SNR} = 10 \log_{10} \frac{s^2}{\sigma^2} = 20 \log_{10} \frac{s}{\sigma} \quad (\text{dB}) \tag{6.23}$$

Consider forming an estimate of the transmitted signal by averaging the samples of the received signal:

$$S_n = \frac{1}{n} \sum_{i=1}^{n} X_i \tag{6.24}$$

This estimate is the *sample mean*.

It has been previously shown that the sample mean is unbiased, i.e., that

$$\mathcal{E}\{S_n\} = s \tag{6.25}$$

and that the variance is given by

$$\sigma_{S_n}^2 = \frac{\text{Var}\,[X_i]}{n} = \frac{\sigma^2}{n} \tag{6.26}$$

Figure 6.5 is a sketch depicting a typical density function for the random variable S_n.

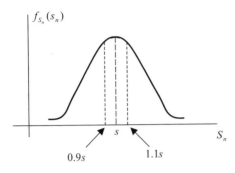

Figure 6.5 Typical density function for the estimate S_n.

The mean of the density is at $s_n = s$ and the variance is a decreasing function of n. It is clear that as $n \to \infty$ the density remains centered about s but becomes narrower and taller so that all of the probability is concentrated at the point $s_n = s$. In other words, the more samples we take to form the estimate, the closer it is likely to be to the true value of the signal. This is a practical illustration of the law of large numbers.

The following example is presented to illustrate the results discussed in this section numerically.

Example 6.9: Consider the signal estimation problem described above. Suppose we want the estimate for s to be within $\pm 10\%$ of its true value (see Fig. 6.5). Since this is a random event let us require that $\Pr[|S_n - s| > 0.1s] = 0.05$. Since the sample mean involves a sum of n IID random variables, we can assume by the central limit theorem that the distribution in Fig. 6.5 is Gaussian with mean and variance given by (6.25) and (6.26). We will determine the number of samples (n) necessary to satisfy this condition as a function of SNR.

First notice from Fig. 6.5 that the probability that S_n lies outside of the $\pm 10\%$ limits is given by

$$\int_{-\infty}^{0.9\,s} f_{S_n}(s_n)ds_n + \int_{1.1\,s}^{\infty} f_{S_n}(s_n)ds_n = 2\int_{-\infty}^{0.9\,s} f_{S_n}(s_n)ds_n = 0.05$$

where f_{S_n} is the PDF for S_n and the probability has been set to the required value of $1 - 0.95 = 0.05$. Since S_n is a Gaussian random variable, the probability can be evaluated using the error function (see Table 3.1) as

$$2\int_{-\infty}^{0.9\,s} f_{S_n}(s_n)ds_n = 2\tfrac{1}{2}\left[1 + \mathrm{erf}\left(\frac{0.9\,s - s}{\sqrt{2}\,\sigma_{S_n}}\right)\right] = 0.05$$

Using (6.26), this equation can be algebraically simplified to

$$\mathrm{erf}\left(0.1(s/\sigma)\sqrt{n/2}\right) = 0.95$$

This equation can be used to plot n as a function of the SNR. Using MATLAB, or tables of the error function, it can be found that the error function is equal to 0.95 when its argument is equal to approximately 1.386. Therefore, we have

$$0.1(s/\sigma)\sqrt{n/2} = 1.386$$

Then by taking logarithms and using the definition (6.23) for SNR, this last equation can be put in the form

$$\log_{10} n + (1/10)\,\mathrm{SNR} = 2.443$$

This function appears as a straight line when plotted in semilog form, as shown in the graph below.

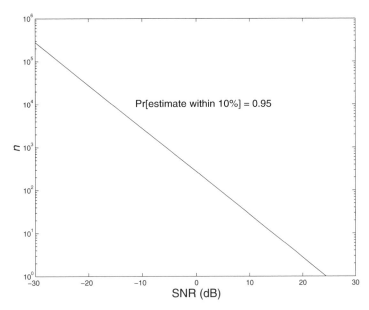

Note that when the SNR is equal to approximately 25 dB, n is equal to 1. This means that the signal with noise stays within $\pm 10\%$ of the true value with probability 0.95, so no averaging is required.[4] As the level of the noise increases (SNR decreases) the number of samples required to estimate the signal increases exponentially. By using a sufficiently large value of n, however, the signal can be estimated to within $\pm 10\%$ for any SNR. This principle was used during the cold war to track submarines at very long distances and very low corresponding SNR values.

□

The previous example is more general than it may seem. First of all, the $\pm 10\%$ criterion is arbitrary and can be changed to any tolerance level without changing the nature of the problem. In addition, the probability required for being within tolerance (0.95 in the example) only serves to set the constant in the equation relating SNR to $\log n$. The straight line form of the curve illustrated in the figure does not change, but only moves up and down as these other conditions are varied.

6.6 Summary

The topics covered in this chapter are useful in the application of probability to practical engineering problems. The Markov and Chebyshev inequalities provide bounds for probabilities involving random variables when only the mean and variance of the distribution are known. While the bounds are typically not tight, they are sufficiently general to state or prove results about random variables that are independent of the actual distribution.

Several important results in the application of probability involve the convergence of a sequence of random variables as the number of terms becomes large. Some types of convergence that are commonly studied are convergence with probability 1, mean-square convergence, convergence in probability, and convergence in distribution. Two important results are discussed that depend on some of these modes of convergence.

---

[4] Strictly speaking, our analysis is only approximate for $n = 1$ unless the noise is Gaussian, because the central limit theorem cannot be invoked for $n = 1$. In many cases of interest, however, the noise *is* Gaussian.

The law of large numbers states that an average of a large number of IID random variables (also known as the "sample mean,") converges to the expectation of these random variables. The central limit theorem states that the distribution of a sum of a large number of random variables converges to the Gaussian distribution. This highlights the Gaussian distribution as a uniquely important distribution because many random variables of practical interest are defined as a sum of other elementary random variables.

In applying probability to real problems, the numerical values of parameters in the distributions must usually be *estimated* from samples of observed or measured data. The properties of such estimates such as 'bias' and 'efficiency' are important characteristics. Moreover, convergence to the true parameter values as the number of observations becomes large (known as 'consistency' of the estimate) is especially important. The *maximum likelihood* (ML) estimate was defined and cited as the most efficient estimate when it exists. Some various examples of how to compute an ML estimate are given in this chapter.

Estimation is a common requirement in communications and signal processing. The chapter concludes with a classical signal estimation problem that demonstrates properties of the estimate as well as application of the central limit theorem.

References

[1] Geoffrey Grimmitt and David Stirzaker. *Probability and Random Processes.* Oxford University Press, New York, third edition, 2001.

[2] O. Hölder. Über einen mittelwertsatz. *Göttingen Nachr.*, 38(7), 1889.

[3] H. Minkowski. *Geometrie der Zahlen - Volume 1.* Leipzig, Germany, 1896.

[4] Alberto Leon-Garcia. *Probability and Random Processes for Electrical Engineering.* Addison-Wesley, Reading, Massachusetts, second edition, 1994.

Problems

Inequalities[5]

6.1 The magnitude of the voltage V across a component in an electronic circut has a mean value of 0.45 volts. Given only this information, find a bound on the probability that $V \geq 1.35$.

6.2 The average time to print a job in the computer laboratory is known to be 30 seconds. The standard deviation of the printing time is 4 seconds. Compute a bound on the probability that the printing time is between 20 and 40 seconds.

6.3 The waiting time K until the first error in a binary sequence is described by the (type 1) geometric PDF. Let the probability of an error be $p = 1/8$.

    (a) Compute the mean and variance of K.

    (b) Apply the Markov inequality to bound the probability of the event $K \geq 16$.

    (c) Use the Chebyshev inequality to bound the probability of this event.

    (d) Use the one-sided Chebyshev inequality to bound the probability.

    (e) Compute the actual probability of the event $K \geq 16$.

---

[5] For the following problems, you may use the formulas in the front of the book for mean and variance of various distributions.

6.4 Let X be an exponential random variable with parameter λ.

    (a) Apply the Markov inequality to bound $\Pr[X \geq 2/\lambda]$.

    (b) Use the Chebyshev inequality to compute the bound.

    (c) Use the one-sided Chebyshev inequality to compute the bound.

    (d) Compute the actual probability of this event.

6.5 Let X be a random variable with mean 0 and variance σ_X^2. It is known that if X is Gaussian, then

$$\Pr[\|X\| > 2\sigma_X] \approx 0.0455$$

    (a) Compute the Chebyshev bound on this probability.

    (b) Compute the probability of this event if X has the following distributions:

        (i) Laplace.

        (ii) Uniform.

Convergence and limit theorems

6.6 A set of speech packets are queued up for transmission in a network. If there are twelve packets queued up, the total delay D until transmission of the twelvth packet is given by the sum

$$D = T_1 + T_2 + \cdots + T_{12}$$

where the T_i are IID *exponential* random variables with mean $\mathcal{E}\{T_i\} = 8$ ms.

    (a) What is the variance $\text{Var}\,[T_i]$? (You may refer to the formulas on the inside cover of the text.)

    (b) Find the mean and variance of the total delay D.

    (c) Speech packets will be lost if the total delay D exceeds 120 ms. Because D is the sum of IID exponential random variables, D is a 12-Erlang random variable. The CDF of this random variable is given by [4, p. 118]

$$F_D(d) = 1 - \sum_{k=0}^{11} \frac{(\lambda d)^k}{k!} e^{-\lambda d}$$

    where λ is the parameter describing the exponential random variables T_i. Using this result, compute the probability that packets are lost, i.e., $\Pr[D > 120]$.

    (d) By applying the central limit theorem and using the Gaussian distribution, compute the approximate value of $\Pr[D > 120]$. Compare this to the exact result computed in part (c).

    (e) Finally evaluate the Markov bound for $\Pr[D > 120]$ and compare it to the two previous results.

    Note: You may use MATLAB or some other program to evaluate the probabilities in parts (c) and (d).

6.7 A simple way to reduce noise is to perform what is called a moving average on a signal. Mathematically, the moving average operation can be written as

$$Y_n = \frac{1}{L} \sum_{k=n-L+1}^{n} X_k$$

where X_k are IID random variables with a mean of 2 and a variance of 3.

(a) Find the mean and the variance of Y_n for $L = 12$.

(b) Based on the central limit theorem, write an expression for the PDF of Y.

(c) How large does L have to be for the variance of Y to be less than 0.1?

(d) For the value of L obtained in (c), what is $\Pr[|Y - m_Y| > 0.1]$?

6.8   The lifetime of a power transistor operating under moderate radiation conditions is known to be an exponential random variable with an average life time of $\lambda^{-1} = 36$ months. Given 16 such transistors, use the central limit theorem to determine the probability that the sum of their life times is 50 years or less.

6.9   With all the handouts in this course, your binder fills up as an exponential random variable on average in a week, i.e., $\lambda^{-1} = 1$ week. What is the minimum number of binders you should buy at the beginning of an 11-week quarter, so the probability that you will not run out of binders during the quarter is 0.9? Use the central limit theorem.

6.10 In a sequence of 800 binary values received at the output of a channel, the values of 1's and 0's are equally likely. Assume that these are IID random variables. Calculate the probability that the number of 1's is between 300 and 500. What is the probability that the number of 1's is between 400 and 440?

6.11 The probability of a bit error in a binary channel is 0.1. Treat each transmission as a Bernoulli random variable; assume that successive transmissions are IID. If you transmit 1000 bits, what is the probability that 115 or fewer errors occur in the channel?

Estimation of parameters

6.12 A random variable Θ is defined by

$$\Theta_n = \sum_{i=1}^{n} \Phi_i$$

where the Φ_i are independent and uniformly distributed from $-\pi$ to π.

(a) What are the mean and variance of Θ_n as a function of n?

(b) What is the PDF of Θ_n for large values of n?

6.13 (a) By expressing the sample variance $\Sigma_n^{2'}$ in terms of Σ_n^2, and using the results of Section 6.3.2, show that $\Sigma_n^{2'}$ is unbiased.

(b) Which of these two estimates is more *efficient*?

6.14 An estimate for the autocorrelation function of a discrete-time random process $X[k]$ is given by

$$\hat{R}_X^{(N)}[l] = \frac{1}{N-l} \sum_{k=0}^{N-l-1} X[k]X[k+l]$$

where N is the number of samples used for the estimate. (Since we already have been using a capital letter for the correlation function, we use a 'hat' here for the estimate and a superscript to indicate its dependence on N.)

(a) Is this estimate unbiased, asymptotically unbiased, or neither?

(b) What about the following estimate?

$$\hat{R}_X[l] = \frac{1}{N} \sum_{k=0}^{N-l-1} X[k]X[k+l]$$

Maximum likelihood estimation

6.15 A random variable X is described by a Gaussian density function with unknown mean μ and variance σ_o^2. (You can assume that σ_o^2 is a *known* parameter.)

$$f_X(x) = \frac{1}{\sqrt{2\pi\sigma_o^2}} e^{-(x-\mu)^2/2\sigma_o^2}$$

Given n independent realizations of the random variable $X_1, X_2, \ldots X_n$, find the maximum likelihood estimate for the mean μ.

6.16 A random variable X is described by the Gaussian density function with mean 0 and unknown variance σ^2.

$$f_X(x) = \frac{1}{\sqrt{2\pi\sigma^2}} e^{-x^2/2\sigma^2}$$

Given n independent realizations of the random variable $X_1, X_2, \ldots X_n$, find the maximum likelihood estimate for the variance σ^2.

6.17 Since the mean of the exponential density function is given by $1/\lambda$, this PDF can be written as

$$f_X(x) = \begin{cases} \frac{1}{\mu} e^{-x/\mu} & x \geq 0 \\ 0 & \text{otherwise} \end{cases}$$

where μ is the mean of the random variable. Using this parametric form of the density, and assuming n independent observations $X_1, X_2, \ldots X_n$, derive the maximum likelihood estimate of the mean μ.

6.18 Repeat Example 6.7, i.e., find the maximum likelihood estimate, using the *log likelihood function*.

Application to signal estimation

6.19 Plot the curve of n vs. SNR in Example 6.9 if it is required that:

(a) the probability that the estimate is within 10% of the true value is 0.98.

(b) the probability that the estimate is within 5% of the true value is 0.95.

(c) the probability that the estimate is within 5% of the true value is 0.98.

6.20 The plot in Example 6.9 shows the value of n needed to ensure $\Pr[|S_n - s| \leq 0.1s]$ is equal to 0.95. Use the Chebyshev inequality to derive a lower bound for n as a function of SNR and compare it to the curve in Example 6.9. (The lower bound means that for *all* distributions of the noise, n must be greater than the value of the bound in order to satisfy the probability 0.95 requirement.)

Computer Projects

Project 6.1

In this project you will empirically study the sums of random variables and demonstrate the central limit theorem. Specifically, you will write computer code to generate Gaussian random variables based on the central limit theorem using uniform and exponential random variables. You will then plot the density (distribution) function of these Gaussian random variables. Also, by successively convolving the Bernoulli and exponential density functions, you will observe that the resulting density function has the shape of a Gaussian density.

1. Generate six sequences of uniform random variables (length $\approx$ 10,000) in the range $[-0.5, 0.5]$. Each element in the ith sequence can be considered to be one realization of the random variable X_i for $i = 1, 2, \ldots, 6$. Note that these six uniform random variables are independent and identically distributed. Compute new random variables as follows:

$$Y_1 = \sum_{i=1}^{6} X_i$$

$$Y_2 = X_1 + 0.5X_2 + 0.8X_3 + 1.8X_4 + 0.3X_5 + 0.5X_6$$

This results in two new sequences, the elements of which can be thought of as $\approx$ 10,000 separate realizations of the random variables Y_1 and Y_2.

(a) Experimentally determine the PDF and CDF of Y_1 and Y_2 by developing a normalized histogram and integrating. (It is recommended that you use about 20 to 30 intervals or "bins" for this work.)

(b) Estimate the mean and variance of Y_1 and Y_2 and compare the estimated values to those from theoretical calculations.

(c) Plot the estimated PDF and CDF of Y_1 and Y_2 and compare their shapes to those of the Gaussian. Do this by plotting a Gaussian curve of the correct mean and variance on top of your PDF result.

2. Generate 20 sequences of IID exponential random variables X_i using parameter $\lambda = 0.5i$ for the ith sequence. Obtain a new random variable Y as follows:

$$Y = \sum_{i=1}^{20} X_i$$

(Note that if you have 10,000 realizations of X_i in the ith sequence, you will have 10,000 realizations of the random variable Y.)

(a) Experimentally determine the PDF and CDF for Y.

(b) Compute the theoretical mean and variance for Y.

(c) Plot the PDF and CDF and compare their shapes to those of the Gaussian.

3. Successively convolve 50 Bernoulli PMFs (not the random variables). Let $p = 0.8$ for all of the PMFs. Plot and compare the shape (only the shape) of the result to that of the Gaussian PDF. What is the theoretical mean and standard deviation of a sum of 50 Bernoulli random variables? How does it compare to what you see in the plot?

MATLAB programming notes

The MATLAB functions 'mean' and 'std' can be used in steps 1(b) and 2(b) to estimate mean and variance. The function 'pdfcdf' from the software package can be used for Step 1(a); the function 'expon' can be used for Step 2.

Project 6.2

The objective in this project is to experimentally investigate the mean and variance of the sample mean statistic using random variables of different kinds. This project builds upon Computer Project 3.1 of Chapter 3 and Computer Project 4.1 of Chapter 4.

1. Generate sequences of random variables as specified in Project 3.1 (a)–(f).
2. For each random variable in Project 3.1 (a)–(f), calculate the sample mean

$$M_n = \frac{1}{n} \sum_{i=1}^{n} X_i$$

   using $n = 10$ variables in the sequence.
3. Repeat the last step a total of $N = 10,000$ times each time using a different set of the X_i so that you have N estimates of the sample mean M_n.
4. Estimate the mean and the variance of the *sample mean* using the following formulas:

$$\mathcal{E}\{M_n\} \approx \frac{1}{N} \sum_{k=1}^{N} M_n^{(k)} \qquad \text{Var}\,[M_n] \approx \frac{1}{N} \sum_{i=1}^{N} X_i^2 - (\mathcal{E}\{M_n\})^2$$

   where $M_n^{(k)}$ represents the kth estimate of the sample mean.
5. Repeat Steps 2, 3, and 4 by increasing the length n to 100, 1,000, and 10,000 random variables. Plot the experimental mean and variance of M_n versus the length n on a logarithmic scale. How well do the experimental results correspond to the theory?

7

Random Processes

This chapter introduces random processes, the autocorrelation function, and the power spectral density function. In the previous chapters, whether it is the case of a single random variable or that of multiple random variables, we discussed the characterization of uncertainty without concern for time. In applications that involve signals, noise, dynamic systems, etc., one needs to account for the fact that the underlying phenomena are a function of time. A *random process* can be thought of as a random variable that is a function of time. While that is a simplistic extension from a random variable to a random process, several new concepts and tools are required to define a random process.

In this connection, we will discuss the concept of an ensemble and the principles of stationarity and ergodicity. While the random processes can be characterized in a number of ways, in order to deal with the random processes of interest, such as electrical signals and queues, we will characterize random processes using the time- and frequency-domain representation of second moment functions, namely, correlation and power spectral density functions. Response of linear time invariant systems to random inputs is briefly treated at an introductory level. Several applications, such as the cross-correlation receiver and system identification, and many examples are also presented throughout this chapter.

7.1 *Random Process*

In Chapter 1, we introduced the probability model, which was then used as basis for defining the random variable in Chapter 2. A random variable X is a mapping that assigns every outcome s in the sample space S to a real number. In other words, a random variable is a function of the outcomes of a random experiment, $X(s)$. The concept of random process is based upon extending the concept of random variable to include time as a parameter. Fig. 7.1 graphically presents this concept of a random process by extending the probability model.

There are many applications in which the outcomes of a random experiment are a function of time. For example, the voltage or current waveform received at the input of a radar or a communication receiver is a function of time and uncertain. In some applications, the outcome of a random experiment can be a function of space; an example is the luminance of a photograph. In applications, involving such signals as video, the outcomes are a function of *both* time and space. Here we limit the discussion to the dependence of the outcomes to either time or space, but not both.

A *random process* $X(t, s)$ is a family of functions for which the underlying mapping takes into account both time t and the outcomes of the random experiment s. Another widely used term for random processes is stochastic processes. A time sample of a random process $X(s, t_1)$ at time $t = t_1$ represents a random variable. Consequently, a random process can be thought of as a continuum of random variables. In the literature, it is common practice to drop the variable s from $X(t, s)$ and denote the

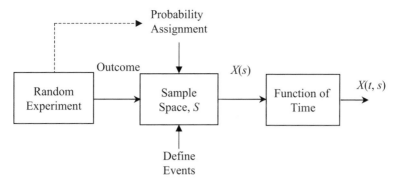

Figure 7.1 The random process model.

random process as $X(t)$. Unless there is a specific reason, here we will use the notation $X(t)$ for the random process.

The sample space can be continuous or discrete; the time index can be continuous or discrete. Consequently, a random process can be of four different types: discrete-magnitude discrete-time, discrete-magnitude continuous-time, continuous-magnitude discrete-time, and continuous-magnitude continuous-time.

Most signals of interest originate in analog form. For example, the audio and video signals are analog in nature. Due to many advantages of digital processing of these signals, we first convert the analog signals into digital form. The digital representation is invariably an approximation of the analog signal, which means that a finite amount of error or noise is introduced in the process of converting an analog waveform into a digital form. Even if the signal being digitized is completely deterministic, the approximation noise introduced during digitization is a random process. Further details of this will be presented in Section 7.5.

Two principal blocks of an analog to digital converter are the sampler and quantizer shown in Figure 7.2. The input $X(t)$ is a continuous-time and continuous-magnitude signal; it has infinite precision in both time and magnitude. At the output of the sampler, we have a discrete-time continuous-magnitude signal. The quantizer takes each of the samples with infinite precision and approximates them to finite values. This step is followed by an encoder that converts each of the quantized samples into a binary sequence, thus creating a discrete-time discrete-magnitude or *digital* signal.

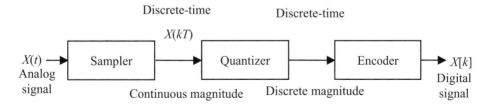

Figure 7.2 Block diagram of an analog to digital converter.

Electrical signals in general can be categorized into deterministic and non-deterministic or random processes (signals). The parameters of a deterministic signal are completely known to the designer and hence there is no uncertainty about the signal values. Random processes (signals) are sometimes further classified into regular random processes and predictable random processes. To explain the difference between these two it suffices to show an example of the latter. In a predictable random process, one or more parameters can be random variables but the form of the signal itself is deterministic. For example, the future values of a predictable random process can be accurately estimated from the knowledge of the past values. A signal of the form

$$X(t) = A\sin(\omega_0 t + \theta)$$

where A and ω_0 are constants, and θ is a random variable is a predictable random process. Figure 7.3(a) illustrates a predictable random process for which ω_0 is a random variable, and Figure 7.3(b) is an example of random amplitude. To construct these, see Figure 7.4 for an example of a regular (non-predictable) random process. The waveform of a regular random process has no definite shape and its values cannot be accurately predicted from the knowledge of the past values. Speech waveforms and electrical noise are examples of a regular random process.

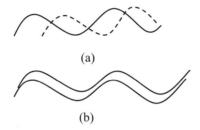

(a)

(b)

Figure 7.3 Examples of a predictable random process.

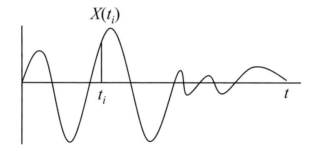

Figure 7.4 A typical regular random process (non-predictable).

7.1.1 The ensemble

In the previous section we defined a random process as a family of functions $X(t, s)$, as a function of both time t and outcome s of a random experiment. This family of functions is known as an *ensemble*. Member functions of the ensemble are also referred to as *sample functions* or *realizations*. Ideally, an ensemble may consist of an infinite number of realizations or sample functions. Graphically, as shown in Fig. 7.5, an ensemble can be considered as having two directions, one corresponding to time and the other corresponding to the realization. Given a sample function of a random process, i.e., for a fixed realization s, one can proceed along the time direction. On the other hand, if the time is fixed, we have a collection of samples corresponding to the various realizations.

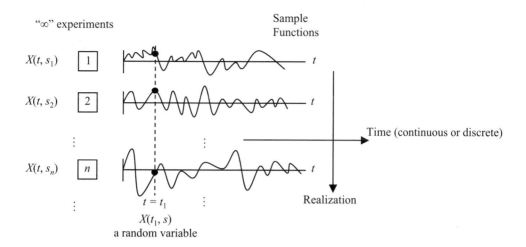

Figure 7.5 Graphical illustration of an ensemble representing a random process.

In this chapter the terms *random process* and *random signal* are used interchangeably. It may seem appropriate to call an electrical signal exhibiting some form of uncertainty in wave shape, frequency, or phase a random process, but one needs to exercise caution in the use of these terms. To that end, we remind the reader that a random process is an abstraction based upon the concept of an ensemble. In other words, a random process truly represents the collection of all the sample functions in the ensemble. On the other hand, a random signal can be considered *a sample function*.

Since any sample of a random process is a random variable, a random process can be considered a continuum of random variables of which theoretically there are an infinite number. In the previous chapters, we learned that a random variable is characterized by a probability density function (PDF) and a cumulative distribution function (CDF). This suggests that every time sample of a random process has a PDF and a CDF.

If we consider a single sample $X(t_i)$ of a random process $X(t)$, then the PDF of the sample, i.e., the PDF of the random process at the time instant $t = t_i$, is $f_{X(t_i)}(x(t_i))$ and the corresponding CDF is $F_{X(t_i)}(x(t_i))$. This notation is comprehensive. For simplicity, however, we denote the PDF and CDF as $f_{X(t)}(x_i; t_i)$ and $F_{X(t)}(x_i; t_i)$, respectively.

Extending this idea to two samples $X(t_i)$ and $X(t_j)$, we now have a joint PDF of these samples $f_{X(t)}(x_i, x_j; t_i, t_j)$ or a joint CDF $F_{X(t)}(x_i, x_j; t_i, t_j)$. This can naturally be generalized to any number of samples of $X(t)$. In essence, the PDF of a random process is a joint PDF of its samples. We will further discuss the PDF and the CDF of a random process in Section 7.3.

Before we proceed further, let us make a quick note about the variable t. In most cases, t represents time and we will refer to it in that way; however, it can also be used to represent other parameters, such as space (for example, signals received at an array of antennas or samples of a photograph).

7.2 First and Second Moments of a Random Process

A given random process can be characterized by an n-dimensional joint PDF, where n is the number of samples. Such a characterization is not always practical and may not be necessary for solving many engineering problems. The first and the second moments of a random process on the other hand provide a partial characterization of a random process, which for most engineering problems is adequate.

7.2.1 Mean

The first moment or the mean of a random process $X(t)$ is defined as

$$m_X(t) = \mathcal{E}\{X(t)\} = \int_{-\infty}^{\infty} x f_{X(t)}(x)dx \tag{7.1}$$

where $f_{X(t)}(x)$ is the PDF of $X(t)$. The mean $m_X(t)$ can be a function of time and is not necessarily a fixed value.

7.2.2 Autocorrelation and autocovariance functions

The second moment of a random process $X(t)$ is expressed as a joint moment between two samples $X_0 = X(t_0)$ and $X_1 = X(t_1)$, where t_0 and t_1 are the sampling instants. Here we define two forms of second moments: the correlation function or the second moment, and the covariance function or the second central moment. The autocorrelation function $R_X(t_1, t_0)$ of a random process $X(t)$ is defined as

$$R_X(t_1, t_0) = \mathcal{E}\{X(t_1)X(t_0)\} = \int_{-\infty}^{\infty}\int_{-\infty}^{\infty} x_1 x_0 f_{X(t_1)X(t_0)}(x_1, x_0)dx_1 dx_0 \tag{7.2}$$

where $f_{X(t_1)X(t_0)}$ is the second order PDF of $X(t)$ or the joint PDF of the samples $X(t_1)$ and $X(t_0)$. It is important to note that the autocorrelation function depends on the sampling time instants t_1 and t_0. In the case of $t_1 = t_0$, we have $X(t_1) = X(t_0)$ and the autocorrelation function becomes the *average power* or the *mean squared value* of the random process

$$R_X(t_1, t_1) = \mathcal{E}\{X^2(t_1)\} = \int_{-\infty}^{\infty} x_1^2 f_{X(t_1)}(x_1)dx_1$$

The autocovariance function $C_X(t_1, t_0)$ of a random process $X(t)$ is defined as the joint central moment between of $X_0 = X(t_0)$ and $X_1 = X(t_1)$

$$\begin{aligned} C_X(t_1, t_0) &= \mathcal{E}\{(X(t_1) - m_X(t_1))(X(t_0) - m_X(t_0))\} \\ &= \int_{-\infty}^{\infty}\int_{-\infty}^{\infty} x_1 x_0 f_{X(t_1)X(t_0)}(x_1, x_0)dx_1 dx_0 \end{aligned} \tag{7.3}$$

where $m_{X(t_1)}$ and $m_X(t_0)$ are the mean values of the random process at sampling time instants t_1 and t_0, respectively. For $t_1 = t_0$, we have $X(t_1) = X(t_0)$ and the autocovariance function becomes the variance of the random process

$$C_X(t_1, t_1) = \sigma_X^2(t_1) = \mathcal{E}\{(X(t_1) - m_X(t_1))^2\} = \int_{-\infty}^{\infty} (x_1 - m_X(t_1))^2 f_{X(t_1)}(x_1)dx_1$$

Note that the variance could, in general, depend on the sampling time t_1.

The autocorrelation function and autocovariance function are related to each other. From (5.28), this relationship can be expressed as

$$C_X(t_1, t_0) = R_X(t_1, t_0) - m_X(t_1)m_X(t_0) \qquad (7.4)$$

The following example illustrates computation of the mean and the autocorrelation function of a random process.

Example 7.1: Consider a random process

$$X(t) = A\sin(\omega t + \theta)$$

where A and ω are known parameters while θ is a uniform random variable in the range $[-\pi, \pi]$.

The first moment of this random process is given by

$$
\begin{aligned}
m_X(t) &= \mathcal{E}\{X(t)\} = \int_{-\pi}^{\pi} A\sin(\omega t + \theta)f_\Theta(\theta)d\theta \\
&= \frac{A}{2\pi}\int_{-\pi}^{\pi} \sin(\omega t + \theta)d\theta
\end{aligned}
$$

where we have used the PDF of θ, $f_\Theta(\theta) = 1/2\pi$ for $-\pi \le \theta \le \pi$. Evaluating the integral, we obtain

$$
\begin{aligned}
m_X(t) &= -\frac{A}{2\pi}\cos(\omega t + \theta)\Big|_{-\pi}^{\pi} = \frac{A}{2\pi}[\cos(\omega t - \pi) - \cos(\omega t + \pi)] \\
&= \frac{A}{2\pi}[-\cos(\omega t) + \cos(\omega t)] = 0
\end{aligned}
$$

The autocorrelation function of $X(t)$ is given by

$$
\begin{aligned}
R_X(t_1, t_0) &= \mathcal{E}\{X(t_1)X(t_0)\} = \int_{-\pi}^{\pi} A\sin(\omega t_1 + \theta)A\sin(\omega t_0 + \theta)f_\Theta(\theta)d\theta \\
&= \frac{A^2}{2\pi}\int_{-\pi}^{\pi}\sin(\omega t_1 + \theta)\sin(\omega t_0 + \theta)d\theta \\
&= \frac{A^2}{4\pi}\int_{-\pi}^{\pi}[\cos\omega(t_1 - t_0) - \cos(\omega(t_1 + t_0) + 2\theta)]d\theta \\
&= \frac{A^2}{2}\cos\omega(t_1 - t_0) - \frac{A^2}{4\pi}\int_{-\pi}^{\pi}\cos(\omega(t_1 + t_0) + 2\theta)d\theta = \frac{A^2}{2}\cos\omega(t_1 - t_0)
\end{aligned}
$$

Since the mean is 0, the autocovariance function is the same as the autocorrelation function, i.e., $C_X(t_1, t_0) = R_X(t_1, t_0)$.

□

On the lines of (5.2a), the correlation coefficient $\rho_X(t_1, t_0)$ of a random process $X(t)$ is defined as the normalized covariance between two samples $X(t_1)$ and $X(t_0)$

$$\rho_X(t_1, t_0) = \frac{C_X(t_1, t_0)}{\sqrt{C_X(t_1, t_1)C_X(t_0, t_0)}} \qquad (7.5)$$

thereby the range of values taken by the correlation function is limited to $-1 \le \rho_X(t_1, t_0) \le 1$.

The above discussion on the first and the second moments of the continuous random processes can be straightforwardly extended to the discrete time case. Table 7.1 summarizes the expressions for the mean, the autocorrelation function, the autocovariance function, and the correlation coefficient of a discrete random process.

quantity	definition
Mean	$m_X[k] = \mathcal{E}\{X[k]\}$
Autocorrelation Function	$R_X[k_1, k_0] = \mathcal{E}\{X[k_1]X[k_0]\}$
Autocovariance Function	$C_X[k_1, k_0] = \mathcal{E}\{(X[k_1] - m_X[k_1])(X[k_0] - m_X[k_0])\}$
Interrelationship	$C_X[k_1, k_0] = R_X[k_1, k_0] - m_X[k_1]m_X[k_0]$
Correlation Coefficient	$\rho_X[k_1, k_0] = \dfrac{C_X[k_1, k_0]}{\sqrt{C_X[k_1, k_1]C_X[k_0, k_0]}}$

Table 7.1 The first and the second moments of a discrete random process.

Example 7.2: Suppose that a discrete-time random process $X[k]$ having a mean and an autocorrelation function

$$m_X[k] = 3$$
$$R_X[k_1, k_0] = 9 + 4e^{-0.2|k_1 - k_0|}$$

is given. Note that the mean is not a function of the time index k. Let us determine the mean, the variance, the covariance, and the correlation coefficient of samples $X_1 = X[8]$ and $X_0 = X[5]$, i.e., $k_1 = 8$ and $k_0 = 5$.

Since the mean is not a function of time, $m_X[8] = m_X[5] = 3$. To determine the variance, we can use (7.4) to write

$$\text{Var}[X[k_1]] = C_X[k_1, k_1] = R_X[k_1, k_1] - m_X^2[k_1]$$

which yields $C_X[8, 8] = C_X[5, 5] = 13 - 9 = 4$.

The covariance function for $k_1 = 8$ and $k_0 = 5$ is given by

$$C_X[8, 5] = R_X[8, 5] - m_X[8]m_X[5]$$
$$= 9 + 4e^{-0.2 \times 3} - 9 = 2.1952$$

For $k_1 = 8$ and $k_0 = 5$, from Table 7.1, the correlation coefficient is

$$\rho_X[8, 5] = \frac{C_X[8, 5]}{\sqrt{C_X[8, 8]C_X[5, 5]}} = \frac{2.1952}{4} = 0.5488$$

□

If the given random process $X(t)$ is a complex random process, the corresponding mean, the autocorrelation function, and the autocovariance function are defined as

$$m_X(t) = \mathcal{E}\{X(t)\}$$
$$R_X(t_1, t_0) = \mathcal{E}\{X(t_1)X^*(t_0)\}$$
$$C_X(t_1, t_0) = \mathcal{E}\{(X(t_1) - m_X(t_1))(X^*(t_0) - m_X^*(t_0))\}$$
$$C_X(t_1, t_0) = R_X(t_1, t_0) - m_X(t_1))m_X^*(t_0) \qquad (7.6)$$

where the asterisk * indicates complex conjugation.

Example 7.3: Let us consider a complex random process

$$X(t) = Ae^{j\omega t}$$

where A is a Gaussian random variable with mean $\mu = 2$ and a variance of $\sigma^2 = 3$ while ω is a deterministic parameter.

The mean of this random process is given by

$$m_X(t) = \mathcal{E}\{X(t)\} = \mathcal{E}\{A\} e^{-j\omega t} = 2e^{j\omega t}$$

Note that the mean retains the dependence on $e^{j\omega t}$.

The autocorrelation function is

$$
\begin{aligned}
R_X(t_1, t_0) &= \mathcal{E}\{X(t_1)X^*(t_0)\} = \mathcal{E}\{Ae^{j\omega t_1} Ae^{-j\omega t_0}\} \\
&= \mathcal{E}\{A^2\} e^{j\omega(t_1 - t_0)} = 7e^{j\omega(t_1 - t_0)}
\end{aligned}
$$

where the second moment $\mathcal{E}\{A^2\} = \sigma_A^2 + m_A^2 = 7$.

The autocovariance function $C_X(t_1, t_0)$ is obtained using the relationship (7.6) as follows

$$C_X(t_1, t_0) = 7e^{j\omega(t_1 - t_0)} - 4e^{j\omega t_1} e^{-j\omega t_0} = 3e^{j\omega(t_1 - t_0)}$$

□

7.2.3 Cross-correlation function

The cross-correlation function of two random processes $X(t)$ and $Y(t)$ is defined as

$$R_{XY}(t_1, t_0) = \mathcal{E}\{X(t_1)Y(t_0)\} = \int_{-\infty}^{\infty} \int_{-\infty}^{\infty} x_1 y_0 f_{X(t_1)Y(t_0)}(x_1, y_0) dx_1 dy_0 \qquad (7.7)$$

where $f_{X(t_1)Y(t_0)}(x_1, y_0)$ is the joint density function of the samples $X_1 = X(t_1)$ and $Y_0 = Y(t_0)$.

The cross-covariance function of $X(t)$ and $Y(t)$ is defined as

$$
\begin{aligned}
C_{XY}(t_1, t_0) &= \mathcal{E}\{(X(t_1) - m_X(t_1))(Y(t_0) - m_Y(t_0))\} \qquad (7.8) \\
&= \int_{-\infty}^{\infty} \int_{-\infty}^{\infty} (x_1 - m_X(t_1))(y_0 - m_Y(t_0)) f_{X(t_1)Y(t_0)}(x_1, y_0) dx_1 dy_0
\end{aligned}
$$

where $m_X(t_1)$ is the mean of random process $X(t)$ at sampling instant $t = t_1$ and $m_Y(t_0)$ is the mean of random process $Y(t)$ at sampling instant $t = t_0$.

From (7.4), we can represent the cross-covariance function in terms of the cross-correlation function and the means of the random processes

$$C_{XY}(t_1, t_0) = R_{XY}(t_1, t_0) - m_X(t_1)m_Y(t_0) \qquad (7.9)$$

Example 7.4: Consider two random processes

$$
\begin{aligned}
X(t) &= A\sin(\omega t + \theta) + B\cos(\omega t) \\
Y(t) &= B\cos(\omega t)
\end{aligned}
$$

where A and ω are known parameters while θ is a uniform random variable in the range $[-\pi, \pi]$, and B is a Gaussian random variable with mean 2 and variance 3. Random variables θ and B are assumed to be independent.

First, let us determine the first moments of these random processes. The mean of $X(t)$ is

$$m_X(t) = \mathcal{E}\{X(t)\} = A\mathcal{E}\{\sin(\omega t + \theta)\} + \mathcal{E}\{B\}\cos(\omega t) = 2\cos(\omega t)$$

where $\mathcal{E}\{\sin(\omega t + \theta)\} = 0$ since θ is uniformly distributed on $[-\pi, \pi]$. The mean of $Y(t)$ is given by

$$m_Y(t) = \mathcal{E}\{Y(t)\} = \mathcal{E}\{B\}\cos(\omega t) = 2\cos(\omega t)$$

We now proceed to determine the cross-autocorrelation function of $X(t)$ and $Y(t)$. From (7.7),

$$
\begin{aligned}
R_{XY}(t_1, t_0) &= \mathcal{E}\{X(t_1)Y(t_0)\} = \mathcal{E}\{(A\sin(\omega t + \theta) + B\cos(\omega t))B\cos(\omega t)\} \\
&= A\cos(\omega t)\mathcal{E}\{B\sin(\omega t_0 + \theta)\} + \mathcal{E}\{B^2\}\cos(\omega t_1)\cos(\omega t_0)
\end{aligned}
$$

Since θ and B are independent, we can write

$$R_{XY}(t_1, t_0) = A\cos(\omega t)\mathcal{E}\{B\}\mathcal{E}\{\sin(\omega t + \theta)\} + \mathcal{E}\{B^2\}\cos(\omega t_1)\cos(\omega t_0)$$

The first term is 0 because $\mathcal{E}\{\sin(\omega t + \theta)\} = 0$ and the second moment of B is given by $\mathcal{E}\{B^2\} = \text{Var}[B] + m_B^2 = 7$. This yields

$$R_{XY}(t_1, t_0) = 7\cos(\omega t_1)\cos(\omega t_0)$$

From (7.9), the cross-covariance function

$$
\begin{aligned}
C_{XY}(t_1, t_0) &= R_{XY}(t_1, t_0) - m_X(t_1)m_Y(t_0) \\
&= 7\cos(\omega t_1)\cos(\omega t_0) - 4\cos(\omega t_1)\cos(\omega t_0) = 3\cos(\omega t_1)\cos(\omega t_0)
\end{aligned}
$$

□

The following example illustrates that the autocorrelation function of a complex random process requires computation of the autocorrelation as well as the cross-correlation functions of the component processes.

Example 7.5: Consider a complex random process

$$Z(t) = X(t) - jY(t)$$

where $X(t) = A\cos\omega t$, $Y(t) = B\sin\omega t$, and A and B are random variables of unknown distribution.

The mean of this random process is

$$m_Z(t) = \mathcal{E}\{Z(t)\} = \mathcal{E}\{A\}\cos\omega t - j\mathcal{E}\{B\}\sin\omega t$$

For complex processes, the autocorrelation function is defined as

$$R_Z(t_1, t_0) = \mathcal{E}\{Z(t_1)Z^*(t_0)\}$$

where the second term is conjugated. We have

$$
\begin{aligned}
R_Z(t_1, t_0) &= \mathcal{E}\{(X(t_1) - jY(t_1))(X(t_0) + jY(t_0))\} \\
&= \mathcal{E}\{X(t_1)X(t_0)\} + j\mathcal{E}\{X(t_1)Y(t_0)\} - j\mathcal{E}\{Y(t_1)X(t_0)\} + \mathcal{E}\{Y(t_1)Y(t_0)\} \\
&= R_X(t_1, t_0) + jR_{XY}(t_1, t_0) - jR_{YX}(t_1, t_0) + R_Y(t_1, t_0)
\end{aligned}
$$

where $R_{YX}(t_1, t_0)$ is the cross-correlation function between $Y(t)$ and $X(t)$, which is not the same as $R_{XY}(t_1, t_0)$, in general.

Let us now determine the component correlation functions in the above equation:

$$
\begin{aligned}
R_X(t_1, t_0) &= \mathcal{E}\{X(t_1)X(t_0)\} = \mathcal{E}\{A^2\}\cos\omega t_1\cos\omega t_0 \\
R_{XY}(t_1, t_0) &= \mathcal{E}\{X(t_1)Y(t_0)\} = \mathcal{E}\{AB\}\cos\omega t_1\sin\omega t_0 \\
R_{YX}(t_1, t_0) &= \mathcal{E}\{Y(t_1)X(t_0)\} = \mathcal{E}\{AB\}\sin\omega t_1\cos\omega t_0 \\
R_Y(t_1, t_0) &= \mathcal{E}\{Y(t_1)Y(t_0)\} = \mathcal{E}\{B^2\}\sin\omega t_1\sin\omega t_0
\end{aligned}
$$

Substituting these into the above equation yields

$$
\begin{aligned}
R_Z(t_1, t_0) &= R_X(t_1, t_0) + jR_{XY}(t_1, t_0) - jR_{YX}(t_1, t_0) + R_Y(t_1, t_0) \\
&= E\{A^2\}\cos\omega t_1 \cos\omega t_0 + jE\{AB\}\cos\omega t_1 \sin\omega t_0 \\
&\quad - jE\{AB\}\sin\omega t_1 \cos\omega t_0 + E\{B^2\}\sin\omega t_1 \sin\omega t_0 \\
&= E\{A^2\}\cos\omega t_1 \cos\omega t_0 + E\{B^2\}\sin\omega t_1 \sin\omega t_0 \\
&\quad + j\left(E\{AB\}\cos\omega t_1 \sin\omega t_0 - E\{AB\}\sin\omega t_1 \cos\omega t_0\right)
\end{aligned}
$$

□

7.3 Properties: Independence, Stationarity, and Ergodicity

A sample $X_0 = X(t_0)$ of a random process $X(t)$ is a random variable. Such a random variable is characterized by a PDF $f_{X(t)}(x_0; t_0)$ or a CDF $F_{X(t)}(x_0; t_0)$. This is referred to as the first-order characterization of the random process. By extension, n samples $\{X(t_0), X(t_1), \dots, X(t_{n-1})\}$ of a random process $X(t)$ are characterized by the nth-order density function, i.e., their nth-order joint PDF

$$
f_{X(t)}(x_0, x_0, \dots, x_{n-1}; t_0, t_1, \dots, t_{n-1})
$$

or their nth-order joint CDF

$$
F_{X(t)}(x_0, x_1, \dots, x_{n-1}; t_0, t_1, \dots, t_{n-1})
$$

The interrelationship between the density and distribution function can be expressed as in Chapter 5.

7.3.1 Statistical independence

Two random processes $X(t)$ and $Y(t)$ are statistically independent if any random variable $\{X(t_0), X(t_1), \dots, X(t_{n-1})\}$ of random process $X(t)$ is statistically independent of any random variable $\{Y(t_0), Y(t_1), \dots, Y(t_{m-1})\}$ of random process $Y(t)$. Stated in equation form, the joint density function between the two groups of random variables must be factorable:

$$
\begin{aligned}
&f_{X(t)Y(t)}(x_0, x_1, \dots, x_{n-1}, y_0, y_1, \dots, y_{m-1}; t_0, t_1, \dots, t_{n-1}, t_0, t_1, \dots, t_{m-1}) \\
&= f_{X(t)}(x_0, x_1, \dots, x_{n-1}; t_0, t_1, \dots, t_{n-1})f_{Y(t)}(y_0, y_1, \dots, y_{m-1}; t_0, t_1, \dots, t_{m-1}) \quad (7.10)
\end{aligned}
$$

This concept is further expanded in Section 7.4 to describe the statistical independence of samples belonging to the same random process.

7.3.2 Strict sense stationarity

A random process is stationary in the *strict sense* if its statistical characterization is independent of the variable t, which usually represents time.

In general, a process is said to be strict sense stationary if $X(t)$ and $X(t + \xi)$ have the same statistical characterization for any ξ or, equivalently, both $X(t)$ and $X(t+\xi)$ have the same density function for any value of ξ. Consequently, the first-order density function

$$
f_{X(t)}(x; t) = f_{X(t)}(x; t + \xi)
$$

for any ξ.

To be more specific, a random process $X(t)$ is said to be *first order* stationary if its first-order density function or distribution function do not change with a shift in t, i.e., samples $X(t_0)$ and $X(t_0 + \xi)$ have the same density and distribution functions

$$
\begin{aligned}
f_{X(t)}(x_0; t_0) &= f_{X(t)}(x_0; t_0 + \xi) = f_{X(t)}(x) \\
F_{X(t)}(x_0; t_0) &= F_{X(t)}(x_0; t_0 + \xi) = F_{X(t)}(x)
\end{aligned}
$$

Thus, the first-order density function of a strict sense stationary random process is independent of the variable t.

Continuing along the same lines as above, the second-order density function

$$
f_{X(t)}(x_0, x_1; t_0, t_1) = f_{X(t)}(x_0, x_1; t_0 + \xi, t_1 + \xi)
$$

for any ξ. Since ξ can have any value, by letting $\xi = -t_0$, we have

$$
f_{X(t)}(x_0, x_1; t_0, t_1) = f_{X(t)}(x_0, x_1; 0, t_1 - t_0) = f_{X(t)}(x_0, x_1; \tau)
$$

where $\tau = t_1 - t_0$. Thus the second-order density function is not a function of the variable t but the difference τ. Similarly, the second-order distribution function is only a function of τ

$$
F_{X(t)}(x_0, x_1; t_0, t_1) = F_{X(t)}(x_0, x_1; \tau)
$$

By extension, for the nth-order case, we have

$$
\begin{aligned}
f_{X(t)}&(x_0, x_1, \ldots, x_{n-1}; t_0, t_1, \ldots, t_{n-1}) \\
&= f_{X(t)}(x_0, x_1, \ldots, x_{n-1}; t_0 + \xi, t_1 + \xi, \ldots, t_{n-1} + \xi) \\
F_{X(t)}&(x_0, x_1, \ldots, x_{n-1}; t_0, t_1, \ldots, t_{n-1}) \\
&= F_{X(t)}(x_0, x_1, \ldots, x_{n-1}; t_0 + \xi, t_1 + \xi, \ldots, t_{n-1} + \xi) \quad (7.11)
\end{aligned}
$$

for any ξ. As in the second-order case, we can show that the density function is a function of time differences, such as $t_1 - t_0$, $t_2 - t_0$, and $t_{n-1} - t_0$, not the absolute time t.

Given two random processes $X(t)$ and $Y(t)$, they are said to be jointly strict sense stationary if their joint statistical characterization is the same as that of $X(t+\xi)$ and $Y(t+\xi)$ for any ξ.

A complex random process $Z(t) = X(t) + jY(t)$ is strict sense stationary if the component random processes $X(t)$ and $Y(t)$ are jointly stationary. In other words the complex random process $Z(t)$ is strict sense stationary if $Z(t)$ and $Z(t + \xi)$ have the same statistics for any ξ.

7.3.3 Wide sense stationarity

Strict sense stationarity is a powerful property that requires the knowledge of the density or distribution functions. Nevertheless, in most problems of practical interest in engineering, we work with first and second order moments (the mean and the autocorrelation function) of a random process. The wide sense stationarity property is useful from a practical standpoint because it is defined in terms of the first and the second moments of a random process.

A random process is said to be *wide sense stationary* if its mean is constant:

$$
\boxed{m_X = \mathcal{E}\{X(t)\}}
\qquad (7.12)
$$

and its autocorrelation function is a function of only the time difference $t_1 - t_0$:

$$R_X(t_1, t_0) = R_X(t_1 - t_0) \tag{7.13}$$

A second-order strict sense random process is also wide sense stationary, but the reverse is not true.

The following examples are illustrations of how to determine if a process is wide sense stationary.

Example 7.6: Consider the random process of Example 7.1:

$$X(t) = A\sin(\omega t + \theta)$$

where A and ω are known parameters while θ is a uniform random variable in the range $[-\pi, \pi]$. Let us examine if this random process is wide sense stationary.

The mean of the random process

$$m_X(t) = \mathcal{E}\{X(t)\} = 0$$

is fixed, thereby satisfying the first of the two required conditions for a process to be wide sense stationary.

The autocorrelation function was shown to be

$$R_X(t_1, t_0) \quad = \quad \mathcal{E}\{X(t_1)X(t_0)\} = \frac{A^2}{2}\cos\omega(t_1 - t_0)$$

This satisfies the second condition that the autocorrelation function can be only a function of the time difference $t_1 - t_0$. Therefore $X(t)$ is a wide sense stationary random process.

□

It is not an implicit requirement that random processes in general be wide sense stationary. For example, speech signals are not wide sense stationary (although they may be considered wide sense stationary over short durations). Let us examine the simple complex random process in Example 7.3 to show that it is not wide sense stationary.

Example 7.7: Consider the complex random process from Example 7.2:

$$X(t) = Ae^{j\omega t},$$

where A is a Gaussian random variable with mean $\mu = 2$ and a variance of $\sigma^2 = 3$.

Its mean and the autocorrelation functions were found to be

$$m_X(t) \quad = \quad \mathcal{E}\{X(t)\} = 2e^{j\omega t}$$
$$R_X(t_1, t_0) \quad = \quad \mathcal{E}\{X(t_1)X^*(t_0)\} = 7e^{j\omega(t_1 - t_0)}$$

Since the mean is a function of t, it does not satisfy the condition in (7.12). As a result, the process is not wide sense stationary.

□

In certain situations, a given random process may be wide sense stationary under certain constraints as illustrated in the following example.

Example 7.8: Consider the complex random process introduced in Example 7.5

$$Z(t) = X(t) - jY(t)$$

where $X(t) = A \cos \omega t$, $Y(t) = B \sin \omega t$, and A and B are random variables of unknown distribution.

The mean and the autocorrelation function of this random process were found to be

$$
\begin{aligned}
m_Z(t) &= \mathcal{E}\{A\} \cos \omega t - j\mathcal{E}\{B\} \sin \omega t \\
R_Z(t_1, t_0) &= \mathcal{E}\{A^2\} \cos \omega t_1 \cos \omega t_0 + j\mathcal{E}\{AB\} \cos \omega t_1 \sin \omega t_0 \\
&= \mathcal{E}\{A^2\} \cos \omega t_1 \cos \omega t_0 + \mathcal{E}\{B^2\} \sin \omega t_1 \sin \omega t_0 \\
&\quad + j\left(\mathcal{E}\{AB\} \cos \omega t_1 \sin \omega t_0 - \mathcal{E}\{AB\} \sin \omega t_1 \cos \omega t_0\right)
\end{aligned}
$$

Clearly, $Z(t)$ is not a wide sense stationary process. Both the mean and the autocorrelation function are a function of t, hence they do not satisfy the conditions (7.12) and (7.13).

Nevertheless, $Z(t)$ can be made wide sense stationary provided the random variables meet certain criteria. For example, if we require the random variables A and B to have 0 mean, then the mean of $Z(t)$ is 0. If we further require that A and B are uncorrelated, i.e., $\mathcal{E}\{AB\} = \mathcal{E}\{A\}\mathcal{E}\{B\}$, the cross terms in the the autocorrelation function become 0 and the above espression simplifies to

$$R_Z(t_1, t_0) = \mathcal{E}\{A^2\} \cos \omega t_1 \cos \omega t_0 + \mathcal{E}\{B^2\} \sin \omega t_1 \sin \omega t_0$$

If we force another assumption that the second moments of A and B are the same, i.e., $\mathcal{E}\{A^2\} = \mathcal{E}\{B^2\}$, we have

$$R_Z(t_1, t_0) = \mathcal{E}\{A^2\}(\cos \omega t_1 \cos \omega t_0 + \sin \omega t_1 \sin \omega t_0) = \mathcal{E}\{A^2\} \cos \omega(t_1 - t_0)$$

Summarizing the above, the random process $Z(t)$ is wide sense stationary if A and B have 0 mean and are uncorrelated random variables with the same second moment.

□

For a wide sense stationary random process, it is conventional to drop the argument t and write the mean as simply m_X since it does not depend on time. Further, in defining the autocorrelation and autocovariance functions it is conventional to replace t_1 with t and t_0 with $t - \tau$, where τ is known as the *lag* variable. The autocorrelation function and the autocovariance function for a wide sense stationary random process and the correlation coefficient ρ are then independent of t and depend only on the lag τ. These definitions and interrelationships are summarized in Table 7.2.

The definition of wide sense stationarity can be easily extended to the discrete-time case. For a discrete-time random process to be wide sense stationary, its mean must be constant and its autocorrelation function must be only a function of $\ell = k_1 - k_0$. The parameter ℓ is the discrete-time lag. Table 7.3 lists the definitions and interrelationships of discrete time random process.

Let us examine the wide sense stationarity of the random process in Example 7.7.

Example 7.9: Consider the discrete-time random process $X[k]$ of Example 7.2 having mean and autocorrelation function

$$
\begin{aligned}
m_X[k] &= 3 \\
R_X[k_1, k_0] &= 9 + 4e^{-0.2|k_1 - k_0|}
\end{aligned}
$$

The mean is constant and the autocorrelation function is dependent only on the lag $\ell = k_1 - k_0$:

$$R_X[\ell] = 9 + 4e^{-0.2|\ell|}$$

quantity	definition
Mean	$m_X = \mathcal{E}\{X(t)\}$
Autocorrelation Function	$R_X(\tau) = \mathcal{E}\{X(t)X(t-\tau)\}$
Autocovariance Function	$C_X(\tau) = \mathcal{E}\{(X(t) - m_X)(X(t-\tau) - m_X)\}$
Interrelationship	$C_X(\tau) = R_X(\tau) - m_X^2$
Correlation Coefficient	$\rho_X(\tau) = \dfrac{C_X(\tau)}{C_X(0)}$

Table 7.2 Wide sense stationary random process: definitions and interrelationships.

Since the mean and the autocorrelation function satisfy the two required conditions, $X[k]$ is a wide sense random process.

□

quantity	definition
Mean	$m_X = \mathcal{E}\{X[k]\}$
Autocorrelation Function	$R_X[\ell] = \mathcal{E}\{X[k]X[k-\ell]\}$
Autocovariance Function	$C_X[\ell] = \mathcal{E}\{(X[k] - m_X)(X[k-\ell] - m_X)\}$
Interrelationship	$C_X[\ell] = R_X[\ell] - m_X^2$
Correlation Coefficient	$\rho_X[\ell] = \dfrac{C_X[\ell]}{C_X[0]}$

Table 7.3 Discrete time wide sense stationary random process: definitions and interrelationships.

Given two random processes $X(t)$ and $Y(t)$, they are said to be jointly wide sense stationary under the following conditions:

1. Each of $X(t)$ and $Y(t)$ is wide sense stationary (see Table 7.2)

2. $R_{XY}(t_1, t_0) = R_{XY}(t_1 - t_0)$

Under these conditions, we can can define

$$R_{XY}(\tau) = \mathcal{E}\{X(t)Y(t-\tau)\} \tag{7.14}$$

where $\tau = t_1 - t_0$. By changing the roles of $X(t)$ and $Y(t)$ in (7.14), we have

$$R_{YX}(\tau) = \mathcal{E}\{Y(t)X(t-\tau)\}$$

By virtue of their definitions, the two cross-correlation functions are interrelated:

$$R_{YX}(\tau) = R_{XY}(-\tau) \qquad\qquad (7.15)$$

The cross-covariance for two jointly wide sense stationary random processes is likewise defined as

$$C_{XY}(\tau) = \mathcal{E}\{(X(t) - m_X)(Y(t-\tau) - m_Y)\}$$

with

$$C_{YX}(\tau) = C_{XY}(-\tau)$$

Analogous results hold for the discrete-time case.

Example 7.10: Consider two random processes

$$
\begin{aligned}
X(t) &= A\sin\omega t + B\cos(\omega t) \\
Y(t) &= A\sin\omega t - B\cos(\omega t)
\end{aligned}
$$

where ω is a constant, and A and B are uncorrelated random variables with 0 mean $\mathcal{E}\{A\} = \mathcal{E}\{B\} = 0$ and equal variance $\mathrm{Var}\,[A] = \mathrm{Var}\,[B] = \sigma^2$. We need to examine if $X(t)$ and $Y(t)$ are jointly wide sense stationary.

Since the mean values are 0, to test wide sense stationarity, we only need to determine if the correlation functions are dependent on only the lag τ as in Table 7.2 and (7.14).

The autocorrelation function of $X(t)$ is given by

$$
\begin{aligned}
R_X(t_1, t_0) &= \mathcal{E}\{X(t_1)X(t_0)\} = \mathcal{E}\{A^2\}\sin\omega t_1 \sin\omega t_0 + \mathcal{E}\{AB\}\sin\omega t_1 \cos\omega t_0 \\
&\quad + \mathcal{E}\{AB\}\cos\omega t_1 \sin\omega t_0 + \mathcal{E}\{B^2\}\cos\omega t_1 \cos\omega t_0 \\
&= \sigma^2(\sin\omega t_1 \sin\omega t_0 + \cos\omega t_1 \cos\omega t_0) \\
&= \sigma^2 \cos\omega(t_1 - t_0) = \sigma^2 \cos\omega\tau = R_X(\tau)
\end{aligned}
$$

where the cross terms are 0 because $\mathcal{E}\{AB\} = \mathcal{E}\{A\}\mathcal{E}\{B\} = 0$. As a result, $X(t)$ is wide sense stationary. Similarly, $Y(t)$ is also wide sense stationary as

$$R_Y(t_1, t_0) = \sigma^2 \cos\omega(t_1 - t_0) = \sigma^2 \cos\omega\tau = R_Y(\tau)$$

The cross-correlation function

$$
\begin{aligned}
R_{XY}(t_1, t_0) &= \mathcal{E}\{X(t_1)Y(t_0)\} = \mathcal{E}\{A^2\}\sin\omega t_1 \sin\omega t_0 - \mathcal{E}\{AB\}\sin\omega t_1 \cos\omega t_0 \\
&\quad + \mathcal{E}\{AB\}\cos\omega t_1 \sin\omega t_0 - \mathcal{E}\{B^2\}\cos\omega t_1 \cos\omega t_0 \\
&= \sigma^2(\sin\omega t_1 \sin\omega t_0 - \cos\omega t_1 \cos\omega t_0) \\
&= -\sigma^2 \cos\omega(t_1 + t_0)
\end{aligned}
$$

is not a function of τ. Consequently, the process is not jointly wide sense stationary.

□

7.3.4 Properties of correlation functions

The autocorrelation function $R_X(\tau)$ and the autocovariance function $C_X(\tau)$ of a wide sense stationary process $X(t)$ are defined in Table 7.2. Both of these functions exhibit some properties that are important in many engineering applications. Some of these properties are listed below without proofs. Even though these properties are stated in terms of $R_X(\tau)$, they are equally applicable to $C_X(\tau)$.

1. For $\tau = 0$, the autocorrelation function yields the second moment or the average power of a wide sense stationary random process

$$R_X(0) = \mathcal{E}\left\{X^2(t)\right\} \geq 0 \qquad (7.16)$$

The value $R_X(0)$ is considered the origin of the autocorrelation function. The second moment is also referred to as the mean squared value. Similarly, for the autocovariance function

$$C_X(0) = \mathcal{E}\left\{(X(t) - m_X)^2\right\} = \sigma^2$$

yields the variance.

2. The autocorrelation function has the highest value at $\tau = 0$. Accordingly, it is bounded by $R_X(0)$ as follows

$$|R_X(\tau)| \leq R_X(0) \qquad (7.17)$$

(This property is implied by Property 4 below.)

3. The autocorrelation function is even symmetric about its origin

$$R_X(\tau) = R_X(-\tau) \qquad (7.18)$$

4. The autocorrelation function is positive semidefinite and is mathematically stated as

$$\int_{-\infty}^{\infty} \int_{-\infty}^{\infty} a(t_1) R_X(t_1 - t_0) a(t_0) dt_1 dt_0 \geq 0 \qquad (7.19)$$

where $a(t)$ is an arbitrary real-valued function.

The above properties are treated for the continuous-time, real-valued case for simplicity but can be easily extended to the discrete-time case and/or the complex-valued case.

Example 7.11: The random process of Example 7.2 satisfies (7.12) and (7.13) and is thus wide sense stationary. The mean and the autocorrelation function can then be written as

$$m_X = 3$$
$$R_X[\ell] = 9 + 4e^{-0.2|\ell|}$$

A plot of the autocorrelation function is shown in the figure below.

From (7.16), the second moment of $X[k]$ is

$$\mathcal{E}\left\{X^2[k]\right\} = R_X[0] = 13$$

From the figure, one can easily verify that any other value of $R_X[\ell]$ is less than 13, thus satisfying (7.17). By observation, it is seen that $R_X[\ell]$ is even symmetric; for example, $R_X[2] = R_X[-2] = 11.6813 \leq 13$.

The autocovariance function is given by

$$C_X[\ell] = R_X[\ell] - m_X^2 = 4e^{-0.2|\ell|}$$

and plotted below.

The variance of $X[k]$ is given by

$$C_X[0] = R_X[0] - m_X^2 = 13 - 9 = 4$$

□

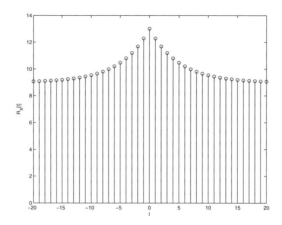

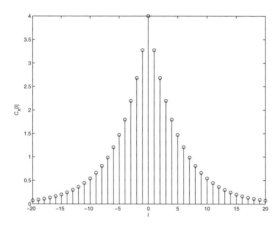

The cross-correlation function of two jointly wide sense stationary processes $X(t)$ and $Y(t)$ is defined in (7.14). The cross-covariance function can be determined in terms of the means and the cross-correlation function as

$$C_{XY}(\tau) \quad = \quad R_{XY}(\tau) - m_X m_Y$$

Some of the properties of the cross-correlation function of two jointly wide sense stationary random processes are listed below without proofs. The same properties apply to the cross-covariance functions as well.

1. If the two random processeses are *statistically jointly independent*, from (7.10) their joint density function can be written as

$$f_{X(t)Y(t)}(x_0, \ldots, x_{n-1}, y_0, \ldots, y_{m-1}; t_0, \ldots, t_{n-1}, t_0, \ldots, t_{m-1})$$
$$= f_{X(t)}(x_0, \ldots, x_{n-1}; t_0, \ldots, t_{n-1}) f_{Y(t)}(y_0, \ldots, y_{m-1}; t_0, \ldots, t_{m-1})$$

Consequently, the cross-correlation function simplifies to

$$R_{XY}(\tau) \quad = \quad \mathcal{E}\{X(t)Y(t-\tau)\} = \mathcal{E}\{X(t)\}\mathcal{E}\{Y(t)\} = m_X m_Y$$

Instead, if the two processes are *uncorrelated*, then we only have

$$R_{XY}(\tau) \quad = \quad \mathcal{E}\{X(t)Y(t-\tau)\} = \mathcal{E}\{X(t)\}\mathcal{E}\{Y(t)\} = m_X m_Y \quad (7.20)$$

Thus independent processes are uncorrelated as well but not vice versa. On the other hand, if the two processes are *orthogonal*, then

$$R_{XY}(\tau) \quad = \quad \mathcal{E}\{X(t)Y(t-\tau)\} = 0 \quad (7.21)$$

2. The values of the cross-correlation function are bounded as follows

$$|R_{XY}(\tau)| \quad \leq \quad \sqrt{R_X(0)R_Y(0)}$$
$$|R_{XY}(\tau)| \quad \leq \quad \frac{1}{2}(R_X(0) + R_Y(0)) \quad (7.22)$$

Example 7.12:   By modifying the random processes given in Example 7.10, let us form two new random processes

$$X(t) \quad = \quad A \sin \omega t + B \cos(\omega t)$$
$$Y(t) \quad = \quad B \sin \omega t - A \cos(\omega t)$$

where ω is a constant, and A and B are uncorrelated random variables with 0 mean $\mathcal{E}\{A\} = \mathcal{E}\{B\} = 0$ and equal variance $\text{Var}[A] = \text{Var}[B] = \sigma^2$. The random processes are given to be jointly wide sense stationary.

The autocorrelation function of $X(t)$ is given by

$$R_X(\tau) = \mathcal{E}\{X(t)X(t-\tau)\} = \sigma^2 \cos \omega \tau$$

while that of $Y(t)$ is

$$R_Y(\tau) = \mathcal{E}\{X(t)X(t-\tau)\} = \sigma^2 \cos \omega \tau$$

From these, we have $R_X(0) = R_Y(0) = \sigma^2$.

The cross-correlation function is given by

$$\begin{aligned} R_{XY}(\tau) \quad &= \quad \mathcal{E}\{X(t)Y(t-\tau)\} \\ &= \quad \mathcal{E}\{AB\}\sin \omega t \sin \omega(t-\tau) - \mathcal{E}\{A^2\}\sin \omega t \cos \omega(t-\tau) \\ &\quad + \mathcal{E}\{B^2\}\cos \omega t \sin \omega(t-\tau) - \mathcal{E}\{AB\}\cos \omega t \cos \omega(t-\tau) \\ &= \quad -\sigma^2(\sin \omega t \cos \omega(t-\tau) - \cos \omega t \sin \omega(t-\tau)) = -\sigma^2 \sin \omega \tau \end{aligned}$$

where we used $\mathcal{E}\{AB\} = \mathcal{E}\{A\}\mathcal{E}\{B\} = 0$.

Let us now check if $R_{XY}(\tau)$ satisfies property (7.22):

$$|R_{XY}(\tau)| \leq \sqrt{R_X(0)R_Y(0)}$$

Substituting on both sides, we have

$$\sigma^2 |\sin \omega \tau| \leq \sigma^2$$

which reduces to

$$|\sin \omega\tau| \le 1$$

and

$$|R_{XY}(\tau)| \le \frac{1}{2}(R_X(0) + R_Y(0))$$
$$|\sin \omega\tau| \le 1$$

Thus the correlation function satisfies both bounds.

$\square$

7.3.5 Time averages and ergodic random processes

The statistical averages based on the expectation operation as detailed thus far can be calculated provided we have a knowledge of the joint PDF of the random process. If such a knowledge is not available, which is common in practice, we determine the mean and the autocorrelation function based on the sample mean or time average computations. It is helpful to think of the expectation and the time based averages in the context of the ensemble as illustrated in Section 7.1. The expectation (statistical average) can be thought of as operating down the ensemble. By this, of course, we consider that the entire ensemble of sample functions is available. On the other hand, a time average is carried out along the time axis. In this case, the assumption is that only one sample function of the ensemble (not the entire ensemble) is available.

A stationary random process can also be ergodic. Ergodicity relates the statistical and time averages. Our interest here is limited to the mean and the correlation ergodicity of a random process. Below we define the time averages followed by the mean and correlation ergodic random processes. For rigorous development of this discussion, the reader is referred to [1, 2, 3].

The time-average based mean $\bar{m}_X$ of a random process $X(t)$ is defined as

$$\bar{m}_X = \langle x(t) \rangle = \lim_{T \to \infty} \frac{1}{2T} \int_{-T}^{T} x(t)dt \qquad (7.23)$$

where $x(t)$ is a sample function of the random process $X(t)$. For a discrete-time random process $X[k]$, the time-averaged mean is given by

$$\bar{m}_X = \langle x[k] \rangle = \lim_{K \to \infty} \frac{1}{2K} \sum_{k=-K}^{K-1} x[k]$$

where $\langle \cdot \rangle$ is the time average operation similar to the statistical average operation of expectation $\mathcal{E}\{\cdot\}$.

The time-averaged autocorrelation function $\bar{R}_X(t_1, t_0)$ is defined as

$$\bar{R}_X(\tau) = \langle x(t)x(t-\tau) \rangle = \lim_{T \to \infty} \frac{1}{2T} \int_{-T}^{T} X(t)X(t-\tau)dt \qquad (7.24)$$

where $\tau = t_1 - t_0$ is the difference between two time instants t_1 and t_0. The corresponding quantity for a discrete case is given by

$$\bar{R}_X[\ell] = \langle x[k]x[k-\ell] \rangle = \lim_{K \to \infty} \frac{1}{2K} \sum_{k=-K}^{K-1} X[k]X[k-\ell] \qquad (7.25)$$

A random process whose first moment is not a function of time, i.e., $\mathcal{E}\{X(t)\} = m_X$, is said to be *mean ergodic* if its time-averaged mean $\bar{m}_X$ converges to its statistical mean m_X:

$$\bar{m}_X = \langle X(t) \rangle \doteq E\{X(t)\} = m_X \tag{7.26}$$

where the symbol $\doteq$ indicates that convergence is in the mean square sense (almost everywhere).

Not all random processes are mean ergodic as shown by the following example.

Example 7.13:   Consider a random process

$$X(t) = A$$

where $A = \pm 1$ is a binary-valued random variable with equally likely probabilities. The mean is

$$m_A = E\{A\} = \frac{1}{2}(+1) + \frac{1}{2}(-1) = 0$$

and the variance is

$$\sigma_A^2 = E\{(A - m_A)^2\} = \frac{1}{2}(+1)^2 + \frac{1}{2}(-1)^2 = 1$$

The sample functions of the random process are "d.c." waveforms of two different magnitudes of $+1$ or -1 as a function of t. Given an ensemble of these sample functions, the statistical mean of this random process is 0.

In order to compute the time-averaged mean of the random process, let us assume that the given sample function is $x(t) = A$. Then the time-averaged mean for this sample function is

$$\bar{m}_X = \langle x(t) \rangle_T = \lim_{T \to \infty} \frac{1}{2T} \int_{-T}^{T} x(t)dt = A \lim_{T \to \infty} \frac{1}{2T} \int_{-T}^{T} dt = A$$

Since A can only be either $+1$ or -1, the time-averaged mean $\bar{m}(t) = A$ and the statistical mean m_A are not the same for different sample functions. As a result, this is not a mean ergodic random process.

□

Many random processes of practical importance are mean ergodic as illustrated by the next example.

Example 7.14:   Consider the random process given in Example 7.1:

$$X(t) = A \sin(\omega t + \theta)$$

where A and ω are known parameters while θ is a uniform random variable in the interval $[-\pi, \pi]$. The statistical mean of the random process was shown to be $m_X(t) = E\{X(t)\} = 0$.

The time-averaged mean for a sample function $x(t)$ is

$$\begin{aligned}
\bar{m}_X &= \langle x(t) \rangle_T = \lim_{T \to \infty} \frac{1}{2T} \int_{-T}^{T} x(t)dt = \lim_{T \to \infty} \frac{A}{2T} \int_{-T}^{T} \sin(\omega t + \theta)dt \\
&= -\lim_{T \to \infty} \frac{A}{2\omega T} \cos(\omega t + \theta)\Big|_{-T}^{T} \\
&= -\lim_{T \to \infty} \frac{A}{2\omega T} (\cos(\omega T + \theta) - A\cos(-\omega T + \theta)) = 0
\end{aligned}$$

Since the two means are 0, the random process is mean ergodic.

□

The definition of mean ergodicity in (7.26) is simple and straightforward. Nevertheless, a more rigorous approach to establish mean ergodicity is through the following theorem [1, 3].

Theorem. A random process $X(t)$ is mean ergodic, i.e.,

$$\langle X(t) \rangle \doteq \mathcal{E}\{X(t)\}$$

in the mean square sense, if and only if its autocovariance function converges to 0 as $T \to \infty$:

$$\lim_{T \to \infty} \frac{1}{T} \int_0^T C_X(\tau)d\tau = 0 \tag{7.27}$$

Example 7.15: Let us apply (7.27) to the random process treated in Examples 7.1 and 7.14. From Example 7.1, the autocovariance function of the random process is

$$C_X(\tau) = \frac{A^2}{2} \cos \omega \tau$$

From (7.27),

$$\lim_{T \to \infty} \frac{A^2}{2T} \int_0^T \cos \omega \tau d\tau = \lim_{T \to \infty} \frac{A^2}{2\omega T} \sin \omega \tau \Big|_0^T = \lim_{T \to \infty} \frac{A^2}{2\omega T} \sin \omega T = 0$$

Hence the random process is mean ergodic.

Now let us apply (7.27) to the random process of Example 7.13. The autocovariance function of this process is given by

$$C_X(\tau) = \mathcal{E}\{X(t)X(t-\tau)\} = \mathcal{E}\{A^2\} = 1$$

From (7.27),

$$\lim_{T \to \infty} \frac{1}{T} \int_0^T d\tau = \lim_{T \to \infty} \frac{1}{T}\tau \Big|_0^T = 1$$

Therefore the random process is not mean ergodic.

□

On the lines of the above discussion for a mean ergodic process, a wide sense stationary random process $X(t)$ is said to be autocorrelation ergodic if the time-averaged autocorrelation function $\bar{R}_X(\tau)$ is equal to the expectation-based autocorrelation function $R_X(\tau)$:

$$R_X(\tau) = \mathcal{E}\{X(t)X(t-\tau)\} \doteq \langle X(t)X(t-\tau)\rangle = \bar{R}_X(\tau) \tag{7.28}$$

in the mean square sense.

Example 7.16: In Example 7.14, we showed that the random process

$$X(t) = A\sin(\omega t + \theta)$$

is mean ergodic. Let us now examine if it is autocorrelation ergodic as well, using (7.28). From Example 7.6, the statistical autocorrelation function is given by

$$R_X(\tau) = \mathcal{E}\{X(t)X(t-\tau)\} = \frac{A^2}{2} \cos \omega \tau$$

From (7.24), the time-averaged autocorrelation function is given by

$$\bar{R}_X(\tau) = \langle x(t)x(t-\tau)\rangle_T = \lim_{T\to\infty} \frac{1}{2T} \int_{-T}^{T} X(t)X(t-\tau)dt$$

$$= \lim_{T\to\infty} \frac{A^2}{2T} \int_{-T}^{T} \sin(\omega t + \theta)\sin(\omega(t-\tau)+\theta)dt$$

$$= \lim_{T\to\infty} \frac{A^2}{4T} \int_{-T}^{T} \cos(\omega\tau)dt - \lim_{T\to\infty} \frac{A^2}{4T} \int_{-T}^{T} \cos(\omega(2t-\tau)+2\theta)dt$$

$$= \frac{A^2}{2}\cos\omega\tau \lim_{T\to\infty} \frac{1}{2T} \, t\Big|_{-T}^{T} - \lim_{T\to\infty} \frac{A^2}{8\omega T} \sin(\omega(2t-\tau)+2\theta)\Big|_{-T}^{T}$$

$$= \frac{A^2}{2}\cos\omega\tau$$

Since the statistical average and the time average are the same, this process is auto-correlation ergodic.

□

The theorem for ergodicity stated in (7.27) can be extended to the autocorrelation ergodicity and is summarized below.

Theorem. A random process $X(t)$ is autocorrelation ergodic, i.e.,

$$\mathcal{E}\{X(t)X(t-\tau)\} \doteq \langle X(t)X(t-\tau)\rangle$$

in the mean square sense, if and only if:

$$\lim_{T\to\infty} \frac{1}{T} \int_0^T C_Y(\tau)d\tau = 0 \qquad (7.29)$$

where $Y(t) = X(t)X(t-\xi)$ and $C_Y(\tau,\xi)$ is the following fourth moment of the random process $X(t)$

$$C_Y(\tau,\xi) = \mathcal{E}\{X(t)X(t-\xi)X(t-\tau)X(t-\xi-\tau)\} - C^2(\xi)$$

Similar discussion on ergodicity can be carried out for discrete-time random processes as well as the complex-valued random processes.

7.4 *Power Spectral Density*

The previous discussion of random processes is focused on the random process as a function of time. For example, the mean and the autocorrelation function provide a time-domain characterization of the random process. For deterministic signals, the signal characterization and system response are undertaken both in the time and the frequency domains for well established reasons. Similarly, we wish to characterize the random processes (signals) in the frequency domain as well. This task is accomplished, not surprisingly, through the application of the Fourier transform, either continuous or discrete as appropriate.

The given random process $X(t)$ is assumed to be wide sense stationary, i.e., its mean m_X is not a function of time t and the autocorrelation function $R_X(\tau)$ is only a function of the time lag τ. The *power spectral density* (PSD) is then defined as the Fourier transform of the autocorrelation function

$$\boxed{S_X(f) = \mathcal{F}\{R_X(\tau)\} = \int_{-\infty}^{\infty} R_X(\tau)e^{-j2\pi f\tau}d\tau} \qquad (7.30)$$

where $\mathcal{F}\{\}$ is the Fourier transform operation and f is the frequency domain variable.

By virtue of the inverse Fourier transform, given the power spectral density, we can obtain the autocorrelation function as follows

$$R_X(\tau) = \mathcal{F}^{-1}\{S_X(f)\} = \int_{-\infty}^{\infty} S_X(f)e^{j2\pi f\tau}\,df \tag{7.31}$$

where $\mathcal{F}^{-1}\{\}$ denotes the inverse Fourier transform. Thus the autocorrelation function and the power spectral density form a Fourier transform pair. The reader may note that for deterministic signals, it is customary to take the Fourier transform of the given signal and then determine the "amplitude" spectral density. On the contrary, for random processes (signals), we determine the "power" spectral density based upon the autocorrelation function, which is a second moment. Consequently, the power spectral density can be considered the second moment of the random process in the frequency domain.

Substituting $R_X(\tau) = C_X(\tau) + m_x^2$ (see Table 7.2) into (7.30), we have

$$S_X(f) = \mathcal{F}\{C_X(\tau) + m_X^2\} = \mathcal{F}\{C_X(\tau)\} + m_X^2\delta(f)$$

where the delta function is a result of taking the Fourier transform of a constant. The power spectral density can thus be divided into two parts: the first part is due to the covariance ("a.c." part of a random signal power) and the second represents the power at 0 frequency ("d.c." power of the signal).

The definition of the power spectral density can be easily extended to the discrete case. The power spectral density of a wide sense stationary, discrete-time random process is given by

$$S_X(e^{j\omega}) = \mathcal{F}\{R_X[\ell]\} = \sum_{\ell=-\infty}^{\infty} R_X[\ell]e^{-j\omega\ell} \tag{7.32}$$

where ω is the radian or "digital" frequency. The inverse Fourier transform has the form

$$R_X[\ell] = \mathcal{F}^{-1}\{S_X(e^{j\omega})\} = \frac{1}{2\pi}\int_{-\infty}^{\infty} S_X(e^{j\omega})e^{j\omega\tau}\,d\omega \tag{7.33}$$

7.4.1 Properties of the power spectral density

Due to the properties of the autocorrelation function listed in Section 7.3.4, the power spectral density exhibits certain useful properties. Some of these properties are briefly presented here without proofs.

1. From (7.31), the second moment of a random process is obtained from the power spectral density

$$R_X(0) = \mathcal{E}\{X^2(t)\} = \int_{-\infty}^{\infty} S_X(f)df \qquad (7.34)$$

which indicates that the average power of the random process is obtained as the area under the power spectral density. While the average power can be obtained from a single value of the autocorrelation function in the time domain, it is spread over all frequencies in the frequency domain. This is why $S_X(f)$ is called *power spectral density*.

2. From (7.18), the autocorrelation function of a real-valued random process is even symmetric, i.e., $R_X(\tau) = R_X(-\tau)$. Correspondingly, the power spectral density is a real-valued even function. From (7.31), we have

$$
\begin{aligned}
S_X(f) &= \int_{-\infty}^{\infty} R_X(\tau)(\cos 2\pi f\tau - j\sin 2\pi f\tau)\, d\tau \\
&= \int_{-\infty}^{\infty} R_X(\tau)\cos 2\pi f\tau\, d\tau \qquad (7.35)
\end{aligned}
$$

where the second term consisting an integral of a product of an even function $R_X(\tau)$ and an odd function $\sin 2\pi f\tau$ is 0. From (7.35), it is easy to see that for real-valued random processes the power spectral density is even symmetric, i.e.,

$$S_X(f) = S_X(-f)$$

3. From the positive semidefinite property of the autocorrelation function (7.19), it can be shown that the corresponding power spectral density is nonnegative [5]:

$$\boxed{S_X(f) \geq 0} \qquad (7.36)$$

Example 7.17: The random process of Example 7.6 has a $m_X = 0$ and the autocorrelation function

$$R_X(\tau) = \frac{A^2}{2}\cos \omega_0 \tau$$

where the constant $\omega_0 = 2\pi f_0$ and we introduced the subscript to keep f_0 separate from the Fourier transform variable f.

The power spectral density of this random process is

$$
\begin{aligned}
S_X(f) &= \mathcal{F}\{R_X(\tau)\} = \frac{A^2}{2}\mathcal{F}\{\cos 2\pi f_0\tau\} \\
&= \frac{A^2}{4}\mathcal{F}\{e^{-j2\pi(f-f_0)\tau}\} + \mathcal{F}\{e^{-j2\pi(f+f_0)\tau}\} = \frac{A^2}{4}\delta(f-f_0) + \frac{A^2}{4}\delta(f-f_0)
\end{aligned}
$$

From the above result, the power spectral density is even symmetric, i.e., $S_X(f) = S_X(-f)$. Also, it is nonnegative, i.e., $S_X(f) \geq 0$ for all values of f.

Let us compute the mean squared value from the time and the freqeuncy domain representations. From the autocorrelation function, $\mathcal{E}\{X^2(t)\} = R_X(0) = A^2/2$. From (7.34), the second moment using the power spectral density

$$\mathcal{E}\{X^2(t)\} = \int_{-\infty}^{\infty} S_X(f)df = \frac{A^2}{4}\int_{-\infty}^{\infty}\delta(f-f_0)df + \frac{A^2}{4}\int_{-\infty}^{\infty}\delta(f-f_0)df = \frac{A^2}{2}$$

□

Here is an example that deals with an exponentially decreasing autocorrelation function.

Example 7.18:   Given that the autocorrelation function of a wide sense stationary random process is

$$R_X(\tau) = 5e^{-2|\tau|}$$

let us determine its power spectral density. From (7.30), we have

$$
\begin{aligned}
S_X(f) &= \mathcal{F}\{R_X(\tau)\} = 5\int_{-\infty}^{\infty} e^{-2|\tau|}e^{-j2\pi f\tau}d\tau \\
&= 5\int_{-\infty}^{\infty} e^{2\tau}e^{-j2\pi f\tau}d\tau + 5\int_{0}^{\infty} e^{-2\tau}e^{-j2\pi f\tau}d\tau \\
&= 5\int_{-\infty}^{\infty} e^{(2-j2\pi f)\tau}d\tau + 5\int_{0}^{\infty} e^{-(2+j2\pi f)\tau}d\tau \\
&= \frac{5}{2-j2\pi f} + \frac{5}{2+j2\pi f} = \frac{5}{1+\pi^2 f^2}
\end{aligned}
$$

which is a second rational polynomial in f. The figure below shows the plots of the autocorrelation function and the power spectral density.

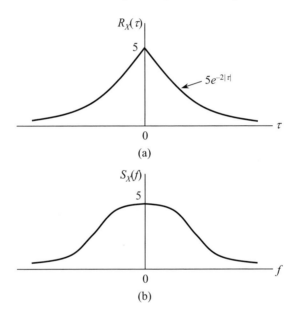

(a)

(b)

□

7.4.2 Cross-power spectral density

The cross-power spectral density of two jointly wide sense stationary random processes $X(t)$ and $Y(t)$ is defined as

$$S_{XY}(f) = \mathcal{F}\{R_{XY}(\tau)\} = \int_{-\infty}^{\infty} R_{XY}(\tau)e^{-j2\pi f\tau}d\tau \qquad (7.37)$$

with a similar definition for $S_{YX}(f)$. The two cross-power spectral densities are, in general, complex valued functions.

We now list some of the properties of the cross-power spectral density of jointly wide sense stationary, real-valued, continuous-time random processes.

1. If the two random processes are orthogonal, then the cross-power spectrum

$$S_{XY}(f) = \mathcal{F}\{R_{XY}(\tau)\} = 0$$

because $R_{XY}(\tau) = 0$. If the two random processes having mean values of m_X and m_Y are uncorrelated, then

$$S_{XY}(f) = \mathcal{F}\{R_{XY}(\tau)\} = m_X m_Y \, \delta(f) \qquad (7.38)$$

because $R_{XY}(\tau) = \mathcal{E}\{X(t)X(t-\tau)\} = \mathcal{E}\{X(t)\}\mathcal{E}\{X(t-\tau)\} = m_X m_Y$ and the Fourier transform of a constant is a delta function.

2. The two cross-power spectral densities are related to each other as follows

$$S_{XY}(f) = S_{YX}^*(f) \qquad (7.39)$$

where $*$ denotes complex conjugation. This results from (7.15).

Example 7.19: From Example 7.12, we have the cross-correlation functions

$$R_{XY}(\tau) = -\sigma^2 \sin \omega_0 \tau$$
$$R_{YX}(\tau) = \sigma^2 \sin \omega_0 \tau$$

where $\omega_0 = 2\pi f_0$.

From (7.37), the cross-power spectral density $S_{XY}(f)$ is given by

$$
\begin{aligned}
S_{XY}(f) &= \mathcal{F}\{R_{XY}(\tau)\} = -\sigma^2 \mathcal{F}\{\sin \omega_0 \tau\} \\
&= j\frac{\sigma^2}{2}\mathcal{F}\{e^{-2\pi(f-f_0)\tau}\} - j\frac{\sigma^2}{2}\mathcal{F}\{e^{-2\pi(f+f_0)\tau}\} = j\frac{\sigma^2}{2}\delta(f-f_0) - j\frac{\sigma^2}{2}\delta(f+f_0)
\end{aligned}
$$

Note that $S_{XY}(f)$ in this case is purely imaginary. Similarly, the cross-power spectral density $S_{YX}(f)$ is given by

$$
\begin{aligned}
S_{YX}(f) &= \mathcal{F}\{R_{YX}(\tau)\} = \sigma^2 \mathcal{F}\{\sin \omega_0 \tau\} \\
&= -j\frac{\sigma^2}{2}\delta(f-f_0) + j\frac{\sigma^2}{2}\delta(f+f_0)
\end{aligned}
$$

Note that $S_{XY}(f)$ and $S_{YX}(f)$ satisfy (7.39) above, i.e., $S_{YX}(f)$ is a conjugate of $S_{XY}(f)$.

$\square$

Similar definitions for the power spectral density and the cross-power spectral density (and the corresponding properties) exist for the discrete-time random processes and for the complex-valued random processes. We are omitting these cases here for brevity.

7.4.3 *White noise*

Consider a random process $X(t)$ with samples $\{X(t_0), X(t_1), \ldots, X(t_{n-1})\}$ sampled at arbitrary time instants $t_0, t_1, \ldots, t_{n-1}$, respectively. Let these samples be independent and identically distributed (IID) random variables having a PDF $f_X(x)$, mean m, and variance σ^2. A random process consisting of such a sequence of samples is called an IID random process. The same definition applies to discrete random processes as well.

The joint PDF of an IID random process can be written as

$$f_{X(t)}(x_0, x_1, \ldots, x_{n-1}; t_0, t_1, \ldots, t_{n-1})$$

$$= f_{X(t)}(x_0; t_0) f_{X(t)}(x_1; t_1) \ldots f_{X(t)}(x_{n-1}); t_{n-1}) = \prod_{i=0}^{n-1} f_X(x_i) \quad (7.40)$$

Example 7.20: Let the given IID random process $X(t)$ be Gaussian with mean m and variance σ^2. The PDF of an arbitrary sample X of the random process is given by

$$f_X(x) = \frac{1}{\sqrt{2\pi}\sigma} e^{-\frac{(x-m)^2}{2\sigma^2}}$$

The joint PDF of three samples $\{X(t_0), X(t_1), X(t_2), X(t_3)\}$ is obtained as follows:

$$
\begin{aligned}
f_{X(t)}(x_0, x_1, x_2, x_3; t_0, t_1, t_2, t_3) &= f_{X(t)}(x_0; t_0) f_{X(t)}(x_1; t_1) f_{X(t)}(x_2; t_2) f_{X(t)}(x_3; t_3) \\
&= \frac{1}{4\pi^2\sigma^4} e^{-\frac{1}{2\sigma^2}\{(x_0-m)^2+(x_1-m)^2+(x_2-m)^2+(x_3-m)^2\}} \\
&= \frac{1}{4\pi^2\sigma^4} e^{-\frac{1}{2\sigma^2}\sum_{i=0}^{3}(x_i-m)^2}
\end{aligned}
$$

□

The mean of an IID random process is constant

$$m_X = \mathcal{E}\{X(t_i)\} = m \quad \text{for all } i \quad (7.41)$$

and the correlation function from (7.2)

$$
\begin{aligned}
R_X(t_j, t_i) &= \mathcal{E}\{X(t_j)X(t_i)\} \\
&= \begin{cases} \mathcal{E}\{X^2(t_i)\} = R_X(0) & \text{if } t_i = t_j \\ \mathcal{E}\{X(t_j)\}\mathcal{E}\{X(t_i)\} = m^2 & \text{if } t_i \neq t_j \end{cases}
\end{aligned}
$$

where if $i \neq j$, the samples $X(t_i)$ and $X(t_j)$ are statistically independent, thus the expectation of their product becomes the product of expectations. From (7.4), the autocorrelation function can be represented in terms of the autocovariance function and the mean

$$
\begin{aligned}
R_X(t_j, t_i) &= C_X(t_j, t_i) + m_X^2 \\
R_X(0) &= C_X(0) + m^2 = \sigma^2 + m^2
\end{aligned}
$$

By substituting this result into the expression for autocorrelation above yields

$$
\begin{aligned}
R_X(t_j, t_i) &= \begin{cases} \sigma^2 + m^2 & \text{if } t_i = t_j \\ m^2 & \text{if } t_i \neq t_j \end{cases} \\
&= \sigma^2 \delta(t_j - t_i) + m^2 \quad (7.42)
\end{aligned}
$$

Clearly an IID random process is wide sense stationary, hence we can write (7.42) as

$$R_X(\tau) = \sigma^2 \delta(\tau) + m^2$$

The corresponding autocovariance function

$$C_X(\tau) = \sigma^2 \delta(\tau)$$

The power spectral density of an IID random process from (7.30)

$$S_X(f) = \mathcal{F}\{R_X(\tau)\} = \mathcal{F}\{\sigma^2\delta(\tau) + m^2\} = \sigma^2 + m^2\delta(f) \quad (7.43)$$

Figure 7.6 illustrates the plots of the autocorrelation function and the power spectral density of an IID random process. The two plots appear similar in their respective domains. In the time domain, the delta function indicates that the samples of the random process are independent; the covariance between any two arbitrary samples is 0 unless they both are sampled at the same time instant, i.e., $X(t_i) = X(t_j)$. The constant part indicates d.c. power of the process (signal). In the frequency domain, the delta function represents the d.c. power at the *0 frequency* while the flat line indicates equal power density covering the entire range of freqeuncies.

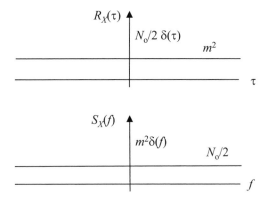

Figure 7.6 The autocorrelation function and the power spectral density of an IID random process with mean m and variance σ^2.

White noise definition

We now present a formal definition of *white noise*. A wide sense stationary random process $W(t)$ having a power spectral density

$$S_W(f) = \frac{N_0}{2} \qquad \text{for all values of } f \tag{7.44}$$

is called white noise. The corresponding autocorrelation function from (7.31)

$$R_W(\tau) = \mathcal{F}^{-1}\{S_W(f)\} = \frac{N_0}{2}\delta(\tau) \tag{7.45}$$

where N_0 is a positive constant. The term $N_0/2$ represents the magnitude of the two-sided power spectral density of white noise. Figure 7.7 illustrates the autocorrelation function and the power spectral density of white noise.

While a zero-mean IID random process is white noise, in general white noise only requires that its samples are uncorrelated for all values of τ. Conceptually, white noise has a flat power spectral density that exists for the entire frequency spectrum. This is analogous to the *white* light, which contains all the natural colors. Random processes that do not have a flat spectrum are typically referred to as *colored* noise. The power spectral density shown in Figure 7.18 is an example of colored noise. In addition to being white, if the random process is also Gaussian, then it is widely known as *white Gaussian noise*.

One of the widely encountered application examples is the case of a sinusoidal signal in additive white Gaussian noise. Typically, the sinusoid is the desired signal while the additive noise is undesired, e.g., communications signals and radar waveforms received

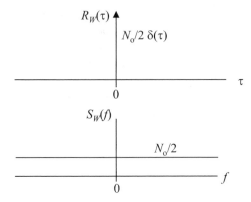

Figure 7.7 The autocorrelation function and the power spectral density of white noise.

at a receiver. The following two examples deal with the cases of sinusoidal signals in noise. The objective here is to extract the desired signal from a signal corrupted with noise.

Example 7.21: Consider the random process consisting of a sinusoid with random phase in additive white noise

$$X(t) = A\sin(\omega_0 t + \theta) + W(t)$$

where A and $\omega_0 = 2\pi f_0$ are constants while θ is a uniform random variable in the range $[-\pi, \pi]$, $W(t)$ is white noise with two-sided power spectral density of $\frac{N_0}{2}$, and θ and $W(t)$ are uncorrelated, i.e.,

$$\mathcal{E}\{A\sin(\omega_0 t + \theta)W(t)\} = \mathcal{E}\{A\sin(\omega_0 t + \theta)\}\,\mathcal{E}\{W(t)\} = 0$$

The autocorrelation function of $X(t)$ is given by

$$
\begin{aligned}
R_X(\tau) &= \mathcal{E}\{X(t)X(t-\tau)\} \\
&= \mathcal{E}\{(A\sin(\omega_0 t + \theta) + W(t))\,(A\sin(\omega_0(t-\tau) + \theta) + W(t-\tau))\} \\
&= A^2\mathcal{E}\{\sin(\omega_0 t + \theta)\sin(\omega_0(t-\tau) + \theta)\} + \mathcal{E}\{W(t)W(t-\tau)\} \\
&\quad + A\mathcal{E}\{\sin(\omega_0 t + \theta)W(t-\tau)\} + A\mathcal{E}\{W(t)\sin(\omega_0(t-\tau) + \theta)\}
\end{aligned}
$$

The third and fourth terms are 0 due to θ and $W(t)$ being uncorrelated. From Example 7.6, the first term yields

$$A^2\mathcal{E}\{\sin(\omega_0 t + \theta)\sin(\omega_0(t-\tau) + \theta)\} = \frac{A^2}{2}\cos\omega\tau$$

From (7.45), the autocorrelation function of white noise is given by

$$\mathcal{E}\{W(t)W(t-\tau)\} = \frac{N_0}{2}\delta(\tau)$$

Sustituting these into the expression for $R_X(\tau)$ yields

$$R_X(\tau) = \frac{A^2}{2}\cos\omega\tau + \frac{N_0}{2}\delta(\tau)$$

The power spectral density of $X(t)$ is obtained as

$$
\begin{aligned}
S_X(f) &= \mathcal{F}\{R_X(\tau)\} = \frac{A^2}{2}\mathcal{F}\{\cos\omega\tau\} + \frac{N_0}{2}\mathcal{F}\{\delta(\tau)\} \\
&= \frac{A^2}{4}\delta(f - f_0) + \frac{A^2}{4}\delta(f + f_0) + \frac{N_0}{2}
\end{aligned}
$$

The figure below illustrates the power spectral density function of $X(t)$. The flat line with a height of $N_0/2$ represents what is called the *noise floor*. As seen, the two impulses representing the sinusoid at frequency f_0 need to exceed the noise floor in order for the sinusoid to be extracted from the received noisy signal.

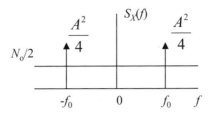

☐

Example 7.22:   Frequently, in certain applications such as digital communication receivers, the receiver has knowledge of the transmitted signal waveforms. What is not known is which of the transmitted waveforms from a predefined set is transmitted during a given symbol period. This scenario can be summarized in the following figure. The received signal is corrupted by additive noise. The reference signal indicates receiver's knowledge of transmitted waveforms. In order to extract the desired sinusoid from the received signal, we compute the cross-correlation function between the received signal and the reference signal.

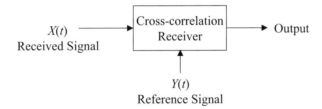

To be more specific, let us assume that the transmitter sends two signals representing a logical 1 and a logical 0:

$$\text{Logical 1} \quad \Rightarrow \quad A\sin(\omega_0 t + \theta)$$
$$\text{Logical 0} \quad \Rightarrow \quad 0 \quad \text{i.e., no signal is transmitted}$$

At the receiver we have the following two random processes

$$X(t) \;=\; A\sin(\omega_0 t + \theta) + W(t)$$
$$Y(t) \;=\; A\sin(\omega_0 t + \theta)$$

to represent a hypothetical case of receiving a logical 1 and

$$X(t) \;=\; W(t)$$
$$Y(t) \;=\; A\sin(\omega_0 t + \theta)$$

for the case of receiving a logical 0. In the above A and $\omega_0 = 2\pi f_0$ are constants while θ is a uniform random variable in the range $[-\pi,\ \pi]$, $W(t)$ is white noise with two-sided power spectral density of $N_0/2$, and θ and $W(t)$ are uncorrelated as in Example 7.21.

First let us consider the case of no sinusoid present (logical 0 received). The corresponding cross-correlation function

$$R_{XY}(\tau) = A\,\mathcal{E}\left\{W(t)\sin(\omega_0(t-\tau)+\theta)\right\} = 0$$

because θ and $W(t)$ are uncorrelated.

For the case of logical 1 received, the cross-correlation function

$$
\begin{aligned}
R_{XY}(\tau) &= A^2 E\left\{\sin(\omega_0 t + \theta)\sin(\omega_0(t-\tau)+\theta)\right\} + A E\left\{\sin(\omega_0 t + \theta)W(t-\tau)\right\} \\
&= \frac{A^2}{2}\cos\omega\tau
\end{aligned}
$$

The power spectral density for this case is given by

$$
\begin{aligned}
S_X(f) &= \mathcal{F}\{R_X(\tau)\} = \frac{A^2}{2}\mathcal{F}\{\cos\omega\tau\} \\
&= \frac{A^2}{4}\delta(f-f_0) + \frac{A^2}{4}\delta(f+f_0)
\end{aligned}
$$

which is similar to that in Example 7.21 with the term corresponding to the noise floor missing.

The figure below illustrates the power spectral density function of $X(t)$. The signal extraction using the crosscorrelation approach is clearly superior to that using the autocorrelation approach. The reason for better performance is due to the availability of the reference signal.

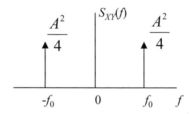

□

For its seemingly universal appeal, white noise is not realizable in practice. From (7.34) and Figure 7.7, it contains infinite power

$$
\int_{-\infty}^{\infty} S_X(f)df = \infty
$$

hence the white noise defined above only has *theoretical* importance. In practice, however, white noise is defined over a narrow band of frequencies

$$
S_W(f) = \begin{cases} \dfrac{N_0}{2} & \text{for } -B \le f \le B \\ 0 & \text{otherwise} \end{cases} \tag{7.46}
$$

where B is typically referred to as the *absolute* bandwidth. Figure 7.8 shows the power spectral density of bandlimited white noise. Since almost all physical systems of practical interest have finite bandwidth, in a given application, and our interest is usually limited to the system bandwidth, the bandlimited white noise can be considered white noise for analysis purposes.

The autocorrelation function corresponding to the bandlimited white noise is

$$
\begin{aligned}
R_W(\tau) &= \mathcal{F}^{-1}\{S_W(f)\} = \frac{N_0}{2}\int_{-B}^{B} e^{j2\pi f\tau}df \\
&= \frac{N_0}{2\pi\tau}\frac{e^{j2\pi B\tau} - e^{-j2\pi B\tau}}{2j} = N_0 B\frac{\sin 2\pi B\tau}{2\pi B\tau}
\end{aligned} \tag{7.47}
$$

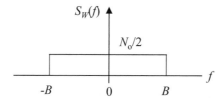

Figure 7.8 The power spectral density of bandlimited white noise.

Figure 7.9 shows the autocorrelation function of bandlimited white noise. The autocorrelation function is maximum at $\tau = 0$ where $R_W(0) = N_0 B$, and is 0 for $\tau = i/2B$ for all i. Since the mean of $W(t)$ is 0 and

$$R_W(\tau) = \mathcal{E}\{W(t)W(t - \tau)\} = 0 \quad \text{for } \tau = \frac{i}{2B} \text{ for all } i,$$

the samples of bandlimited white noise are uncorrelated only for specific values of τ. Compare this to the case of *theoretical* white noise with infinite bandwidth whose samples are uncorrelated for all values of τ. As $B \to \infty$ in Figure 7.8, the $(\sin x)/x$ shape of the autocorrelation function approaches an impulse (see Figure 7.7), i.e., the bandlimited white noise approaches theoretical white noise.

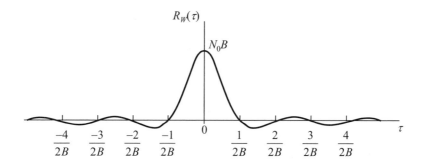

Figure 7.9 The autocorrelation function of bandlimited white noise.

7.4.4 An application

In baseband transmission of binary data, the binary digits logical 1 and logical 0 are represented using electrical pulses. A variety of pulse shapes are used for this purpose; to name a few: polar non-return to 0, bipolar return to 0, and Manchester. In general these waveforms are referred to as semi-random binary waveforms. Since in a given sequence the binary digits occur randomly, the corresponding pulse waveform is also random, hence the name. In the following example, we determine the autocorrelation and the power spectral density of a semi-random binary waveform for polar signaling.

Example 7.23: In polar signaling, a logical 1 is represented by a positive voltage of magnitude A and a logical 0 is represented by a negative voltage of magnitude $-A$. The duration of each binary digit is T_b. The figure below shows the waveform for a binary sequence 011010010.

Assuming no pulse synchronization, the starting time of the first complete pulse with respect to a reference time point (say, $t = 0$) is uniformly distributed in the range

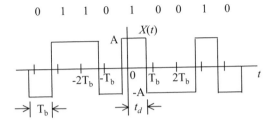

[0, T_b] and is represented as

$$f_{T_d}(t_d) = \begin{cases} \dfrac{1}{T_b}, & 0 \le t_d \le T_b \\ 0, & \text{otherwise} \end{cases}$$

Let us assume that the occurrence of 1's and 0's is equally likely, i.e., the pulse magnitude is $\pm A$ during any bit period with equal probability. Also, the occurrences are independent from bit period to bit period. Consequently, the mean of the process

$$m_X(t) = \mathcal{E}\{X(t)\} = 0$$

The autocorrelation function

$$R_X(t_1, t_0) = \mathcal{E}\{X(t_1)X(t_0)\}$$

We evaluate the autocorrelation function for two cases: the time difference $t_1 - t_0 > T_b$ and $t_1 - t_0 < T_b$.

For $|t_1 - t_0| > T_b$, the random variables $X(t_1)$ and $X(t_0)$ occur in different bit periods and are independent. As a result

$$R_X(t_1, t_0) = \mathcal{E}\{X(t_1)X(t_0)\} = \mathcal{E}\{X(t_1)\}\mathcal{E}\{X(t_0)\} = 0$$

For $t_1 - t_0 < T_b$ and $t_d > T_b - |t_1 - t_0|$, the random variables occur in different bit periods and as above, the autocorrelation function is 0.

For $0 \le |t_1 - t_0| < T_b$ and $0 \le t_d < T_b - |t_1 - t_0|$, the random variables occur in the same bit period. We thus have a conditional expectation

$$\begin{aligned} R_X(t_1, t_0, t_d) &= \mathcal{E}\{X(t_1)X(t_0) \mid t_d < T_b - |t_1 - t_0|\} \\ &= \begin{cases} A^2, & 0 < |t_1 - t_0| < T_b \text{ and } 0 < t_d < T_b - |t_1 - t_0| \\ 0, & \text{otherwise} \end{cases} \end{aligned}$$

To remove the effect of t_d on the autocorrelation function, we take the expectation of the above term with respect to t_d

$$\begin{aligned} R_X(t_1, t_0) &= \mathcal{E}\{R_X(t_1, t_0, t_d)\} = \int_{-\infty}^{\infty} \mathcal{E}\{R_X(t_1, t_0, t_d)\} f_{T_d}(t_d) \, dt_d \\ &= \frac{1}{T_b} \int_0^{T_b - |t_1 - t_0|} A^2 \, dt_d \quad \text{for } 0 \le |t_1 - t_0| < T_b \\ &= \frac{A^2}{T_b} (T_b - |t_1 - t_0|) \quad \text{for } 0 \le |t_1 - t_0| < T_b \end{aligned}$$

Since the autocorrelation function is only a function of $t_1 - t_0$, we can simplify the above into

$$R_X(\tau) = \begin{cases} A^2 \left(1 - \frac{|\tau|}{T_b}\right) & \text{for } 0 \le |\tau| < T_b \\ 0 & \text{otherwise} \end{cases}$$

A plot of the autocorrelation function is shown in the figure below. The reader may note that the autocorrelation function of a rectangular waveform is triangular in shape.

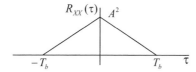

The power spectral density of $X(t)$ is then obtained by taking the Fourier transform of its autocorrelation function and is given by (see Problem 7.12.)

$$S_X(f) = \mathcal{F}\{R_X(\tau)\} = A^2 T_b \left(\frac{\sin \pi f T_b}{\pi f T_b} \right)^2$$

The $(\sin z/z)^2$ shaped power spectral density of $X(t)$ is shown in the figure below. In baseband communication applications, the first null width of the power spectral density is used as a figure of merit in evaluating different pulses for transmission. From the figure, the *first null bandwidth* of polar signaling is $1/T_b$.

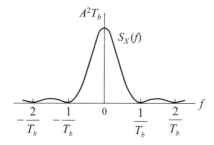

If the bit period of the polar signal is 1 msec, its first null bandwidth is 1 kHz. In other words, to transmit 1000 bits/sec using polar signaling requires a first null bandwidth of 1 kHz.

□

7.5 *Noise Sources*

Typically noise is an unwanted random process (signal). Noise that interferes with the operation of electronic systems is originated by a variety of sources. Most sources produce noise in a limited spectral range. As a result, the noise of interest depends on the frequency range of operation of the electronic system. Analog as well as digital systems are hampered by noise in different ways. In the following, we will first detail the sources of noise for analog systems followed by a discussion of quantization noise, a source of digital noise.

Whether or not a given noise waveform is correlated depends on the nature of its power spectral density. If the power spectral density is relatively flat over the frequency band of interest, then for all practical purposes we call the noise uncorrelated. Recall that we have *white* noise if in addition to being uncorrelated it also has 0 mean. There are essentially two types of uncorrelated noise: external and internal. Noise generated outside of the system under study is termed external noise. We have no control over the generation or the power spectral density of external noise. We typically attempt to shield the system from external noise. Atmospheric noise, extraterrestrial noise, and man-made noise are some of the externally produced noise. Depending on the frequency band of operation of the system, each of these can become an issue in the system design and performance analysis.

Noise generated by devices and circuits within the system under study is termed internal noise. Internal noise cannot be completely eliminated, but through proven design techniques and by choosing appropriate devices its effects can be somewhat mitigated. Shot noise, excess noise, transit-time noise, and thermal noise are predominant internal noise. All four types of noise are a result of the mobility of carriers of electric current in an electronic device. An electronic device, such as a transistor, operates based on the movement of subatomic carriers known as electrons and holes. Excess noise and transit-time noise are little understood while shot noise and thermal noise have been extensively studied. These internal noise sources are a hindrance in analog systems.

In digital systems, quantization noise is the predominent source of system performance degradation. The quantization noise occurs due to rounding errors during the analog to digital conversion process.

In this section, we will examine the thermal noise and quantization noise as representative sources of noise in analog and digital systems, respectively.

7.5.1 Thermal noise

Thermal noise is due to random motion of electrons in a conductor caused by thermal agitation. Thermal noise is also referred to as Brownian noise, Johnson noise, or white noise.

Electric current in a conductor is a result of electron movement. The flow of electrons is, however, not deterministic but random. As the electrons move through the conductor, they collide with molecules, thus generating short pulses of electric disturbances. The root mean square (rms) velocity of electrons is known to be proportional to the absolute temperature and, the *mean square* voltage density generated across the terminals of a resistor due to thermal noise is given by

$$\mathcal{E}\{V_T^2\} = 2\kappa T R \ \text{volts}^2/\text{Hz} \tag{7.48}$$

where T is temperature in degrees Kelvin, R is the resistance in ohms, and κ is is Boltzmann's constant equal to 1.38×10^{-23} joules per degrees Kelvin [4].

Example 7.24: Consider an electronic device operating at a temperature of 300 K over a bandwidth of 5 MHz with an internal resistance of 1000 Ω. We can determine the rms noise voltage across the terminals of this device as follows.

From (7.48), the mean square voltage density across a 1000-Ω is given by

$$\mathcal{E}\{V_T^2\} = 2\kappa T R = 2 \times 1.38^{-23} \times 300 \times 1000 = 8.28 \times 10^{-18} \ \text{volts}^2/\text{Hz}$$

The rms voltage across the terminals of this resistor is then given by

$$\sqrt{\mathcal{E}\{V_T^2\}(2B)} = \sqrt{8.28 \times 10^{-18} \times 2 \times 5 \times 10^6} = 9.0995 \times 10^{-6} \ \text{volts}$$

where B is the bandwidth in Hz. Commonly, the definition of bandwidth indicates only the postive side of the spectral range. Thus, the term $2B$ above accounts for the positive and the negative side of the spectrum.

□

Consider the circuit diagram in Figure 7.10 in which the thermal noise source is the resistance R. The mean square voltage spectral density so generated is transferred to a load resistor R_L. The power spectral density of the source is given by

$$S_S(f) = \frac{\mathcal{E}\{V_T^2\}}{R + R_L} = \frac{2\kappa TR}{R + R_L}$$

It is well known that the maximum power spectral density is delivered to a load resistor when the load resistance is matched to the source resistance, i.e., $R = R_L$. As a consequence, the source resistor R and the load resistor R_L share the power spectral density equally. Under these conditions, the power spectral density delivered to the load resistor (also known as available power spectral density) is one half of that generated by the source

$$S_W(f) = \frac{1}{2}S_S(f) = \frac{1}{2}\frac{2\kappa TR}{2R} = \frac{\kappa T}{2} = \frac{N_0}{2}\text{watts/Hz} \qquad (7.49)$$

Since it is customary to represent the power spectral density of white noise in terms of N_0, we have $N_0 = \kappa T$.

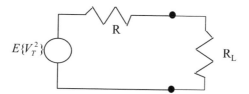

Figure 7.10 Thermal noise source with a mean square voltage spectral density $\mathcal{E}\{V_T^2\}$ and an internal resistance R.

Example 7.25: Let us determine the available power spectral density and the available average noise power generated for the electronic device in Example 7.24.

The power spectral density of thermal noise generated in the device from (7.49)

$$S_W(f) = \frac{\kappa T}{2} = \frac{1.38 \times 10^{-23} \times 300}{2} = 2.07 \times 10^{-21} \text{ watts/Hz}$$

From this, we have $N_0 = 4.14 \times 10^{-21}$.

The available average noise power

$$P_N = \int_{-B}^{B} S_W(f)df = 2.07 \times 10^{-21}(2B) = 2.07 \times 10^{-21} \times 10^7 = 2.07 \times 10^{-14} \text{ watts}$$

where the available average power is not dependent on the value of resistance; the temperature and the bandwidth are relevant parameters.

□

Suppose that we send a perfectly clean sinusoid into an electronic device with an internal thermal noise source as depicted in Figure 7.11. At the output of the device, the signal now has additive thermal noise. Let us define the *signal to noise ratio* (SNR) as a ratio of signal power to noise variance, given by

$$\text{SNR}_{dB} = 10\log_{10}\frac{P_S}{\sigma_W^2} \qquad (7.50)$$

where P_S is the signal power of the sinusoid

$$P_S = \frac{A^2}{2}$$

and from (7.45) and (7.49)

$$\sigma_W^2 = R_W(\tau)|_{\tau=0} = \mathcal{F}^{-1}\{S_W(f)\}|_{\tau=0)} = \frac{\kappa T}{2}\delta(\tau)|_{\tau=0} = \frac{\kappa T}{2}$$

Substituting these into (7.50) yields

$$\mathrm{SNR_{dB}} = 10\log_{10}\frac{A^2}{\kappa T}$$

The SNR at the input of the system is infinite while the output SNR is finite, indicating that there is signal quality degradation.

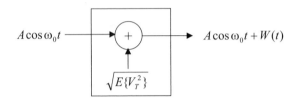

Figure 7.11 Depiction of noise added by a thermal noise source in an electronic device.

Example 7.26:   Let us continue from Example 7.25 by considering that the input signal to the device is $3\cos\omega_0 t$. The SNR at the output of the device from (7.50) is

$$\mathrm{SNR_{dB}} = 10\log_{10}\frac{A^2}{kT} = 10\log_{10}\frac{9}{1.38\times10^{-23}\times300} = 213.37\mathrm{dB}$$

This signal quality degradation is fairly small, but it could become significant if the system has high gain and there is a cascade of systems with internal thermal noise sources.

□

7.5.2 Quantization noise

Most signals of interest originate in analog form and are converted to digital form due to many advantages of digital processing, storage, and transmission of these signals. Digitization requires rounding or truncation of the analog signal samples. Rounding (truncation) causes loss of precision, which is equivalent to adding noise to the signal. This approximation error is termed *quantization noise* or digitization noise.

Figure 7.2 shows the schematic diagram of the digitization process in which the quantizer block introduces quantization noise. Figure 7.12 shows the transfer characteristic of a typical quantizer. The continuous magnitude input to the quantizer is denoted by $X_c[kT]$ and the corresponding approximated value is $X[nT]$.

Given an input sample $X_c[kT]$, we can represent the quantized (approximated) sample $X[kT]$ as

$$X[kT] = X_c[kT] + W_q[kT]$$

where $W_q[kT]$ is a random variable representing the quantization (rounding) error. For the input range $\frac{-\Delta}{2} \leq X_c < \frac{\Delta}{2}$, the corresponding output value is 0, where Δ is called the *step* size. Given that the input dynamic range is $2A$, i.e., it extends from $-A$ to A, the step size is given by

$$\Delta = \frac{2A}{N}$$

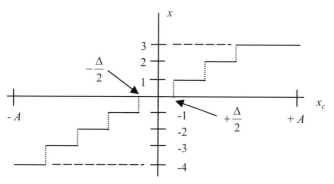

Figure 7.12 A quantizer with an input range of $-A$ to $+A$.

where N is the number of steps (levels) in the quantizer and is invariably a power of 2, i.e.,

$$N = 2^b$$

where b is the number of bits used by the encoder. The quantization errors $W_q[kT]$ are assumed to be IID and uniformly distributed

$$f_{W_q}(w_q) = \begin{cases} \dfrac{1}{\Delta} & \text{for } \frac{-\Delta}{2} \le X_c < \frac{\Delta}{2} \\ 0 & \text{otherwise} \end{cases}$$

The mean of $W_q[kT]$ is

$$m_{W_q} = \mathcal{E}\{W_q[kT]\} = 0$$

and its variance (see Problem 4.15)

$$\sigma_{W_q}^2 = \mathcal{E}\{W_q^2[kT]\} = \frac{\Delta^2}{12}$$

Substituting $\Delta = \frac{2A}{2^b}$ into the above yields

$$\sigma_{W_q}^2 = \frac{A^2}{3 \times 2^{2b}} \tag{7.51}$$

Since the quantization noise is assumed to be white, the power spectral density is

$$S_{W_q}(e^{j\omega}) = \frac{A^2}{3 \times 2^{2b}}$$

with the corresponding autocorrelation function

$$R_{W_q}[\ell] = \frac{A^2}{3 \times 2^{2b}} \delta[\ell]$$

Let us assume that the signal being digitized is a pure sinusoid (i.e., no additive noise) with amplitude A and power $A^2/2$. After digitizing, the signal is now equivalent to sinusoid in additive white noise. The corresponding signal to quantization noise ratio (SQNR) along the lines of (7.50) is given by

$$\begin{aligned}
\text{SQNR}_{\text{dB}} &= 10\log_{10}\frac{P_S}{P_q} = 10\log_{10}\frac{P_S}{\sigma_{W_q}} = 10\log_{10} 1.5 \times 2^{2b} \\
&= 10\log_{10} 1.5 + 2\,b\log_{10} 2 = 1.7609 + 6.0206\,b \tag{7.52}
\end{aligned}$$

which is commonly approximated to

$$\text{SQNR}_{\text{dB}} \approx 6b \text{ dB}$$

and used as a rule of thumb: "*6 dB per bit.*"

Example 7.27: A given analog to digital converter has a resolution of 8 bits. The upper
bound on the signal to quantization noise ratio in dB is then given by $\text{SQNR}_{\text{dB}} = 48$
dB, i.e., in the absence of any other noise, the SNR degrades from infinity to 48 dB
due to digitization.

Suppose that we require about 72 dB of SQNR in a certain application. We then need
to use at least a 12-bit analog to digital converter.

□

7.6 *Response of Linear Systems*

The discussion of random processes has thus far centered around their characterizaton
in terms of a joint PDF, their properties such as stationarity, and their representation
using the first and the second moments. The second moment analysis based on the
autocorrelation function in the time domain and the power spectral density in the
frequency domain provide powerful techniques for analysis of random waveforms. From
a traditional standpoint of signals and systems in electrical engineering, all of this deals
with signal representation and analysis.

In deterministic signal and system studies, one is typically interested in the response
of a system to such inputs as an impulse, a step waveform, or a sinusoid. Such studies
yield very useful tools of system representation, e.g., impulse response, step response,
and the freqeuncy response. Our interest here is the response of linear time-invariant
(LTI) systems to random inputs. For these systems, we may argue that the system
when driven by a random input will not introduce any uncertainty but will alter the
nature of the randomness of the input process. With that in mind, it is reasonable to
state that a random input $X(t)$ produces a random output $Y(t)$. Assuming that we
have a probabilistic description of the input random process, we wish to determine
the PDF and the first and the second moments of the output random process.

The discussion here has some parallels to the topic of *functions of random variables*
discussed in Section 3.6 in which the function (or mapping) was memoryless but can be
linear or nonlinear. In this chapter, the system maps a given input random process $X(t)$
into a new random process $Y(t)$. The system is linear and time invariant but possesses
memory. Due to the system memory, for example, if the input is an uncorrelated
random process, the output, which is a linear combination of present and past input
samples, is a correlated process. In the frequency domain, this means that a white
noise input process produces a non-white output spectrum. We will examine this and
other related issues in the following.

Our focus here will be the correlation function and the power spectral density for
the continuous time case. The discussion for the discrete case is very similar and is
summarized at the end of the section. Since the development of the covariance functions
is similar to that of correlation functions, we will only summarize the final expressions
for the covariance functions.

7.6.1 *Linear time-invariant system*

Figure 7.13 illustrates a typical LTI system with $X(t)$ and $Y(t)$ as input and ouput
random processes, respectively. An LTI system is completely characterized by its im-

pulse response $h(t)$, which is the system output when $X(t) = \delta(t)$. For a general input $X(t)$, the output of an LTI system can be obtained as a convolution of the system impulse response and the input process:

$$Y(t) = h(t) \circledast X(t) = \int_{-\infty}^{\infty} h(\lambda) X(t - \lambda) \, d\lambda \tag{7.53}$$

For the sake of the following definition, assume that the system input is deterministic. Taking the Fourier transform on both sides of (7.53) yields

$$Y(f) = H(f) X(f) \tag{7.54}$$

where $H(f)$ is called the system frequency response or *transfer function*.

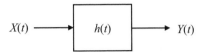

Figure 7.13 A linear time invariant system driven by a random process.

Example 7.28: The impulse response of an LTI system is given by

$$h(t) = e^{-2t} u(t)$$

Its transfer function is given by

$$
\begin{aligned}
H(f) &= \mathcal{F}\{h(t)\} = \mathcal{F}\{e^{-2t} u(t)\} = \int_0^{\infty} e^{-2t} e^{-j2\pi ft} \, dt \\
&= -e^{\frac{-(2+j2\pi f)t}{2+j2\pi f}} \Big|_{t=0}^{t=\infty} = \frac{1}{2+j2\pi f}
\end{aligned}
$$

The system output in the frequency domain is given by

$$Y(f) = \frac{1}{2 + j2\pi f} X(f)$$

$\square$

In the following discussion, we assume that the input $X(t)$ is a random process and is wide sense stationary. By extension, since the system is LTI, the output is a random process, which is also wide sense stationary. We now determine the mean value and the autocorrelation function of the system output followed by the ouput power spectral density.

7.6.2 *Output mean*

The mean value of the system output can be easily obtained from (7.53). In particular

$$
\begin{aligned}
m_Y &= \mathcal{E}\{Y(t)\} = \mathcal{E}\{h(t) \circledast X(t)\} = \mathcal{E}\left\{ \int_{-\infty}^{\infty} h(\lambda) X(t - \lambda) \, d\lambda \right\} \\
&= \int_{-\infty}^{\infty} h(\lambda) \mathcal{E}\{X(t - \lambda)\} \, d\lambda = m_X \int_{-\infty}^{\infty} h(\lambda) \, d\lambda
\end{aligned}
$$

is a constant.

The following example illustrates computation of the output mean.

Example 7.29: The LTI system in Example 7.28 is driven by a random process

$$X(t) = 2 + 5\sin(\omega_0 t + \theta)$$

where θ is a uniform random variable in the range $[-\pi, \pi]$. The mean of this random process

$$m_X = E\{2 + 5\sin(\omega_0 t + \theta)\} = 2 + 5E\{\sin(\omega_0 t + \theta)\} = 2$$

where we observe that the second term is 0 (see Example 7.1).

The mean of the system output, from (7.55), is

$$m_Y \;=\; 2\int_0^\infty e^{-2\lambda}\,d\lambda = 2\frac{e^{-2\lambda}}{-2}\bigg|_0^\infty = 1$$

□

7.6.3 Cross-correlation functions and cross-power spectra

The cross-correlation function of the system output and the input random processes is given from (7.12)

$$R_{YX}(t, t - \tau) = E\{Y(t)X(t - \tau)\}$$

Substituting for $Y(t)$ from (7.53) results in

$$R_{YX}(t, t - \tau) \;=\; E\left\{\left(\int_{-\infty}^\infty h(\lambda)X(t - \lambda)\,d\lambda\right)X(t - \tau)\right\}$$

$$= \int_{-\infty}^\infty h(\lambda)E\{X(t - \lambda)X(t - \tau)\}\,d\lambda$$

Then using (7.11) in the above equation yields

$$R_{YX}(\tau) = \int_{-\infty}^\infty h(\lambda)R_X(\tau - \lambda)\,d\lambda = h(\tau) \circledast R_X(\tau) \tag{7.55}$$

i.e., the cross-correlation function $R_{YX}(\tau)$ is a *convolution* of the system response and the input autocorrelation function. Observe that $R_{XY}(\tau)$ is only a function of τ, and not of absolute time t.

From (7.37), the cross-spectral density is given by

$$S_{YX}(f) = \mathcal{F}\{R_{YX}(\tau)\} = \mathcal{F}\{h(\tau) \circledast R_X(\tau)\} = H(f)S_X(f) \tag{7.56}$$

The relationships in (7.55) and (7.56) are illustrated in Figure 7.14.

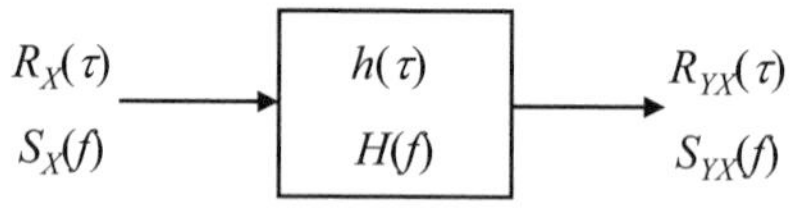

Figure 7.14 Computation of the cross-correlation function $R_{YX}(\tau)$ and the cross-power spectral density $S_{YX}(f)$.

Due to the interrelationship (7.15) between the two cross-correlation functions, we have

$$R_{XY}(\tau) = R_{YX}(-\tau) = h(-\tau) \circledast R_X(-\tau) = h(-\tau) \circledast R_X(\tau) \tag{7.57}$$

Similarly, since from (7.41) $S_{XY}(f) = S_{YX}^*(f)$, the cross-power spectral density

$$S_{XY}(f) = \mathcal{F}\{h(-\tau) \circledast R_X(\tau)\} = H^*(f)S_X(f) \tag{7.58}$$

where $*$ denotes complex conjugation.

The following example determines both cross-correlation functions when the system input is white noise.

Example 7.30: Consider the LTI system of Example 7.28. Suppose that the input is white noise with an autocorrelation function given by

$$R_X(\tau) = \frac{N_0}{2}\delta(\tau)$$

The system impulse response and the input autocorrelation function are illustrated in the figure below.

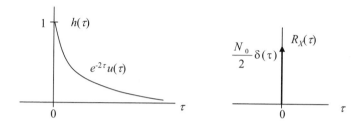

From (7.55), the cross-correlation function $R_{YX}(\tau)$ is

$$R_{YX}(\tau) = h(\tau) \circledast R_X(\tau) = e^{-2\tau}u(\tau) \circledast \frac{N_0}{2}\delta(\tau) = \frac{N_0}{2}e^{-2\tau}u(\tau)$$

and from (7.57), the cross-correlation function $R_{XY}(\tau)$ is

$$R_{XY}(\tau) = R_{YX}(-\tau) = \frac{N_0}{2}e^{2\tau}u(-\tau)$$

The figure below shows plots of these two cross-correlation functions.

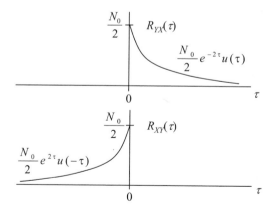

☐

In the last example, you can observe that $R_{YX}(\tau)$ turns out to be just a scaled version of the system impulse response. This is an important result that can be used to determine the impulse response of an unknown system. We consider the general precedure below.

System identification

System identification is a method of determining the impulse response $h(t)$ of an unknown system through the output-input cross-correlation function defined in (7.55). Figure 7.15 depicts a block diagram of a system identification scheme. The unknown system is driven by a white noise having $N_0/2 = 1$, i.e., its autocorrelation function

$$R_X(\tau) = \delta(\tau)$$

The cross-correlation function between the system output $Y(t)$ and the input $X(t)$ is determined from (7.55)

$$R_{YX}(\tau) = h(\tau) \circledast R_X(\tau) = h(\tau) \circledast \delta(\tau) = h(\tau) \qquad (7.59)$$

The corresponding relationship in the frequency domain from (7.56) yields $S_{YX}(f) = H(f)S_X(f) = H(f)$ since $S_X(f) = 1$ in this case. This is a useful and a more practical

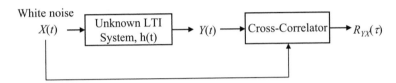

Figure 7.15 Schematic representation of system identification of an unknown system.

method than driving a system with an impulse for determining the impulse response. Ideally, an impulse has infinite magnitude and 0 width, which make generating an impulse impractical. For the sake of this argument, suppose that we are able to generate a waveform having a narrow pulse width and a large magnitude in order to closely resemble an impulse. Such waveform may cause undue stress on the system when applied at the input. On the other hand, using white noise as in the above procedure alleviates this problem. Nevertheless, the reader may note that ideal white noise does not exist and one has to choose bandlimited white noise that has flat spectrum over the frequency band of interest.

7.6.4 Autocorrelation function and power spectral density of system output

The autocorrelation function of the system output $R_Y(\tau)$ is

$$R_Y(\tau) = \mathcal{E}\{Y(t)Y(t-\tau)\}$$

Since $Y(t) = h(t) \circledast X(t)$, we have

$$R_Y(\tau) = \mathcal{E}\left\{ \left(\int_{-\infty}^{\infty} h(\lambda)X(t-\lambda)d\lambda \right) Y(t-\tau) \right\} = \int_{-\infty}^{\infty} h(\lambda)\mathcal{E}\{X(t-\lambda)Y(t-\tau)\}\,d\lambda$$

Substituting $R_{XY}(\tau) = \mathcal{E}\{X(t)Y(t-\tau)\}$ into the above equation yields

$$R_Y(\tau) = \int_{-\infty}^{\infty} h(\lambda)R_{XY}(\tau-\lambda)d\lambda = h(\tau) \circledast R_{XY}(\tau) \qquad (7.60)$$

Then, substituting (7.57) into (7.60) yields

$$\boxed{R_Y(\tau) = h(\tau) \circledast h(-\tau) \circledast R_X(\tau)} \qquad (7.61)$$

which can be written in the integral form as

$$R_Y(\tau) = \int_{-\infty}^{\infty} \int_{-\infty}^{\infty} h(\lambda)h(\mu)R_X(\tau - \lambda + \mu) \, d\lambda \, d\mu$$

This operation is depicted in Figure 7.16. The purpose of the figure is not to suggest a way of computing the output autocorrelation function but to illustrate the concept.

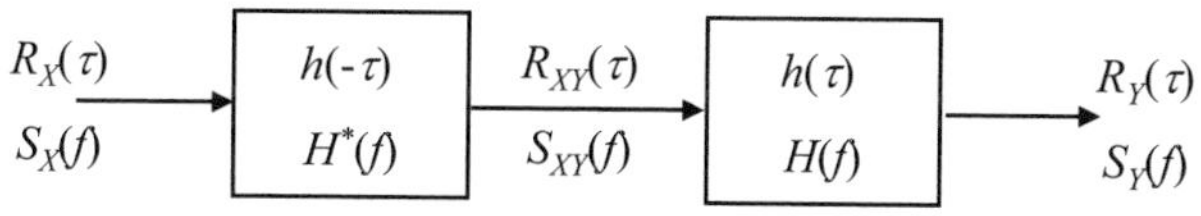

Figure 7.16 Conceptual schematic representation of output autocorrelation function (power spectral density) in terms of the system and the input autocorrelation (power spectral density).

The output power spectral density is obtained by taking the Fourier transform of the output autocorrelation function

$$S_Y(f) = \mathcal{F}\{h(\tau) \circledast h(-\tau) \circledast R_X(\tau)\} = H(f)H^*(f)S_X(f)$$

or

$$S_Y(f) = |H(f)|^2 S_X(f) \qquad (7.62)$$

where the term $|H(f)|$ is the magnitude response of the system transfer function. In other words, the output power spectral density is a product of the input power spectral density and the square of the system magnitude response (see Figure (7.16)).

The expressions for the cross-covariance functions and the ouput autocovariance function are summarized in Table 7.4. The development of these expressions parallels those of the correlation functions discussed above.

quantity	*definition*
Cross-covariance Function	$C_{YX}(\tau) = h(\tau) \circledast C_X(\tau)$
	$C_{XY}(\tau) = h(-\tau) \circledast C_X(\tau)$
Output Autocovariance Function	$C_Y(\tau) = h(\tau) \circledast C_{XY}(\tau)$
	$C_Y(\tau) = h(\tau) \circledast h(-\tau) \circledast C_X(\tau)$
Interrelationships	$C_{YX}(\tau) = R_{YX}(\tau) - m_Y m_X$
	$C_{XY}(\tau) = R_{XY}(\tau) - m_X m_Y$
	$C_Y(\tau) = R_Y(\tau) - m_Y^2$

Table 7.4 Covariance functions.

Example 7.31: From Example 7.28, consider the system having impulse response

$$h(t) = e^{-2t}u(t)$$

The system input is an IID random process having an autocorrelation function

$$R_X(\tau) = 2 + 3\delta(\tau)$$

Determine the ouput mean, the cross-correlation functions, the output autocorrelation function, and the outuput power spectral density. The impulse response and the input autocorrelation function are shown below.

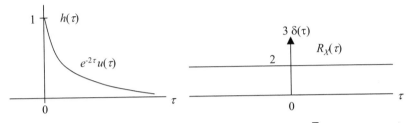

From the autocorrelation function, the input mean $m_X = \sqrt{2}$ (see Table 7.2). The output mean is

$$m_Y = \sqrt{2} \int_0^\infty e^{-2t}\,dt = \frac{1}{\sqrt{2}}$$

The input autocovariance function is given by

$$C_X(\tau) = R_X(\tau) - m_X^2 = 3\delta(\tau)$$

The cross-covariance function is thus

$$C_{YX}(\tau) = h(\tau) \circledast C_X(\tau) = 3\left(e^{-2\tau}u(\tau)\right) \circledast \delta(\tau) = 3e^{-2\tau}u(\tau)$$

and the corresponding cross-correlation function becomes

$$R_{YX}(\tau) = C_{YX}(\tau) + m_Y m_X = 3e^{-2\tau}u(\tau) + 1$$

which is shown in the figure below.

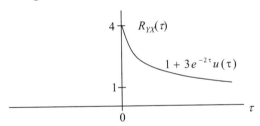

By symmetry, we then have

$$C_{XY}(\tau) = C_{YX}(-\tau) = 3e^{2\tau}u(-\tau)$$

$$R_{XY}(\tau) = C_{XY}(\tau) + m_X m_Y = 3e^{2\tau}u(-\tau) + 1$$

and a plot of the cross-correlation function is shown below.

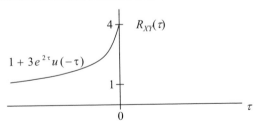

We now turn to computing the second moment quantities associated with the output $Y(t)$. For ease of computation, we first determine the output autocovariance function

$$C_Y(\tau) = h(\tau) \circledast C_{XY}(\tau) = 3e^{-2\tau}u(\tau) \circledast e^{2\tau}u(-\tau)$$

The above convolution is carried out for two cases:

Case 1: $\tau \geq 0$

The figure below provides a graphical illustration of this case. The integral is evaluated

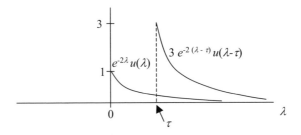

as follows:

$$C_Y(\tau) = 3\int_\tau^\infty e^{-2\lambda}e^{-2(-\tau+\lambda)}d\lambda = 3e^{2\tau}\left.\frac{e^{-4\lambda}}{-4}\right|_\tau^\infty = \frac{3}{4}e^{-2\tau}$$

Case 2: $\tau \leq 0$

The figure below provides a graphical illustration of this case. The corresponding

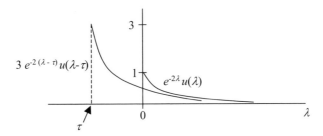

integral is evaluated as follows:

$$C_Y(\tau) = 3\int_0^\infty e^{-2\lambda}e^{-2(-\tau+\lambda)}d\lambda = 3e^{2\tau}\frac{e^{-4\lambda}}{-4}|_0^\infty = \frac{3}{4}e^{2\tau}$$

Combination of these two results yields the output autocovariance function:

$$C_Y(\tau) = \frac{3}{4}e^{-2|\tau|}$$

The output autocorrelation function is then given by

$$R_Y(\tau) = C_Y(\tau) + m_Y^2 = \frac{3}{4}e^{-2|\tau|} + \frac{1}{2}$$

Finally, a plot of the output autocorrelation function is shown in the figure at the top of the next page. Clearly, the output autocorrelation function exhibits the required even symmetry.

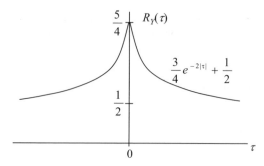

Let us now determine the output power spectral density. From Example 7.28, the transfer function of the system and its complex conjugation are

$$H(f) = \frac{1}{2 + j2\pi f} \quad H^*(f) = \frac{1}{2 - j2\pi f}$$

The input power spectral density is

$$S_X(f) = \mathcal{F}\{R_X(\tau)\} = \mathcal{F}\{3\delta(\tau) + 2\} = 3 + 2\delta(f)$$

Thus from (7.62), we have the output power spectral density

$$S_Y(f) = |H(f)|^2 S_X(f) = H(f)H^*(f)S_X(f) = \frac{1}{1 - \pi^2 f^2}\left(\frac{3}{4} + \frac{1}{2}\delta(f)\right)$$

□

Equation 7.62 is one of the most important and widely used results in the analysis of random signals and linear systems. We close this section with one final example of using 7.62. This example determines the output power spectral density graphically.

Example 7.32: The magnitude response of a bandpass LTI system is given by

$$H(f) = \begin{cases} 1 & \text{for} \quad 200 \text{ Hz} \leq f \leq 300 \text{ Hz} \\ 0 & \text{otherwise} \end{cases}$$

and is sketched in the top part of the figure below.

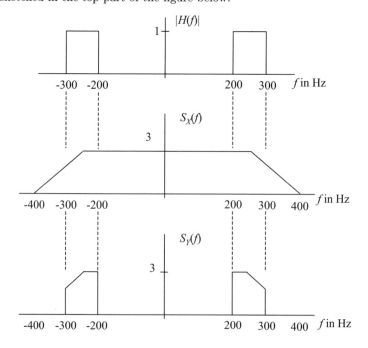

The power spectral density of a random process applied at the input of the system, and the resulting output power spectral density are also shown in the figure. The system response has a magnitude of 1, so in squaring it we still have the same shape. By multiplying the squared system magnitude response with the input power spectral density $S_X(f)$, we obtain the output power spectral density, as shown in the figure.

□

7.6.5 Response of linear systems: discrete case

The above discussion can be easily extended to the discrete system case. Let $h[k]$ be the impulse response of a discrete linear shift-invariant system. The system output is given by

$$Y[k] = h[k] \circledast X[k] = \sum_{i=-\infty}^{\infty} h[i]X[k-i] \tag{7.63}$$

Assuming that the system input is deterministic, the system response in the frequency domain is

$$Y(e^{j\omega}) = H(e^{j\omega})X(e^{j\omega}) \tag{7.64}$$

where $H(e^{j\omega})$ is the transfer function of the discrete system defined by

$$H(e^{j\omega}) = \sum_{k=-\infty}^{\infty} h[k]e^{-j\omega k}$$

The output mean of a discrete system is (see (7.55))

$$m_Y = E\{Y[k]\} = \sum_{i=-\infty}^{\infty} h[i]E\{X[k-i]\} = m_X \sum_{i=-\infty}^{\infty} h[i] \tag{7.65}$$

The following example illustrates the output mean computation.

Example 7.33: Given a finite impulse response system

$$Y[k] = 0.5X[k] + 0.2X[k-1]$$

with a white noise input, let us determine the output mean value.

The impulse response of the system is

$$h[k] = 0.5\delta[k] + 0.2\delta[k-1]$$

By definition, the mean of the input white noise is 0, i.e., $m_X = 0$.

From (7.65), the mean value of the system output is

$$m_Y = m_X (0.5 + 0.2) = 0 \times 0.7 = 0$$

□

The cross-correlation function between the system output and the input is given by

$$R_{YX}[\ell] = h[\ell] \circledast R_X[\ell] \tag{7.66}$$

By changing the roles of $Y[k]$ and $X[k]$ in the above equation (see (7.15)), we have

$$R_{XY}[\ell] = h[-\ell] \circledast R_X[\ell]$$

The corresponding cross-power spectral densities are

$$
\begin{aligned}
S_{YX}(e^{j\omega}) &= H(e^{j\omega})S_X(e^{j\omega}) \\
S_{XY}(e^{j\omega}) &= H^*(e^{j\omega})S_X(e^{j\omega})
\end{aligned}
\tag{7.67}
$$

where * indicates complex conjugation and ω is the digital frequency. The following example illustrates the computation of these cross-correlation functions.

Example 7.34: Consider the system in Example 7.33. Assume that this system is driven by an IID random process having

$$R_X[\ell] = \delta[\ell+1] + 2\delta[\ell] + \delta[\ell-1]$$

These two functions are graphically illustrated in the figure below.

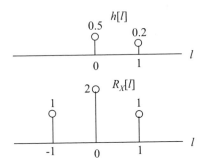

The cross-correlation function $R_{YX}[\ell]$ is

$$R_{YX}[\ell] = h[\ell] \circledast R_X[\ell] = 0.5\delta[\ell+1] + 1.2\delta[\ell] + 0.9\delta[\ell-1] + 0.2\delta[\ell-2]$$

A plot is shown below.

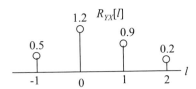

The corresponding cross-correlation function $R_{XY}[\ell]$ is

$$R_{XY}[\ell] = R_{YX}[-\ell] = 0.2\delta[\ell+2] + 0.9\delta[\ell+1] + 1.2\delta[\ell] + 0.5\delta[\ell-1]$$

and its plot is shown below.

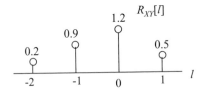

☐

Following the development in Section 7.6.4, the output autocorrelation of a discrete system is

$$
\begin{aligned}
R_Y[\ell] &= h[\ell] \circledast R_{XY}[\ell] \\
&= h[\ell] \circledast h[-\ell] \circledast R_X[\ell] = \sum_{i=-\infty}^{\infty} \sum_{j=-\infty}^{\infty} h[i]h[j]R_X[\ell-i+j] \qquad (7.68)
\end{aligned}
$$

The output power spectral density is defined as

$$S_Y(e^{j\omega}) = \mathcal{F}\{h[\ell] \circledast h[-\ell] \circledast R_X[\ell]\} = H(e^{j\omega})H^*(e^{j\omega})S_X(e^{j\omega}) = |H(e^{j\omega})|^2 S_X(e^{j\omega})$$

(7.69)

where $|H(e^{j\omega})|$ is the magnitude frequency response of the discrete system.

Given the system impulse response and the cross-correlation function $R_{XY}[\ell]$, the example below computes the output autocorrelation function.

Example 7.35: Continuing from Example 7.34, let us determine the output autocorrelation function:

$$R_Y[\ell] = h[\ell] \circledast R_{XY}[\ell] = \sum_{i=-2}^{1} R_{XY}[i]h[\ell - i]$$

The sum is evaluated for $\ell = -2, -1, 0, 1, 2$. The figure below illustrates discrete convolution for $\ell = 0$ and the corresponding output autocorrelation value

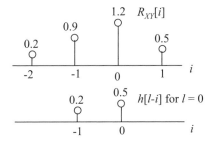

$$R_Y[0] = 0.9 \times 0.2 + 1.2 \times 0.5 = 0.78$$

The figure below illustrates the two functions in the sum for $\ell = -2$ and $\ell = 1$

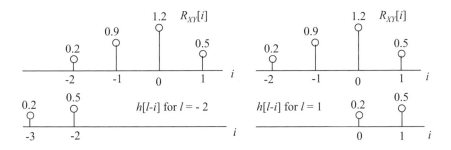

and the corresponding output autocorrelation values are

$$R_Y[-2] = 0 \times 0.2 + 0.2 \times 0.5 = 0.10 \qquad R_Y[1] = 1.2 \times 0.2 + 0.5 \times 0.5 = 0.49$$

The remaining output autocorrelation values can be computed similarly. A plot of the output autocorrelation function is shown below. By observation, $R_Y[\ell]$ is even

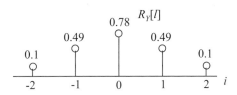

symmetric and $R_Y[\ell] \leq R_Y[0]$ for all values of ℓ.

The output power spectral density can be obtained by taking the discrete-time Fourier transform of the output autocorrelation function

$$
\begin{aligned}
S_Y(e^{j\omega}) &= \sum_{i=-2}^{2} R_Y[\ell]e^{-j\omega i} \\
&= 0.78 + 0.49\left[e^{j\omega} + e^{-j\omega}\right] + 0.10\left[e^{j2\omega} + e^{-j2\omega}\right] \\
&= 0.78 + 0.98\cos\omega + 0.20\cos 2\omega
\end{aligned}
$$

A plot of $S_Y(e^{j\omega})$ is shown below. The output power spectral density is real, positive, and even symmetric.

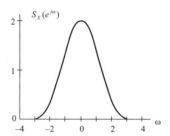

$\square$

A summary of expressions for the cross-covariance functions and the output auto-covariance function is provided in Table 7.5 [5, 6].

quantity	definition
Cross-covariance Function	$C_{YX}[\ell] = h[\ell] \circledast C_X([\ell]$
	$C_{XY}[\ell] = h(-\tau) \circledast C_X[\ell]$
Output Autocovariance Function	$C_Y[\ell] = h[\ell] \circledast C_{XY}[\ell]$
	$C_Y[\ell] = h[\ell] \circledast h[-\ell] \circledast C_X[\ell]$
Interrelationships	$C_{YX}[\ell] = R_{YX}[\ell] - m_Y m_X$
	$C_{XY}[\ell] = R_{XY}[\ell] - m_X m_Y$
	$C_Y[\ell] = R_Y[\ell] - m_Y^2$

Table 7.5 Covariance functions: discrete case.

7.7 *Summary*

This chapter introduces the concept of random process as a random variable that is a function of time. The time index can be continuous or discrete. Even though the discussion here is focused on the continuous-time random processes, we extend the treatment to the discrete-time random processes where applicable. In both cases, a sample of a random process is a random variable; consequently, a random process can be considered as a time sequence of random variables. Since a single random variable can be characterized by its PDF, a random process can be completely characterized by a joint PDF.

Since the PDF of a random process may not be readily available in engineering applications, the random process can be partially characterized by its first and second moments. Higher moments are also important in characterizing a random process, but their discussion is beyond the scope of this book. The first moment is called the mean while the second moment quantities are called the autocorrelation function and the autocovariance function. Where two random processes are involved, we also have cross-correlation and cross-covariance functions.

The properties of stationarity and ergodicity are important in the treatment of random processes and their application to engineering problems. A wide sense stationary random process has a constant mean and an autocorrelation function that is a function of only the time difference (i.e., independent of the absolute time instants). In essence, the first and the second moments of a wide sense stationary random process do not change with the time origin. For a given random process, the first and second moments can be determined based on the expectation operation or using time averages. When these moments are the same regardless of whether they are based upon the expectation or the time averages, we have an ergodic random process. In other words, if a random process is ergodic, its first and second moments can be computed using the time averages. This is advantageous since in most problems of practical interest precise knowledge of PDF is frequently not available.

Second moment analysis in the frequency domain is beneficial in many engineering problems. The power spectral density function is obtained as a Fourier transform of the autocorrelation function. The power spectral density is a representation of the spread of the total average power of a random process over all frequencies in the frequency domain. The power spectral density function for real-valued random processes is even symmetric and non-negative. White noise is a special random process whose power spectral density is constant over all frequencies; consequently, white noise has an autocorrelation function that is an impulse.

Thermal noise and quantization noise are presented as examples of random processes occurring in analog and digital systems, respectively. Thermal noise occurs due to the random fluctuations of electrons caused by thermal agitation. The power spectral density function of thermal noise is assumed to be flat (white noise) and is proportional to the absolute temperature of the conducting medium. As a consequence, thermal noise is omnipresent in electronic devices; it is particularly bothersome in high-gain radio frequency power amplifiers. Quantization noise, on the other hand, is due to approximation of continuous magnitude values to finite levels and occurs when a signal is being converted into the digital form. The degradation due to quantization noise is inversely proportional to the number of bits used to represent the signal in the digital form; hence, the effect of quantization noise can be mitigated by increasing the bit resolution.

The response of linear time-invariant systems to random inputs is of interest in electrical engineering problems and is presented using both time and frequency domain approaches. The cross-correlation function between the system output and input is obtained as a convolution of the input autocorrelation function and the system impulse response. If the input is white noise with a variance of 1, this cross-correlation function is equal to the system impulse response. This is exploited in the system identification problem. Expressions for the output autocorrelation function as a function of the input autocorrelation function and the system impulse response and the output power spectral density as a product of the input power spectral density and the system frequency response are obtained.

References

[1] Alberto Leon-Garcia. *Probability and Random Processes for Electrical Engineering.* Addison-Wesley, Reading, Massachusetts, second edition, 1994.

[2] Carl W. Helstrom. *Probability and Stochastic Processes for Engineers.* MacMillan, New York, second edition, 1991.

[3] Athanasios Papoulis. *Probability, Random Variables, and Stochastic Processes.* McGraw-Hill, New York, third edition, 1991.

[4] Wilbur B. Davenport, Jr. and William L. Root. *An Introduction to the Theory of Random Signals and Noise.* McGraw-Hill, New York, 1958.

[5] Charles W. Therrien. *Discrete Random Signals and Statistical Signal Processing.* Prentice Hall, Inc., Englewood Cliffs, New Jersey, 1992.

[6] Henry Stark and John W. Woods. *Probability, Random Processes, and Estimation Theory for Engineers.* Prentice Hall, Inc., Englewood Cliffs, New Jersey, 1986.

Problems

First and second moments of a random process

7.1   A *complex*-valued continuous random signal is defined by

$$X(t) = Ae^{\jmath(\omega t + \Phi)}$$

where A and ω are constant parameters (i.e., not random variables) and Φ is a random variable uniformly distributed between $-\pi$ and π.

(a) What is the mean, $\mathcal{E}\{X(t)\}$?

(b) The covariance function for a complex random process is defined as $C(t_1, t_0) = \text{Cov}\,[X(t_1), X^*(t_0)]$. What is the covariance function of this process?

(c) What is $\text{Cov}\,[X(t_1), X(t_0)]$ (*without* the conjugate)?

7.2   Given the following random signal $X(t) = A + \cos(\omega t + \Phi)$, where Φ is a uniform random variable in the range $[-\pi, \pi]$ and A is a constant, determine the following:

(a) The mean of $X(t)$.

(b) The correlation function of $X(t)$.

(c) The covariance function of $X(t)$.

Properties: Independence, stationarity and ergodicity

7.3 A discrete-time random process known as a *Bernoulli process* is defined by

$$X[k] = \begin{cases} +1 & \text{with probability } p \\ -1 & \text{with probability } 1-p \end{cases}$$

Successive values of the process $X[k], X[k+1], X[k+2]\ldots$ are *independent*.

(a) What are the mean and variance of the process, $\mathcal{E}\{X[k]\}$ and $\mathrm{Var}\,[X[k]]$? Do they depend on k?

(b) What are the correlation function $R[k_1, k_0]$ and covariance function $C[k_1, k_0]$ of the process? Is the random process stationary?

7.4 $X[k]$ is a random process defined as an independent sequence of Bernoulli random variables taking on values of $+1$ and -1 with equal probability.

(a) What is the mean of this random process?

(b) What is the correlation function?

(c) What is the covariance function?

(d) A new random process is defined as the difference

$$Y[k] = X[k] - X[k-1]$$

What is the mean of the process $Y[k]$?

(e) What is the correlation function for the process $Y[k]$?

7.5 A discrete-time random process $Y[k]$ is defined by

$$Y[k] = \sum_{i=0}^{k} X[i]$$

where $X[k]$ is the Bernoulli process defined in Prob. 7.3. This process $Y[k]$ is known as a *random walk*.

(a) Sketch some typical realizations of the Bernoulli process.

(b) Sketch the corresponding realizations of the random walk.

(c) Compute the mean and variance of $Y[k]$. Do they depend on the time index k?

(d) For the special case $p = \frac{1}{2}$, do the mean and variance depend on k?

7.6 A certain signal is modeled by a continuous Gaussian random process $X(t)$ with mean $\mathcal{E}\{X(t)\} = 2$ and correlation function $R(t_1, t_0) = 4 + e^{-0.5|t_1 - t_0|}$.

(a) Compute the covariance function $C(t_1, t_0)$.

(b) Compute $\mathrm{Var}\,[X(t)]$.

(c) Write the PDF for a single sample $X(t_0)$.

(d) Write the joint PDF for two samples $X(t_1)$ and $X(t_0)$ for $t_0 = 3$ and $t_1 = 7$.

7.7 A *complex* continuous random process is defined by

$$X(t) = Ae^{j(\omega_o t + \Phi)} + W(t)$$

where A and Φ are real parameters, ω_o is a known radian frequency, and $W(t)$ is a stationary zero-mean white noise process with correlation function $R_W(\tau) = \frac{1}{2}\delta(\tau)$. Assume that $W(t)$ is independent of the parameters A and Φ. Find the autocorrelation function for $X(t)$ under the following conditions:

(a) $\Phi = 0$ and A is a Rayleigh random variable with parameter $\alpha = \frac{1}{2}$.

(b) Φ is uniformly distributed between $-\pi$ and π and $A = 1/\sqrt{2}$.

(c) Φ is uniformly distributed between $-\pi$ and π and A is a Rayleigh rv with $\alpha = \frac{1}{2}$. A and Φ are independent.

(d) Same conditions as Part (b) but $W(t)$ has a correlation function $R_W(\tau) = \frac{1}{2}e^{-|\tau|}$. (When the noise $W(t)$ is not white, it is said to be *colored*.)

Note: The autocorrelation function for a stationary complex random process can be defined as $R_X(\tau) = \mathcal{E}\{X(t)X^*(t-\tau)\}$ where the $*$ denotes the complex conjugate. Note also that the process is of the form $X(t) = S(t)+W(t)$ and since $W(t)$ is independent of $S(t)$ and has mean 0, the expectation of crossproducts involving $S(t)$ and $W(t)$ vanishes.

7.8   The mean and the autocovariance function of a random process are given by

$$m(t) \quad = \quad 0$$

$$C_X(t_1, t_0) \quad = \quad \cos(\pi(t_1 - t_0)).$$

(a) For $t_1 = 3.1$ and $t_0 = 1.8$, compute the values of mean and correlation.

(b) Compute the correlation coefficient.

(c) Determine the average total power.

(d) What is the rms value of $X(t)$?

7.9   Two random processes $X(t)$ and $Y(t)$ are independent and wide sense stationary.

(a) If we define, $Z(t) = X(t) + Y(t)$, is $Z(t)$ wide sense stationary?

(b) If we define, $Z(t) = X(t)Y(t)$, is $Z(t)$ wide sense stationary?

Power spectral density

7.10   Consider the autocorrelation function of the Gaussian random process in Problem 6 and determine the power spectral density $S_X(f)$.

7.11   Determine the autocorrelation function of the random process in Problem 8 and compute the corresponding power spectral density.

7.12   Show that the Fourier transform of the triangular-shaped autocorrelation function in Example 7.23 is given by

$$S_X(f) = A^2 T_b \left(\frac{\sin \pi f T_b}{\pi f T_b} \right)^2$$

Hint: Consider the Fourier transform of a square pulse $p(t)$ with width T_b. The convolution $p(t) \circledast p(-t)$ is a triangular waveform similar to $R_X(\tau)$ in the example. Use the properties of the Fourier transform as it relates to convolution to devise $S_X(f)$.

7.13   The autocorrelation function of a discrete random process is given by

$$R_X[\ell] = 0.3\delta[\ell + 2] + 2\delta[\ell + 1] + 5\delta[\ell] + 2\delta[\ell - 1] + 0.3\delta[\ell - 2]$$

(a) Determine and sketch the power spectral density $S_X(e^{j\omega})$.

(b) Is this is a valid power spectral density?

7.14 The cross-correlation function of a random process is given by

$$R_{XY}(\tau) = 5e^{-\tau}u(\tau) + 2$$

Find the cross-power spectral density $S_{XY}(f)$.

Noise sources

7.15 In a power amplifier operating at $17°C$ with a bandwidth of 10 kHz:

(a) What is the thermal noise power in dB?

(b) Find the rms noise voltage given that the internal resistance is 100 ohms and a load resistance of 100 ohms.

(c) If the signal being amplified is a perfect sinusoid with a magnitude of 3 mV, what is the SNR as the signal leaves the amplifier?

7.16 Consider an N-step quantizer. The input to the quantizer is a sinusoidal waveform with a peak magnitude of A. The quantization error due to rounding is characterized by the following PDF

$$f_E(e) = \left\{ \frac{2}{\Delta}\left(1 - \frac{2}{\Delta}|e|\right) \qquad -\frac{\Delta}{2} \leq e \leq \frac{\Delta}{2}, \right.$$

(a) Determine the mean and the variance of the quantization error.

(b) Obtain the expression for the signal to quantization noise in dB.

(c) If the SQNR desired is at least 80 dB, how many (binary) bits are required to achieve this?

Response of linear systems

7.17 The impulse response $h(\tau)$ of an LTI system and the autocorrelation function of a random process $X(t)$ at the input of this system, respectively, are

$$h(\tau) = e^{-2t}u(t)$$
$$R_X(\tau) = \begin{cases} 2 - |\tau| & \text{for } -2 \leq \tau \leq 2 \\ 0 & \text{otherwise} \end{cases}$$

Determine and sketch the cross-correlation function $R_{YX}(\tau)$.

7.18 The input to a continuous-time linear system is a white noise process with correlation function $R_X(\tau) = 5\delta(\tau)$. The impulse response of the system is

$$h(t) = e^{-at}u(t)$$

where a is a fixed positive deterministic parameter and $u(t)$ represents the unit step function.

(a) Find the output autocorrelation function under the assumption that the mean of the input is 0. Be sure to specify it for $-\infty < \tau < \infty$.

(b) If the expected value of the input is 2, what is the expected value (mean) of the output?

7.19 Refer to Prob. 7.18 and answer these questions: (Assume the input has 0 mean.)

(a) What is the frequency response $H(f)$ of the system?

(b) Take the Fourier transform of $R_Y(\tau)$ as computed in Prob. 7.18 to find $S_Y(f)$.

(c) Use the formula $S_Y(f) = |H(f)|^2 S_X(f)$ to check the result of part (b).

(d) Find the average output power $E\{Y^2(t)\}$ from the autocorrelation function $R_Y(\tau)$ that you computed for the output.

(e) Compute the average output power by integrating the power spectral density function $S_Y(f)$ over frequency from $-\infty$ to ∞. You may use a table of definite integrals in this part.

7.20 The input to an ideal bandpass filter is white Gaussian with a variance of 3. The bandpass filter response is shown below:

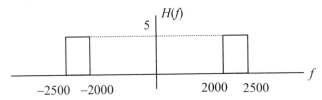

(a) Obtain and plot the input signal power spectral density function.

(b) What is the output power spectral density? (Show it graphically.)

(c) Compute the output power.

7.21 Consider the filter shown in Problem 7.20. The input to the filter is given by

$$X(t) = 2\cos(4400\pi t + \Phi) + W(t)$$

where $W(t)$ is white Gaussian with a variance of 4, Φ is uniform in the range $[-\pi, \pi]$. Assume $W(t)$ and Φ are independent.

(a) Determine and plot the input power spectral density function.

(b) Plot the output power spectral density function.

(c) Determine the output SNR in dB.

7.22 A linear system is described by a first-order differential equation

$$\frac{dy}{dt} + ay(t) = x(t).$$

The impulse response of this system is $h(t) = e^{-at}u(t)$. The input is a white noise process with a correlation function

$$R_X(\tau) = \frac{N_0}{2}\delta(\tau).$$

(a) Determine the correlation function of the output random process $R_Y(\tau)$.

(b) What is the power spectral density function of the input?

(c) What is the power spectral density function of the output?

(d) Find the total average power of the output.

7.23 A continuous white noise process is applied to the continuous-time system shown below. The input $W(t)$ has 0 mean and $E\{W(t_1)W(t_0)\} = \delta(t_1 - t_0)$.

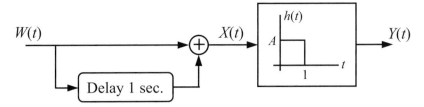

(a) What is $X(t)$ and what is the correlation function $R_X(t_1, t_0)$? Is $X(t)$ stationary?

(b) What is the correlation function of the output signal $Y(t)$?

(c) What is the frequency response $H(f)$ corresponding to the filter with impulse response $h(t)$?

(d) What is the power spectral density function $S_X(f)$?

(e) What is the output power spectral density function $S_Y(f)$? Express it in the simplest form possible.

7.24 Show that the output power spectral density in Example 7.31 given by

$$S_Y(f) = \frac{1}{1 - \pi^2 f^2} \left(\frac{3}{4} + \frac{1}{2}\delta(f) \right)$$

can be obtained by directly computing the Fourier transform of the autocorrelation function given by

$$R_Y(\tau) = \frac{3}{4}e^{-2|\tau|} + \frac{1}{2}$$

7.25 Show that cross-covariance function $C_{YX}(\tau)$ is given by

$$C_{YX}(\tau) = R_{YX}(\tau) - m_Y m_X$$

as listed in Table 7.4.

7.26 The autocorrelation functions for two discrete-time random processes $X[k]$ and $Y[k]$ are sketched below:

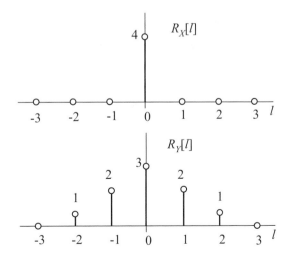

(a) Are both of these correlation functions legitimate?

(b) Which process has higher correlation?

(c) What would be the impulse response $h[k]$ of a causal time-invariant linear system such that if $X[k]$ were the input, $Y[k]$ would be the output of this system?

7.27 A discrete-time random process $X[k]$ is defined by

$$X[k] = 2W[k] - 3W[k-1] + W[k-2]$$

where $W[k]$ is a wide-sense stationary white noise process with mean 0 and correlation function $R_W[l] = \delta[l]$. (Note that this means that $\mathcal{E}\{W[k_1]W[k_0]\} = \delta[k_1 - k_0]$.)

(a) Find the mean and autocorrelation function for the process $X[k]$. Is it stationary? If so, express the correlation function in terms of just the lag variable l, and plot $R_X[l]$ versus l.

(b) Another random process is defined by

$$Y[k] = W[k] - 2W[k-1]$$

Find the autocorrelation function for $Y[k]$ and the cross-correlation function for X and Y. Are the two random processes *jointly stationary*? If so, express all correlation functions in terms of the lag variable $l = k_1 - k_0$.

7.28 $X[k]$ is the input to a discrete-time linear shift-invariant system. $X[k]$ has mean 0 and correlation function $R_X[l] = \delta[l]$. A zero-mean process with this form of autocorrelation function is referred to as discrete-time white noise.

The linear system has an impulse response given by

$$h[k] = 2\delta[k] - 3\delta[k-1] + \delta[k-2]$$

The output of the system is a random process $Y[k]$.

(a) Find the cross-correlation function between input and output $R_{XY}[l]$.

(b) Find the autocorrelation function of the output $R_Y[l]$.

(c) If the mean of the input is not 0, but $m_X = \mathcal{E}\{X[k]\} = 2.5$, then what is the mean of the output m_Y?

7.29 Refer to Prob. 7.28 and answer the following questions: (Assume the input has 0 mean.)

(a) What is the frequency response of the system $H(e^{j\omega})$?

(b) Compute the power spectral density function $S_Y(e^{j\omega})$ of the output as the discrete-time Fourier transform of the autocorrelation function $R_Y[l]$.

(c) Confirm the result in part (b) by using the formula $S_Y(e^{j\omega}) = |H(e^{j\omega})|^2 S_X(e^{j\omega})$.

(d) Show that $S_Y(e^{j\omega})$ is real and positive for all values of ω.

7.30 Given that the cross-covariance function $C_{XY}[\ell]$ is

$$C_{XY}[\ell] = R_{XY}[\ell] - m_X m_Y$$

show that the output autocovariance function $C_Y[\ell]$ is given by

$$C_Y[\ell]) = R_Y[\ell] - m_Y^2$$

as listed in Table 7.5.

Computer Projects

Project 7.1

The data sets s00.dat, s01.dat, s02.dat, and s03.dat from the data package represent realizations of four different random signals. In this project you will examine the autocorrelation and power spectral density functions for these signals.

1. Plot each of the signals versus time. It is best to use two plots per page and orient the plot in landscape mode. Tell whether you think each random signal exhibits high correlation, low correlation, or negative correlation.

2. Estimate the mean for each process; it should be close to 0.

3. Estimate the autocorrelation function $R[l]$ for each process and plot it over the interval $-128 \leq l \leq 127$. Compare the correlation of each of the signals based on the estimated correlation function. How well does this compare to your guesses about correlation in Step 1?

4. Compute the power spectral density of each process as the Fourier transform of $R[l]$ using a 256-point discrete Fourier transform (DFT). How do your estimates about correlation relate to the spectrum of each process?

MATLAB programming notes

You may use the MATLAB function 'getdata' from the software package to retrieve the data.

The MATLAB function 'xcorr' can be used to estimate the autocorrelation function. The MATLAB 'fft' function can be used for the DFT.

Project 7.2

In this project, you will compute the autocorrelation and cross-correlation functions for two given sequences as well as their power spectral density and cross-power spectral density functions. You will also compare the experimental results with theoretical results.

1. Generate the following discrete-time sequences:

$$y[k] = 5\cos(0.12\pi k + 0.79)$$
$$x[k] = y[k] + 7w[k]$$

for $0 \leq k \leq 199$. Here $w[k]$ is a zero-mean white Gaussian noise sequence with variance one. What is the SNR for $x[k]$?

2. Compute the autocorrelation function $R_X[l]$ for $-100 \leq l \leq 99$. Use the *unbiased* form of the estimate.

3. Now compute an estimate for the power density spectrum as the discrete Fourier transform (DFT) of the autocorrelation function. Plot $y[k]$, $x[k]$, $R_X[l]$, and power spectral density function (it should be real). Plot all of these quantities beneath one another on the same page. In your plots, clearly label the axes with appropriate values.

4. Compute the cross-correlation function $R_{XY}[l]$ for $-100 \leq l \leq 99$ and an estimate for the cross-power spectral density function as the DFT of $R_{XY}[l]$. Plot $y[k]$, $x[k]$, $R_{XY}[l]$, and the *magnitude* of the cross-power spectral density function on the same page.

5. Compare all of the computed quantities to theoretical values.

MATLAB programming notes

The MATLAB function 'xcorr' can be used to estimate both the autocorrelation function and the cross-correlation function. Use the 'unbiased' option. The MATLAB 'fft' function can be used for the DFT.

Project 7.3

This project explores the identification of an unknown linear time-invariant system using the cross-correlation approach. The discrete-time system to be identified has the transfer function

$$H(z) = \frac{0.50 - 1.32z^{-1} + 0.82z^{-2}}{1 + 1.75z^{-1} + 0.89z^{-2}}$$

1. Compute the impulse response $h[k]$ of the above system for $0 \le k \le 127$. Also compute the magnitude frequency response $|H(e^{j\omega})|$. Plot both $h[k]$ and $|H(e^{j\omega})|$. Use at least 256 frequency samples to plot the frequency response.

2. Generate 10,000 samples of the input $X[k]$ as a unit-variance white Gaussian sequence. Plot the first 100 samples of $X[k]$.

3. Compute the autocorrelation function of the input sequence and plot the autocorrelation function $R_X[l]$ for $-128 \le l \le 127$. Verify that the input sequence has the autocorrelation function you expect for white noise.

4. With $X[k]$ as input, generate the output of the "unknown" system $Y[k]$. Compute the cross-correlation function $R_{YX}[l]$ between $Y[k]$ and $X[k]$.

5. Plot the cross-correlation function $R_{YX}[l]$ for $0 \le l \le 127$ and compare it to the previously computed impulse response of the system. Also compute and plot the magnitude of the cross-spectral density as the Fourier transform of the cross-correlation function and compare it to the magnitude frequency response found in Step 1.

6. Repeat Steps 2 through 5 for input sequences of lengths 1,000 and 100,000.

MATLAB programming notes

You may use the MATLAB functions 'filter' and 'freqz' to compute the impulse response and frequency response in Step 1.

The MATLAB function 'xcorr' can be used to estimate both the autocorrelation function and the cross-correlation function.

8 Markov and Poisson Random Processes

This chapter deals with two special types of random processes, the Markov and Poisson processes. You have seen aspects of both of these processes in earlier chapters. This final chapter, however brings this knowledge together in one place and develops it further. In particular, the theory and techniques needed for the analysis of systems characterized by random requests for service and random service times are developed here. Such systems include multi-user and multi-process computer operating systems, and the traffic on computer networks, switches, and routers.

We begin by developing the Poisson process from some basic principles. You will see that many types of random variables encountered earlier are related to this process. We then move on to discuss the discrete random process known as a Markov chain. This process is characterized by "states" which determine its future probabilistic behavior, and is a useful model for many physical processes that evolve in time. The Markov and Poisson models are then combined in the *continuous-time Markov chain*. This random process is also characterized by states, but state transitions may occur randomly at *any* time (i.e., not at just discrete epochs of time) according to the Poisson model. This combined model is what is necessary to represent the computer and network problems cited above.

We end the chapter with an introduction to queueing theory, the branch of probability and statistics that deals with analysis of these combined systems. Among other things, queueing models provide ways to describe message traffic in a system, to estimate service times and predict delays, and to estimate needed resources (or predict catastrophe) under various operating conditions. The study of modern computer networks, requires at least a basic understanding of these types of systems.

8.1 The Poisson Model

In earlier chapters of this book we have encountered several types of distributions for random variables dealing with the arrival of "events." Let us just review these ideas.

1. The *Poisson* PMF (Chapter 3) was introduced as a way to characterize the number K of discrete events (telephone calls, print jobs, "hits" on a webpage, etc.) arriving within some continuous time interval of length t. The arrival rate of the events was assumed to be λ, and the parameter in the Poisson PMF is then given by $\alpha = \lambda t$.

2. The *exponential* PDF (also Chapter 3) was introduced as characterizing the *waiting time T* to the next event or equivalently, the interarrival time *between* two successive events. Thus T is a continuous random variable.

3. In Chapter 5 the Erlang PDF was derived as the total waiting time $T = T_1 + T_2 + \ldots + T_K$ to the arrival of the Kth event. It was assumed here that the first order interarrival times T_i are exponentially distributed.

As you may have guessed, all of these distributions are related and are different aspects of of a single probabilistic model for the arrival of events. This model is called the *Poisson model*.

8.1.1 Derivation of the Poisson model

A Poisson model is defined by considering very small increments of time of length Δ_t. Events in these time intervals are considered to be independent. Further, as $\Delta_t \to 0$, the probability of one event occuring in the time interval is proportional to the length of the time interval, i.e., the probability is $\lambda\Delta_t$. The probability that more than one event occurs in the small time interval is essentially zero. These postulates are summarized in Table 8.1.

---

1. Events in non-overlapping time intervals are independent.

2. For $\Delta_t \to 0$:

   (a) $\Pr[1 \text{ event in } \Delta_t] = \lambda\Delta_t$

   (b) $\Pr[> 1 \text{ event in } \Delta_t] \approx 0$

   (c) $\Pr[0 \text{ events in } \Delta_t] = 1 - \lambda\Delta_t$

Table 8.1 Postulates for a Poisson model.

To proceed with developing this model, let us define

$$P[K;T] \overset{\text{def}}{=} \Pr[K \text{ arrivals in time } T] \tag{8.1}$$

The mixed parentheses are used to show that this quantity is a function of both a discrete and a continuous random variable. When considered as a function of K, $P[K;T]$ has a discrete character; when plotted versus T, it is a continuous function.

Now consider the time interval depicted in Fig. 8.1. Using the principle of total

$$0 \qquad\qquad\qquad t \quad t+\Delta_t$$

Figure 8.1 Time interval for arrival of events.

probability and the fact that events in non-overlapping intervals are independent, we can write

$$\Pr[k \text{ events in } t + \Delta_t] = \Pr[k-1 \text{ in } t] \cdot \Pr[1 \text{ in } \Delta_t] + \Pr[k \text{ in } t] \cdot \Pr[0 \text{ in } \Delta_t]$$

With the notation of (8.1) and the probabilities from Table 8.1 this equation becomes

$$P[k; t + \Delta_t] = P[k-1; t] \cdot \lambda\Delta_t + P[k; t] \cdot (1 - \lambda\Delta_t)$$

Rearranging, dividing through by Δ_t, and taking the limit as $\Delta_t \to 0$ yields

$$\lim_{\Delta_t \to 0} \frac{P[k; t + \Delta_t] - P[k; t]}{\Delta_t} + \lambda P[k; t] = \lambda P[k-1; t]$$

or

$$\frac{dP[k; t]}{dt} + \lambda P[k; t] = \lambda P[k-1; t] \tag{8.2}$$

This is an interesting equation because it is both a differential equation in the continuous variable t and a difference equation in the discrete variable k. You can verify by direct substitution that the solution to (8.2) is given by

$$P[k; t] = \frac{(\lambda t)^k e^{-\lambda t}}{k!} \qquad t \geq 0, \ k = 0, 1, 2, \ldots \tag{8.3}$$

and that it satisfies the required initial condition[1]

$$P[k; 0] = \begin{cases} 1 & k = 0 \\ 0 & k > 0 \end{cases}$$

Equation (8.3) is the essence of the Poisson model. Some of its implications are discussed below.

Distribution for number of arrivals

For any fixed time interval t, the probability of k arrivals in that time interval is given by (8.3). If K is considered to be a random variable, then its PMF is thus

$$f_K[k] = \frac{(\lambda t)^k e^{-\lambda t}}{k!} \qquad k = 0, 1, 2, \ldots \tag{8.4}$$

This is the same as the Poisson PMF given in Chapter 3 (see (3.8)) with parameter $\alpha = \lambda t$. Thus the Poisson PMF first introduced in Chapter 3 is based on the Poisson model just derived.

Distribution of interarrival times

The PDF for the kth order interarrival times of the Poisson model can be derived as follows. Let T be the random variable that represents the waiting time to the kth event. The probability that T is within a small interval of time $(t, t + \Delta_t]$ is given by

$$\Pr[t < T \leq t + \Delta_t] = f_T(t)\Delta_t$$

where $f_T(t)$ is the PDF for the interarrival time. Now refer to Fig. 8.1 once again. If the interarrival time is equal to t, that means the kth event falls in the interval of length Δ_t depicted in the figure. In other words, $k-1$ events occur in the interval $(0, T]$ and *one* event occurs in the interval of length Δ_t. The probability of this situation is

$$\Pr[k-1 \text{ in } t] \cdot \Pr[1 \text{ in } \Delta_t] = P[k-1; t] \cdot \lambda\Delta_t = \frac{(\lambda t)^{(k-1)} e^{-\lambda t}}{(k-1)!} \lambda\Delta_t$$

where we have used (8.3) in the last step. Since these last two equations represent the probability of the same event, we can set them equal to each other and cancel the common term Δ_t. This provides the PDF for the waiting time as

$$f_T(t) = \frac{\lambda^k t^{k-1} e^{-\lambda t}}{(k-1)!} ; \qquad t \geq 0 \tag{8.5}$$

which is the Erlang density function introduced in Chapter 5 (see (5.56)).

To complete discussion of the Poisson model, notice that for $k = 1$ (8.5) reduces to the exponential density function

$$f_T(t) = \lambda e^{-\lambda t} ; \qquad t \geq 0 \tag{8.6}$$

---

[1] Recall that most differential and difference equations are solved by "guessing" the solution and then verifying that the form is correct. Needed constants are then obtained by applying the initial conditions. In this case the initial condition states that the probability of 0 arrivals in a time interval of length 0 is 1, and the probability of more than 0 arrivals is 0.

Therefore the first order interarrival times of the Poisson model are in fact exponentially distributed. The mean interarrival time is given by $\mathcal{E}\{T\} = 1/\lambda$ (see Example 4.7).

8.1.2 The Poisson process

The random process known as a Poisson process counts the number of events occurring from some arbitrary time point designated as $t = 0$.[2] The process is homogeneous in time, so it does not matter where the time origin is taken. If the Poisson process is denoted by $N(t)$, then a typical realization is as shown in Fig. 8.2. The process

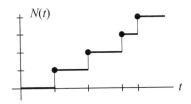

Figure 8.2 Realization of a Poisson process.

is continuous in time but discrete in its values. The random process is completely described by the fact that events in non-overlapping intervals are independent and that for any point in time the probability that $N(t) = k$ is given by the Poisson PMF (8.4).

The first and second moment properties of the Poisson process can be easily derived. The mean of the process is given by

$$m_N(t) = \mathcal{E}\{N(t)\} = \lambda t \tag{8.7}$$

This follows since the mean of the Poisson process is just the mean of the Poisson PMF (8.4). It was shown earlier (see Chapter 4 (Example 4.7)) that the mean and variance of this distribution are both equal to the parameter $\alpha = \lambda t$.

The covariance function for the Poisson process is defined as

$$C_N(t_1, t_0) = \text{Cov}\,[N(t_1), N(t_0)]$$

and can be derived as follows. Since t_1 and t_0 can be any time points, first assume that $t_1 > t_0$. Then the number of arrivals up to t_1 is equal to the number of arrivals up to t_0 plus the number of arrivals from t_0 to t_1, i.e.,

$$N(t_1) = N(t_0) + N(t_1 - t_0)$$

Now the following identity holds for any three random variables X, Y, and Z. (You should check this out using the definition of covariance.)

$$\text{Cov}\,[X + Y, Z] = \text{Cov}\,[X, Z] + \text{Cov}\,[Y, Z]$$

This can be used to write

$$\begin{aligned}
\text{Cov}\,[N(t_1), N(t_0)] &= \text{Cov}\,[N(t_0) + N(t_1 - t_0), N(t_0)] \\
&= \text{Cov}\,[N(t_0), N(t_0)] + \text{Cov}\,[N(t_1 - t_0), N(t_0)]
\end{aligned}$$

The last term on the right is 0, however, since $N(t_1 - t_0)$ and $N(t_0)$ are defined on non-overlapping intervals. Thus for $t_1 > t_0$ the last equation states

$$\text{Cov}\,[N(t_1), N(t_0)] = \text{Cov}\,[N(t_0), N(t_0)] = \text{Var}\,[N(t_0)] = \lambda t_0$$

---

[2] In the following we consider 't' to be the time index of a continuous random process rather than the realization of a random variable T.

If the opposite assumption is made about the ordering of the time points (i.e., $t_1 < t_0$), then an identical argument shows that $\mathrm{Cov}\,[N(t_1), N(t_0)] = \lambda t_1$. Thus we have the covariance function for the Poisson process as:

$$C_N(t_1, t_0) = \lambda \min(t_1, t_0) \tag{8.8}$$

The correlation function for the Poisson process can be obtained by starting with the covariance and adding the product of the means (see (...)). The result is

$$R_N(t_1, t_0) = \lambda \min(t_1, t_0) + \lambda^2 t_1 t_0 \tag{8.9}$$

Notice that the Poisson process is *not* a stationary random process. The mean is a (linear) function of time and the correlation and covariance functions do not depend on just the time difference $t_1 - t_0$.

8.1.3 An application of the Poisson process: the random telegraph signal

A classic example of a stationary process related to the Poisson process is known as the *random telegraph signal*. A typical realization of this signal is shown in Fig. 8.3. The

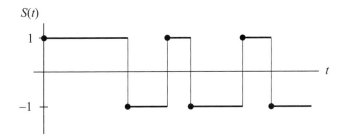

Figure 8.3 Random telegraph signal.

signal changes from $+1$ to -1 and back at random times; those change times or "zero crossings" are described by a Poisson process with parameter λ. In these days where digital modulation schemes predominate, this process is less important than it used to be as a model for communications. Nevertheless it arises in many other applications; for example, the digits of a binary register that counts the arrival of random events follow this model. It is traditional however, to assume this process takes on values of ± 1 rather than 0 and 1.

Probabilistic description

Assume that the process starts at time $t = 0$ and that the initial probabilities of a plus 1 or a minus 1 at that time are given by

$$\begin{array}{ll} \Pr[S(0){=}{+}1] = p & \text{(a)} \\ \Pr[S(0){=}{-}1] = 1 - p & \text{(b)} \end{array} \tag{8.10}$$

For any other time we can write

$$\begin{aligned} \Pr[S(t){=}{+}1] \;=\; & \Pr[\text{even}\,|S(0){=}{+}1] \cdot \Pr[S(0){=}{+}1] \\ & + \Pr[\text{odd}\,|S(0){=}{-}1] \cdot \Pr[S(0){=}{-}1] \end{aligned} \tag{8.11}$$

where the terms "even" and "odd" represent an even and odd number of arrivals (zero-crossings) in the interval $(0, t]$, and likewise,

$$\Pr[S(t) = -1] \quad = \quad \Pr[\text{odd} \,|\, S(0) = +1] \cdot \Pr[S(0) = +1]$$
$$+ \Pr[\text{even} \,|\, S(0) = -1] \cdot \Pr[S(0) = -1] \qquad (8.12)$$

Now, the probability of k arrivals in the interval is given by the Poisson PMF (8.4), which does *not* depend on the value of $S(0)$. Therefore

$$\Pr[\text{even} \,|\, S(0) = \pm 1] \quad = \quad \Pr[\text{even}]$$
$$= \quad \sum_{k \text{ even}} f_K[k] = e^{-\lambda t} \sum_{k \text{ even}} \frac{(\lambda t)^k}{k!} \qquad (8.13)$$

To evaluate the infinite sum, consider the Taylor series for the exponentials

$$e^{\lambda t} = \sum_{k=0}^{\infty} \frac{(\lambda t)^k}{k!} \quad \text{and} \quad e^{-\lambda t} = \sum_{k=0}^{\infty} \frac{(-\lambda t)^k}{k!}$$

Observe that the sum of these two expressions produces twice the *even* terms in the series. Equation 8.13 can thus be evaluated as

$$\Pr[\text{even} \,|\, S(0) = \pm 1] \quad = \quad \Pr[\text{even}]$$
$$= \quad e^{-\lambda t} \left(\frac{e^{\lambda t} + e^{-\lambda t}}{2} \right) = \tfrac{1}{2} \left(1 + e^{-2\lambda t} \right) \qquad (8.14)$$

Similarly, the difference of the two Taylor series results in twice the odd terms; therefore

$$\Pr[\text{odd} \,|\, S(0) = \pm 1] \quad = \quad \Pr[\text{odd}]$$
$$= \quad e^{-\lambda t} \left(\frac{e^{\lambda t} - e^{-\lambda t}}{2} \right) = \tfrac{1}{2} \left(1 - e^{-2\lambda t} \right) \qquad (8.15)$$

Substituting (8.10), (8.14), and (8.15) in (8.11) and (8.12) produces the two equations

$$\Pr[S(t) = +1] \quad = \quad \tfrac{1}{2} \left(1 + e^{-2\lambda t} \right) \cdot p + \tfrac{1}{2} \left(1 - e^{-2\lambda t} \right) \cdot (1 - p)$$
$$\Pr[S(t) = -1] \quad = \quad \tfrac{1}{2} \left(1 - e^{-2\lambda t} \right) \cdot p + \tfrac{1}{2} \left(1 + e^{-2\lambda t} \right) \cdot (1 - p)$$

or

$$\Pr[S(t) = +1] = \tfrac{1}{2} + (p - \tfrac{1}{2}) e^{-2\lambda t} \quad \text{(a)}$$
$$\Pr[S(t) = -1] = \tfrac{1}{2} - (p - \tfrac{1}{2}) e^{-2\lambda t} \quad \text{(b)} \qquad (8.16)$$

Notice that if $p \neq \tfrac{1}{2}$ the probabilities for $S(t)$ are time-varying and the process is therefore not stationary. This time variation is transient however, and dies out as t gets large. Therefore both $\Pr[S(t) = +1]$ and $\Pr[S(t) = -1]$ approach steady-state values of $\tfrac{1}{2}$. In fact, if $p = \tfrac{1}{2}$ and/or $t \to \infty$ the process becomes stationary in the strict sense. In this case for any two distinct points t_1 and t_0, the samples are independent with $\Pr[+1] = \Pr[-1] = \tfrac{1}{2}$.

Mean and second moments

To further describe the random telegraph signal, let us compute its mean, autocorrelation, and covariance function. Since $S(t)$ can take on only two possible values, computation of the mean is straightforward using (8.16):

$$m_S(t) = \mathcal{E}\{S(t)\} = (+1) \cdot \Pr[S(t) = +1] + (-1) \cdot \Pr[S(t) = -1]$$

$$= (+1) \cdot \left[\tfrac{1}{2} + (p - \tfrac{1}{2})e^{-2\lambda t}\right] + (-1) \cdot \left[\tfrac{1}{2} - (p - \tfrac{1}{2})e^{-2\lambda t}\right]$$

$$= (2p - 1)e^{-2\lambda t}$$

Since the autocorrelation function is defined as $R_S(t_1, t_0) = \mathcal{E}\{S(t_1)S(t_0)\}$ we could compute this expectation as the sum of

$$(+1)(+1)\Pr[S(t_1) = +1, S(t_0) = +1]$$

and three similar terms. Although straightforward, this method is quite tedious.

A simpler procedure is as follows. Assume first that $t_1 \geq t_0$. Let us think of the product $S(t_0)S(t_1)$ as a random variable X, so that $\mathcal{E}\{S(t_1)S(t_0)\} = \mathcal{E}\{X\}$. Since X can take on only two possible values (± 1), we can write

$$\mathcal{E}\{X\} = (+1) \cdot \Pr[X = 1] + (-1) \cdot \Pr[X = -1]$$

Now notice that X is equal to 1 when there is an even number of events in the interval $(t_0, t_1]$ and X equals -1 otherwise. Thus the last equation can be restated as

$$\mathcal{E}\{S(t_1)S(t_0)\} = \mathcal{E}\{X\} = (+1) \cdot \Pr[\text{even}] + (-1) \cdot \Pr[\text{odd}]$$

The probabilities of an even number and odd number of events in the interval $(t_0, t_1]$ are given by the expressions (8.14) and (8.15) with $t = t_1 - t_0$. Substituting these equations yields

$$R_S(t_1, t_0) = \mathcal{E}\{S(t_1)S(t_0)\}$$

$$= (+1)\tfrac{1}{2}\left(1 + e^{-2\lambda(t_1 - t_0)}\right) + (-1)\tfrac{1}{2}\left(1 - e^{-2\lambda(t_1 - t_0)}\right) = e^{-2\lambda(t_1 - t_0)}$$

Finally, recall that it was assumed that $t_1 \geq t_0$. If the reverse were true, t_1 and t_0 would simply be interchanged. Therefore the autocorrelation function for *any* combination of t_1 and t_0 can be written as

$$R_S(t_1, t_0) = e^{-2\lambda|t_1 - t_0|} \tag{8.17}$$

The covariance function can be computed using the relation (7.4) between autocorrelation and covariance:

$$C_S(t_1, t_0) = R_S(t_1, t_0) - m_S(t_1)m_S(t_0)$$

$$= e^{-2\lambda|t_1 - t_0|} - (2p - 1)^2 e^{-2\lambda(t_1 + t_0)}$$

Observe that if $p = \tfrac{1}{2}$ or as t_1 and $t_0 \to \infty$ the mean approaches 0 and the autocorrelation and covariance functions depend only on the difference $\tau = t_1 - t_0$. This indicates that under these conditions the random telegraph signal is wide sense stationary. In fact, we have observed that under the same conditions this process is also strict sense stationary. A sketch of the autocorrelation function is shown in Fig. 8.4.

8.1.4 Additional remarks about the Poisson process

Sum of independent Poisson processes

One of the remarkable facts about the Poisson model is that the sum of two independent Poisson processes is also a Poisson process. Consider the following example.

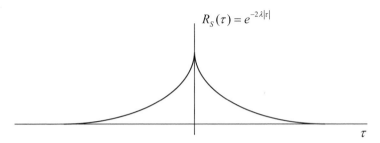

Figure 8.4 Autocorrelation function for the random telegraph signal.

Example 8.1:　Assume that customers at the website "badticket.com" are from two distinct populations: those wanting to buy rock concert tickets and those wanting to buy classical music tickets. The two groups are described by independent Poisson processes with arrival rates λ_1 and λ_2. We can show that the entire set of customers (independent of their musical tastes) forms a Poisson process with arrival rate $\lambda = \lambda_1 + \lambda_2$.

Since the two processes are independent and each is a Poisson process, events defined on non-ovelapping intervals remain independent. This fulfills the first requirement in Table 8.1. Now consider a small time interval Δ_t. If just one event occurs in the interval, this means there was either one rock customer and no classical customer, or one classical customer and no rock customer. The probability of this event is

$$
\begin{aligned}
\Pr[\text{1 customer}] \;&=\; \lambda_1\Delta_t \cdot (1 - \lambda_2\Delta_t) + \lambda_2\Delta_t \cdot (1 - \lambda_1\Delta_t) \\
&=\; (\lambda_1 + \lambda_2)\Delta_t + \mathcal{O}(\Delta_t^2)
\end{aligned}
$$

where $\mathcal{O}(\Delta_t^2)$ represents terms proportional to Δ_t^2, which become negligible as $\Delta_t \to 0$.

The probability of two or more customers in the interval Δ_t is 0 because of the properties of Poisson processes, unless a rock *and* a classical customer should both arrive in the interval Δ_t. Since the two Poisson processes are independent, the probability of this event is $(\lambda_1\Delta_t)(\lambda_2\Delta_t)$ which is also of the order of Δ_t^2 and becomes negligible as $\Delta_t \to 0$. Therefore we are left with:

$$
\begin{aligned}
\Pr[\text{1 customer}] \;&=\; (\lambda_1 + \lambda_2)\Delta_t \\
\Pr[> \text{1 customer}] \;&\approx\; 0 \\
\Pr[\text{0 customers}] \;&=\; 1 - (\lambda_1 + \lambda_2)\Delta_t
\end{aligned}
$$

This defines a Poisson process with rate $\lambda_1 + \lambda_2$.

□

Relation to the Bernoulli process

The Poisson process can also be described as a limiting case of a Bernoulli process. Recall that a Bernoulli process is a discrete random process taking on values of 1 with probability p and 0 with probability $1 - p$. Time samples $X[n_0]$ and $X[n_1]$ for $n_1 \neq n_0$ are independent.

Now consider a *continuous* time interval of length t and assume that the interval is divided into N short segments of length $\Delta_t = t/N$. Suppose that events arrive according to the postulates in Table 8.1. If we associate the arrival of an event with a 1 and no arrival with a 0, then we can define a Bernoulli process over the N time segments with probability parameter

$$p = \lambda \Delta_t = \frac{\lambda t}{N}$$

The probability of k events occuring in the N time intervals is given by the Binomial PMF (see Chapter 3 Eq. (3.5)) with p as above:

$$\Pr[k \text{ events}] = \binom{N}{k} \left(\frac{\lambda t}{N}\right)^k \left(1 - \frac{\lambda t}{N}\right)^{N-k}$$

Let us rewrite this expression and take the limit as $N \to \infty$:

$$\lim_{N \to \infty} \binom{N}{k} \cdot \left(\frac{\lambda t/N}{1 - \lambda t/N}\right)^k \cdot \left(1 - \frac{\lambda t}{N}\right)^N \qquad (8.18)$$

The first term can be written as

$$\binom{N}{k} = \frac{N!}{k!(N-k)!} \to \frac{1}{k!} N^k$$

where the right hand side denotes the behavior of the term as N gets very large. The second term becomes

$$\left(\frac{\lambda t/N}{1 - \lambda t/N}\right)^k \to (\lambda t/N)^k$$

The last term in (8.18) has a well known limit[3], namely $e^{-\lambda t}$. Therefore the limit in (8.18) becomes

$$\frac{1}{k!} N^k \cdot (\lambda t/N)^k \cdot e^{-\lambda t} = \frac{(\lambda t)^k e^{-\lambda t}}{k!}$$

which is the same as (8.3).

The relation between the Poisson and the Bernoulli processes is intuitively satisfying since we can think of a Poisson process as a "microscopic" Bernoulli process with the fixed time increments between events approaching 0.

8.2 Discrete-Time Markov Chains

In Section 5.5, we introduced the concept of a Markov process as a way to model a sequence of dependent random variables. A special class of Markov processes, known as a Markov chain is now discussed in more detail.

8.2.1 Definitions and dynamic equations

A Markov chain can be thought of as a random process that takes on a discrete set of values, which (for simplicity) we denote by the integers $0, 1, 2, \ldots q - 1$ (q may be infinite). A typical realization of a discrete-time Markov chain is shown in Fig. 8.5. At each time point 'k' the Markov chain takes on a value i, which is a non-negative integer. When $X[k] = i$, the Markov chain is said to be "in state i."

The Markov chain can also be represented by a state diagram such as the one shown in Fig. 8.6. The possible states of the process are denoted by "bubbles" and possible transitions between states are indicated by branches with arrows.

The Markov chain is defined by a set of *state probabilities*

$$p_i[k] \stackrel{\text{def}}{=} \Pr[X[k] = i] = \Pr[\text{state } i \text{ at time } k] \qquad (8.19)$$

---

[3] The general form is $\lim_{N \to \infty} \left(1 + \frac{a}{N}\right)^N = e^a$. See e.g., [1] p. 471.

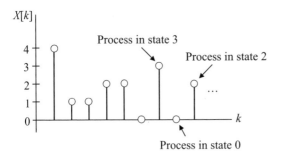

Figure 8.5 Typical realization of a Markov chain.

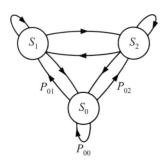

Figure 8.6 State diagram for a Markov chain.

and a set of state *transition probabilities*

$$P_{ij} \overset{\text{def}}{=} \Pr\big[X[k+1] = j \,|\, X[k] = i\big]$$
$$= \Pr[\text{transition from state } i \text{ to state } j] \qquad (8.20)$$

The P_{ij} are assumed to be constant in time.

The time behavior of the Markov chain can be described using these two types of probabilities. For example, referring to Fig. 8.6 we can write

$$p_0[k+1] = P_{00} \cdot p_0[k] + P_{10} \cdot p_1[k] + P_{20} \cdot p_2[k]$$

This equation articulates that the probability of being in state 0 at time $k+1$ is the sum of probabilities of being in state i at time k and transferring from state i to state 0. (Notice that this is just a statement of the *principle of total probability*!) The corresponding equations for all of the states can be written as a matrix equation

$$\begin{bmatrix} p_0[k+1] \\ p_1[k+1] \\ p_2[k+1] \end{bmatrix} = \begin{bmatrix} P_{00} & P_{10} & P_{20} \\ P_{01} & P_{11} & P_{21} \\ P_{02} & P_{12} & P_{22} \end{bmatrix} \begin{bmatrix} p_0[k] \\ p_1[k] \\ p_2[k] \end{bmatrix}$$

By following this procedure, the set of equations for any Markov chain with finite number of states q can be expressed in matrix notation as

$$\mathbf{p}[k+1] = \mathbf{P}^T \mathbf{p}[k] \qquad (8.21)$$

where

$$\mathbf{p}[k] = \begin{bmatrix} p_0[k] \\ p_1[k] \\ \vdots \\ p_{q-1}[k] \end{bmatrix} \qquad (8.22)$$

is the vector of state probabilities, and $\mathbf{P}$ is the *state transition matrix*

$$
\mathbf{P} = \begin{bmatrix}
P_{00} & P_{01} & \cdots & P_{0,q-1} \\
P_{10} & P_{11} & \cdots & P_{1,q-1} \\
\vdots & \vdots & \cdots & \vdots \\
P_{q-1,0} & P_{q-1,1} & \cdots & P_{q-1,q-1}
\end{bmatrix}
\tag{8.23}
$$

Notice that the transition matrix appears *transposed* in (8.21)[4]. Notice also that each *row* of the transition matrix contains probabilities conditioned on one particular state (see 8.20); thus *each row of the transition matrix sums to 1.*

8.2.2 Higher-order transition probabilities

Suppose that at time k a discrete-time Markov chain has a set of probabilities $\mathbf{p}[k]$ associated with its states. Then according to (8.21), the vector of state probabilities after ℓ transitions is

$$
\mathbf{p}[k+\ell] = \underbrace{\mathbf{P}^T\mathbf{P}^T\cdots\mathbf{P}^T}_{\ell \text{ times}}\mathbf{p}[k] = (\mathbf{P}^\ell)^T\mathbf{p}[k]
$$

The matrix $\mathbf{P}^\ell$ is called the ℓth order transition matrix. The elements of this matrix, denoted by $P_{ij}^{(\ell)}$ are interpreted as

$$
P_{ij}^{(\ell)} = \Pr\left[X[k+\ell] = j \mid X[k] = i\right]
\tag{8.24}
$$

In other words, $P_{ij}^{(\ell)}$ is the probability of transition from state i to state j in ℓ transitions. Notice that $P_{ij}^{(\ell)} \neq (P_{ij})^\ell$. That is, the ℓth order transition probabilities are *not* obtained by raising P_{ij} to the ℓth power. Rather, they are the elements of the matrix obtained by multiplying the transition matrix $\mathbf{P}$ by itself ℓ times.

To further understand higher order transition probabilities, let us consider the state diagram of Fig. 8.6 and compute $P_{01}^{(2)}$, the probability of starting in state 0 and arriving in state 1, after two transitions. The possible sequences of states and their probabilities are given by

state sequence	probability
$0 \rightarrow 0 \rightarrow 1$	$P_{00} \cdot P_{01}$
$0 \rightarrow 1 \rightarrow 1$	$P_{01} \cdot P_{11}$
$0 \rightarrow 2 \rightarrow 1$	$P_{02} \cdot P_{21}$

The probability of starting in state 0 and arriving in state 1 after two transitions is therefore the sum

$$
P_{01}^{(2)} = P_{00} \cdot P_{01} + P_{01} \cdot P_{11} + P_{02} \cdot P_{21}
$$

Note that this is just the element in the first row and second column of the product

---

[4] It would seem simpler if we were to just define the transition matrix as the transpose of (8.23). The current definition has become standard, however, due to the practice in the statistical literature to define the vector of state probabilities as a row vector and write (8.21) as the product of a row vector on the left times a matrix $(\mathbf{P})$ on the right[2, 3, 4, 5].

matrix

$$\mathbf{P}^2 = \mathbf{P} \cdot \mathbf{P} = \begin{bmatrix} P_{00} & P_{01} & P_{02} \\ \times & \times & \times \\ \times & \times & \times \end{bmatrix} \begin{bmatrix} \times & P_{01} & \times \\ \times & P_{11} & \times \\ \times & P_{21} & \times \end{bmatrix}$$

which is obtained by multiplying the first row in the first matrix by the second column in the second matrix. (The $\times$ denotes other entries.) The entire set of second order transition probabilities is given by the terms in the matrix $\mathbf{P}^2$.

The higher-order transition matrix can be used to derive a fundamental result for Markov chains known as the *Kolmogorov equation*. Suppose the matrix $\mathbf{P}^\ell$ is factored as

$$\mathbf{P}^\ell = \mathbf{P}^r \, \mathbf{P}^{(\ell-r)} \tag{8.25}$$

where r is any integer between 0 and ℓ; then $\mathbf{P}^r$ and $\mathbf{P}^{(\ell-r)}$ are the rth order and $(\ell-r)^{th}$ order transition matrices, respectively. Writing out the matrix product in terms of its elements produces

$$P_{ij}^{(\ell)} = \sum_{m=1}^{q-1} P_{im}^{(r)} P_{mj}^{(\ell-r)} \qquad 0 < r < l \tag{8.26}$$

This equation states that the probability of moving from state 'i' to 'j' in ℓ transitions is a sum of products of the probabilities of moving from state 'i' to *each* of the other states in r transitions and the probability of moving from each of these intermediate states to state 'j' in $\ell - r$ transitions. This fundamental principle is elegantly expressed in matrix form by (8.25).

8.2.3 Limiting state probabilities

For most Markov chains that are of interest, the state probabilities approach limiting values as the time 'k' gets large. Figure 8.7 illustrates this effect for a simple two-

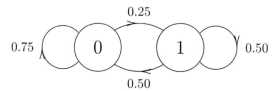

k	0	1	2	3	4	5	6
$p_0[k]$	0.500	0.625	0.656	0.664	0.666	0.667	0.667
$p_1[k]$	0.500	0.375	0.344	0.336	0.334	0.333	0.333

Figure 8.7 Evolution of state probabilities for a two-state Markov chain.

state Markov chain. The table in the figure lists the state probabilities for $k = 0$ through 6. Starting with the initial state probabilities $p_0[0] = p_1[0] = 0.5$, after just five transitions, the system reaches a steady-state condition with probabilities $\bar{p}_0 = 0.667$ and $\bar{p}_1 = 0.333$. These are the *limiting state probabilities* and their values are *independent of the initial conditions* $p_0[0]$ and $p_1[0]$. Note that

$$\bar{\mathbf{p}} = \lim_{k \to \infty} \mathbf{p}[k] = \lim_{k \to \infty} (\mathbf{P}^k)^T \mathbf{p}[0] \tag{8.27}$$

Thus the existence of limiting state requires that the ℓth order transition matrix $\mathbf{P}^\ell$ reaches a limit. While it would appear that $\bar{\mathbf{p}}$ depends on the initial probability vector $\mathbf{p}[0]$, it in fact does not because all of the rows of $\mathbf{P}^\ell$ become identical as $\ell \to \infty$. (It is an interesting excercise to show that this must be the case.) In our example, the sequence of transition matrices is

$$\mathbf{P} = \begin{bmatrix} 0.75 & 0.25 \\ 0.5 & 0.5 \end{bmatrix} \; ; \quad \mathbf{P}^2 = \begin{bmatrix} 0.688 & 0.312 \\ 0.625 & 0.375 \end{bmatrix} \; ; \quad \mathbf{P}^3 = \begin{bmatrix} 0.672 & 0.328 \\ 0.656 & 0.344 \end{bmatrix} \; ;$$

$$\mathbf{P}^4 = \begin{bmatrix} 0.668 & 0.332 \\ 0.664 & 0.336 \end{bmatrix} \; ; \quad \mathbf{P}^5 = \begin{bmatrix} 0.667 & 0.333 \\ 0.666 & 0.334 \end{bmatrix} \; ; \quad \mathbf{P}^6 = \begin{bmatrix} 0.667 & 0.333 \\ 0.667 & 0.333 \end{bmatrix} \cdots$$

where it is obvious that the matrix approaches a limit in just a few transitions and that the rows become identical with elements equal to the limiting state probabilities $\bar{p}_0$ and $\bar{p}_1$. Under this condition $\bar{\mathbf{p}}$ computed from (8.27) is the same for any initial probability vector $\mathbf{p}[0]$ that appears on the right.

To find the limiting state probabilities, it is only necessary to note that when the corresponding limits exist,

$$\mathbf{p}[k+1] \to \mathbf{p}[k] \to \bar{\mathbf{p}}$$

as $k \to \infty$. Therefore (8.21) becomes

$$\bar{\mathbf{p}} = \mathbf{P}^T \bar{\mathbf{p}} \quad \text{or} \quad (\mathbf{P}^T - \mathbf{I})\bar{\mathbf{p}} = \mathbf{0} \tag{8.28}$$

Equation 8.28 determines $\bar{\mathbf{p}}$ only to within a constant; that is, if $\bar{\mathbf{p}}$ is a solution to this equation, then $c\bar{\mathbf{p}}$ is also a solution, where c is any constant. The remaining condition, however, that the elements of $\bar{\mathbf{p}}$ sum to 1, insures a unique solution:

$$\sum_{i=0}^{q-1} \bar{p}_i = 1 \tag{8.29}$$

The following example illustrates the procedure of solving for the limiting state probabilities.

Example 8.2:   A certain binary sequence, with values of 0 and 1, is modeled by the Markov chain shown in Fig. 8.7. The transition matrix is

$$\mathbf{P} = \begin{bmatrix} 0.75 & 0.25 \\ 0.5 & 0.5 \end{bmatrix}$$

It is desired to find the probability of a run of five 0's and a run of five 1's.

The probabilities of the first binary digit (0 or 1) when the sequence is observed at a random time are the limiting state probabilities. These are found from (8.28)

$$\begin{bmatrix} -0.25 & 0.50 \\ 0.25 & -0.50 \end{bmatrix} \begin{bmatrix} \bar{p}_0 \\ \bar{p}_1 \end{bmatrix} = \begin{bmatrix} 0 \\ 0 \end{bmatrix}$$

and (8.29)

$$\bar{p}_0 + \bar{p}_1 = 1$$

The solution of these two equations is $\bar{p}_0 = 2/3$ and $\bar{p}_1 = 1/3$. The probability of a run of five 0's is then

$$(\tfrac{2}{3})(0.75)^4 = 0.211$$

and the probability of a run of five 1's is

$$(\tfrac{1}{3})(0.5)^4 = 0.021$$

Now compare these results to the probability of runs in a Bernoulli process with the same limiting-state probabilities. The probability of five 0's for the Bernoulli process is

$$\left(\tfrac{2}{3}\right)^5 = 0.132$$

while the probability of five 1's is

$$\left(\tfrac{1}{3}\right)^5 = 0.004$$

As you might expect, these probabilities are significantly smaller than the corresponding probabilities for the Markov chain, since successive values of the Bernoulli process are independent and runs are therefore less likely to occur.

□

8.3 Continuous-Time Markov Chains

A basic model for systems characterized by random traffic and random service times can be obtained by combining the concept of a Markov chain with that of the Poisson process. The result, a continuous-time Markov chain, is the basis for many standard models of traffic in computer networks and other systems.

8.3.1 Simple server system

Consider the simple server system shown in Fig. 8.8. For purposes of providing a

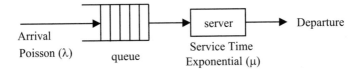

Figure 8.8 A simple queueing system.

familiar application, think of this as a model for customers buying tickets at a box office. Customers line up and are served one at a time. New customers arrive and enter at the back of the line, or queue. Such a system is commonly referred to as a *queueing system* or a *queueing model*, and the mathematics that deals with these systems is known as *queueing theory*. Clearly this model applies to a host of other situations besides the box office scenario. For example, Fig. 8.8 can be a model for print requests sent to a single printer on a computer or computer network. Here requests for service are placed in an electronic queue similar to the line of customers and are processed one at a time on a first-in first-out basis. In any such system, both the arrivals and the server are frequently described by a Poisson model. The arrivals are Poisson with rate λ; thus, the interarrival times are exponential random variables. The service times are likewise assumed to be exponential random variables with parameter μ (mean service time is $1/\mu$); thus departures also comprise a Poisson process with rate μ.

The simple queueing model of Fig. 8.8 can be described by a Markov chain with state diagram shown in Fig. 8.9; here the state represents the number of customers in the system (those in the queue plus the one being served). Unlike the discrete-time Markov chain, however, state transitions may occur *at any time*. According to the Poisson model, in any small interval of time Δ_t, a new customer will arrive with probability $\lambda\Delta_t$; thus the state will change from j to $j+1$ with probability $\lambda\Delta_t$. Likewise, the

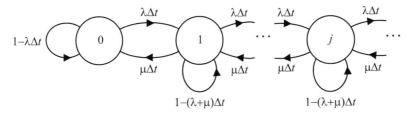

Figure 8.9 State diagram for the simple queueing system.

service may end within any small interval Δ_t so that the state may change from j to $j-1$ with probability $\mu\Delta_t$. Since the probabilities of state transitions must sum to 1, for all states except the 0 state, the probability of remaining in the state is given by $1 - (\lambda + \mu)\Delta_t$.

A Markov chain with state diagram of the form shown in Fig. 8.9 is known as a "birth and death" process. Since state transitions occur only between left or right adjacent states, this model can be used to represent population changes due to births or deaths in a community. The birth and death model also describes the queue of most service systems.

The random process taking on a discrete set of states but where transitions can occur at any point in time is known as a *continuous-time Markov chain*. The number of customers $N(t)$ in the simple server system is such a process; a typical realization is depicted in Fig. 8.10. The number of customers in just the queue and the number

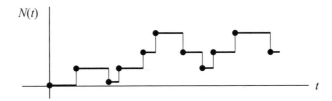

Figure 8.10 Random process representing the number of customers in the system (continuous-time Markov chain).

of customers in service (state of the server) are also continuous-time Markov chains. For the single server system discussed here, the server state may be only 0 or 1. When there are no customers in the queue the server process is as shown in Fig. 8.11.

8.3.2 Analysis of continuous-time Markov chains

Having introduced the concept of a continuous-time Markov chain, let us move on to describe some of its general characteristics and the tools used for analysis. The process $N(t)$ is characterized by *transition rates* r_{ij} which cause it to change from state i to state j with probability $r_{ij}\Delta_t$. In the case of Fig. 8.9 all of these transition rates are equal to either λ or μ; however, in a more general situation the state diagram and the transition rates could be quite different. It is conventional to represent the continuous time Markov chain $N(t)$ by a modified state diagram known as a *transition diagram*. In such a diagram only the *rates* are represented (not the interval Δ_t) and self loops are eliminated. A transition diagram for the single server process is shown in Fig. 8.12. (Compare this to Fig. 8.11 (a).)

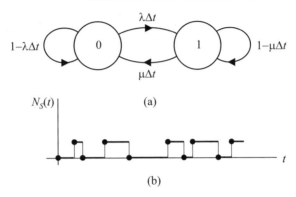

(a)

(b)

Figure 8.11 Random process representing a server. (a) State diagram. (b) Typical Markov chain realization.

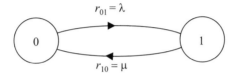

Figure 8.12 Transition diagram for the server process.

Let us now develop the equations that describe the evolution of a general Markov chain $N(t)$ with q states denoted by $0, 1, \ldots, q - 1$. A typical state j has multiple transitions to and from other states, as shown in Fig. 8.13. Notice that the rates r_{ij}

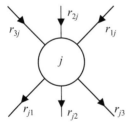

Figure 8.13 Typical state in a general Markov chain.

represent transitions *into* state j while the rates r_{ji} represent transitions *out* of state j. Let us define the time-varying state probabilities $p_i(t)$ for the random process as

$$p_i(t) = \Pr[N(t) = i] \quad i = 0, 1, \ldots, q - 1 \qquad (8.30)$$

Then for a particular state j we can write

$$p_j(t + \Delta_t) = \sum_{\substack{i=0 \\ i \neq j}}^{q-1} r_{ij}\Delta_t \cdot p_i(t) + \left(1 - \sum_{\substack{i=0 \\ i \neq j}}^{q-1} r_{ji}\Delta_t\right) p_j(t)$$

Observe again that the terms in the first sum represent transitions into state j while the terms in the second sum represent transitions out of state j. Thus the term in parentheses represents the "self loop" probability, which is not explicitly represented in the transition diagram. Rearranging this equation and dividing through by Δ_t

produces

$$\frac{p_j(t + \Delta_t) - p_j(t)}{\Delta_t} = \sum_{\substack{i=0 \\ i \neq j}}^{q-1} r_{ij} p_i(t) - \left(\sum_{\substack{i=0 \\ i \neq j}}^{q-1} r_{ji} \right) p_j(t)$$

Then taking the limit as $\Delta_t \to 0$ yields

$$\frac{dp_j(t)}{dt} = \sum_{i=0}^{q-1} r_{ij} p_i(t) \qquad j = 0, 1, \ldots, q-1 \tag{8.31}$$

where the "rate" r_{jj} has been defined as

$$r_{jj} \overset{\text{def}}{=} -\sum_{\substack{i=0 \\ i \neq j}}^{q-1} r_{ji} \tag{8.32}$$

Equations 8.31 are the *Chapman-Kolmogorov* equations for continuous-time Markov chains. These are a coupled set of first order differential equations that can be solved for the probabilities $p_i(t)$. Rather than solve these equations, which requires advanced linear algebra techniques, it is more useful to examine the limiting steady-state condition, where the probabilities approach constant values that do not change with time. As in the case of discrete-time Markov chains, the existence of such limiting state probabilities depends on the topology of the state diagram, or equivalently, the eigenvalues in the matrix form of the Chapman-Kolmogorov equations. In the following, it will be assumed that conditions are such that these limiting-state probabilities do in fact exist.

In the limiting condition we have

$$\lim_{t \to \infty} p_j(t) \to \bar{p}_j \text{ and } \frac{dp_j(t)}{dt} \to 0$$

Therefore, (8.31) becomes

$$\sum_{i=0}^{q-1} r_{ij} \bar{p}_i = 0 \qquad j = 0, 1, \ldots, q-1$$

Now, bringing the jth term out of the sum and substituting (8.32) for r_{jj} produces

$$\sum_{\substack{i=0 \\ i \neq j}}^{q-1} r_{ij} \bar{p}_i - \left(\sum_{\substack{i=0 \\ i \neq j}}^{q-1} r_{ji} \right) \bar{p}_j = 0$$

or

$$\bar{p}_j \sum_{\substack{i=0 \\ i \neq j}}^{q-1} r_{ji} = \sum_{\substack{i=0 \\ i \neq j}}^{q-1} \bar{p}_i r_{ij} \qquad j = 0, 1, \ldots, q-1 \tag{8.33}$$

Equations 8.33 are known as the *global balance equations*. These equations show that in the limiting-state condition, the probability of state j times the sum of transition rates *out* of the state must be equal to the sum of probabilities for the other states times the transitions *into* state j. This principle must hold for all states j. The global balance principle is illustrated in Fig. 8.14 where the directions of arrows on the branches indicate transitions into and out of the states.

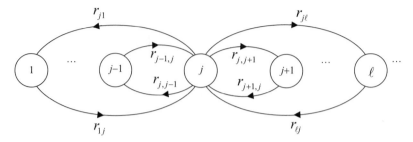

Figure 8.14 Transition rate diagram illustrating the global balance principle.

The global balance equations (8.33) are not quite enough to solve for the limiting-state probabilities. In fact, as illustrated in the example below, one equation is always redundant. (This set of equations is analogous to (8.28) in the discrete-time case.) One more condition is needed to solve for the $\bar{p}_i$, namely

$$\sum_{i=0}^{q-1} \bar{p}_i = 1 \tag{8.34}$$

Let us now illustrate the procedure with an example.

Example 8.3: Consider the 4-state birth and death process shown below.

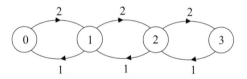

The transition rates in and out of adjacent states are identical.

The global balance equations (8.33) can be written for this process as follows:

$$\begin{aligned}
\text{state 0:} \quad \bar{p}_0 \cdot 2 &= \bar{p}_1 \cdot 1 \\
\text{state 1:} \quad \bar{p}_1(2+1) &= \bar{p}_0 \cdot 2 + \bar{p}_2 \cdot 1 \\
\text{state 2:} \quad \bar{p}_2(2+1) &= \bar{p}_1 \cdot 2 + \bar{p}_3 \cdot 1 \\
\text{state 3:} \quad \bar{p}_3 \cdot 1 &= \bar{p}_2 \cdot 2
\end{aligned}$$

These equations can then be written as a single matrix equation

$$\begin{bmatrix} 2 & -1 & 0 & 0 \\ -2 & 3 & -1 & 0 \\ 0 & -2 & 3 & -1 \\ 0 & 0 & -2 & 1 \end{bmatrix} \begin{bmatrix} \bar{p}_0 \\ \bar{p}_1 \\ \bar{p}_2 \\ \bar{p}_3 \end{bmatrix} = \begin{bmatrix} 0 \\ 0 \\ 0 \\ 0 \end{bmatrix}$$

The matrix on the left has determinant 0. (You can tell this since all of the columns add to 0.) Hence the probabilities $\bar{p}_i$ can only be determined to within a constant, and one of the global balance equations is redundant. The condition

$$\bar{p}_0 + \bar{p}_1 + \bar{p}_2 + \bar{p}_3 = 1$$

corresponding to (8.34) is needed to provide a unique solution. If we replace one of the redundant global balance equations (say the first equation) with this, the foregoing matrix equation becomes

$$\begin{bmatrix} 1 & 1 & 1 & 1 \\ -2 & 3 & -1 & 0 \\ 0 & -2 & 3 & -1 \\ 0 & 0 & -2 & 1 \end{bmatrix} \begin{bmatrix} \bar{p}_0 \\ \bar{p}_1 \\ \bar{p}_2 \\ \bar{p}_3 \end{bmatrix} = \begin{bmatrix} 1 \\ 0 \\ 0 \\ 0 \end{bmatrix}$$

This can now be solved to find

$$\bar{p}_0 = 1/15 \quad \bar{p}_1 = 2/15 \quad \bar{p}_2 = 4/15 \quad \bar{p}_3 = 8/15$$

□

8.3.3 Special condition for birth and death processes

The procedure for finding the limiting-state probabilities discussed in the previous subsection is general for any circumstances where these limiting values exist. In the case of a birth and death process (such as the model for a queue) the global balance equations can be reduced to a simpler set of equations. In this special case, let the variables λ and μ represent transitions into or out of an adjacent state as shown in Fig. 8.15. Note that the rates λ and μ need not be constant but can depend on the state.

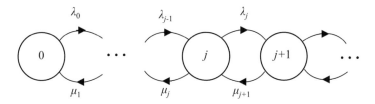

Figure 8.15 Transitions for a birth and death process.

The global balance equations for the birth and death process can then be replaced by the conditions

$$\bar{p}_j \lambda_j = \bar{p}_{j+1} \mu_{j+1} \qquad j = 0, 1, \dots, q - 2 \tag{8.35}$$

which say that in steady state, the probabilities of transitioning up and down the chain are equal (see Fig. 8.15). To see how conditions (8.35) arise, let us write the global balance equations (8.33) for the first three states in Fig. 8.15:

$$\begin{aligned} \bar{p}_0 \lambda_0 &= \bar{p}_1 \mu_1 \\ \bar{p}_1 (\lambda_1 + \mu_1) &= \bar{p}_0 \lambda_0 + \bar{p}_2 \mu_2 \\ \bar{p}_2 (\lambda_2 + \mu_2) &= \bar{p}_1 \lambda_1 + \bar{p}_3 \mu_3 \end{aligned}$$

$$\vdots$$

The first equation is already in the form (8.35). If we reverse the first equation left to right and subtract it from the second equation we obtain

$$\bar{p}_1 \lambda_1 = \bar{p}_2 \mu_2$$

which is of the form (8.35) and can replace the second equation. Then this equation can be reversed and subtracted from the third equation above to obtain

$$\bar{p}_2 \lambda_2 = \bar{p}_3 \mu_3$$

This procedure is continued until all the equations are of the form (8.35).

By solving (8.35) for $\bar{p}_{j+1}$ we see that the limiting-state probabilities for a birth and death process can be computed recursively using

$$\bar{p}_{j+1} = \frac{\lambda_j}{\mu_{j+1}} \bar{p}_j \qquad j = 0, 1, \ldots, q - 2 \tag{8.36}$$

and the relation (8.34). The procedure is illustrated in the following example.

Example 8.4: Consider the problem of Example 8.3. From the transition rate diagram in that example, (8.36) becomes

$$\begin{aligned}
\bar{p}_1 &= 2\bar{p}_0 \\
\bar{p}_2 &= 2\bar{p}_1 = 4\bar{p}_0 \\
\bar{p}_3 &= 2\bar{p}_2 = 8\bar{p}_0
\end{aligned}$$

Substituting these equations into (8.34) then yields

$$\bar{p}_0 + 2\bar{p}_0 + 4\bar{p}_0 + 8\bar{p}_0 = 1$$

or

$$\bar{p}_0 = \frac{1}{1 + 2 + 4 + 8} = \frac{1}{15}$$

The other probabilities then become

$$\bar{p}_1 = 2/15 \qquad \bar{p}_2 = 4/15 \qquad \bar{p}_3 = 8/15$$

as before.

□

8.4 Basic Queueing Theory

This section provides an introduction to the systems known as queueing systems. While these systems occur in many applications, they are of particular interest to electrical and computer engineers to model traffic conditions on computer systems and computer networks. The most general form of queueing system to be treated in this chapter is shown in Fig. 8.16. This type of system is known in the literature as an M/M/K/C queueing system. The M's stand for "Markov," and describe the arrival and service processes; K is the number of servers, and C stands for the *capacity* of the system, i.e., the number of servers plus the maximum allowed length of the queue.

In the figure $A(t)$ and $D(t)$ represent the Poisson arrival and departure processes. We also define the following additional random variables and Poisson random processes:

W	Waiting time (in the queue)	$N_q(t)$	Number of requests in the queue
S	Service time	$N_s(t)$	Number of requests in service
T	Total time spent in the system	$N(t)$	Number of requests in the system

Table 8.2 Definition of random variables and processes for a queueing system.

These variables satisfy the relations (see Fig. 8.16)

$$T = W + S \tag{8.37}$$

and

$$N(t) = N_q(t) + N_s(t) \tag{8.38}$$

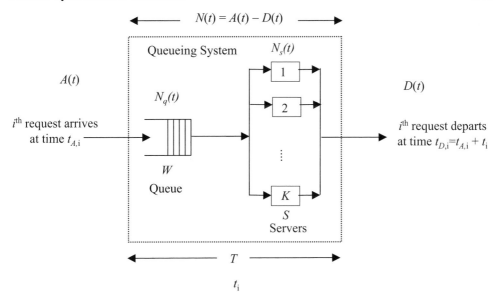

Figure 8.16 An $M/M/K/C$ queueing system.

The following subsections provide an introduction to the topic of queueing systems. Several texts devoted to this topic provide a more extensive and advanced treatment (e.g., [2, 3, 6]).

8.4.1 The single-server system

We consider first a single-server system with no restriction on the length of the queue. It is assumed that the arrival process is a Poisson process with parameter λ and that the service process is a Poisson process with parameter μ. In other words, the service time is an exponential random variable with mean service time of $1/\mu$. Such a system is referred to as $M/M/1/\infty$ or simply an $M/M/1$ queueing system.

The number of requests in the system can be described by an infinitely long birth and death process as shown in Fig. 8.17.

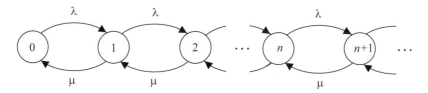

Figure 8.17 Birth and death process model for the number of requests in an $M/M/1/\infty$ system.

Let us develop the PMF for the random variable N which represents the number of requests in the system at some time t. We can define

$$f_N[n] = \Pr[N(t) = n] = \bar{p}_n \tag{8.39}$$

where $\bar{p}_n$ represents the limiting-state probability for state n. To find the limiting-state probabilities, we follow the procedure in Section 8.3.3. Define the new parameter

$$\rho = \frac{\lambda}{\mu} \tag{8.40}$$

This parameter is called the traffic intensity or the *utilization factor*; it relates to how busy the system is. With this definition in (8.36), the limiting-state probabilities satisfy the relation

$$\bar{p}_{j+1} = \rho\,\bar{p}_j \qquad j = 0, 1, 2, \ldots$$

Then applying this equation recursively, starting with $j = 0$ leads to the general expression

$$\bar{p}_n = \rho^n \bar{p}_0 \qquad n = 0, 1, 2, \ldots \tag{8.41}$$

In order to find the probability $\bar{p}_0$ of the 0 state, we observe that the probabilities must sum to 1 and apply the condition (8.34) to obtain

$$\sum_{n=0}^{\infty} \rho^n \bar{p}_0 = \bar{p}_0 \sum_{n=0}^{\infty} \rho^n = \bar{p}_0 \frac{1}{1-\rho} = 1 \tag{8.42}$$

where we have used the formula for an infinite geometric series, assuming that $\rho < 1$. (This last assumption is an important one because if $\rho \geq 1$, the arrival rate λ is equal to or greater than the service rate μ and the system breaks down mathematically as well as physically.) Solving the last equation yields

$$\bar{p}_0 = 1 - \rho$$

Finally, applying this to (8.41) and substituting the result in (8.39) produces the desired expression

$$\boxed{f_N[n] = \bar{p}_n = (1 - \rho)\rho^n \qquad n = 0, 1, 2, \ldots} \tag{8.43}$$

Notice that N is a (type 0) *geometric* random variable with parameter $p = 1 - \rho$. The distribution is sketched in Fig. 8.18.

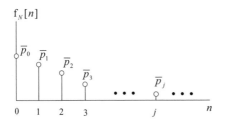

Figure 8.18 PMF for the number of requests in an M/M/1/∞ system.

Two important results regarding the number of requests in the system are as follows.

Probability of long queues.

Using the PMF just derived, it is possible to calculate the probability that the number of requests in the system will reach or exceed some chosen value n. We can write

$$\Pr[N < n] = \sum_{j=0}^{n-1}(1 - \rho)\rho^j = (1 - \rho)\sum_{j=0}^{n-1}\rho^j$$

$$= (1 - \rho)\frac{1 - \rho^n}{1 - \rho} = 1 - \rho^n$$

where we have used the formula for a finite geometric series. The probability that the number of requests is greater than or equal to a chosen value n is therefore

$$\Pr[[N \geq n] = \rho^n \tag{8.44}$$

Average number of requests in the system

The average number of requests in the system is given by the mean of the Geometric PMF. Thus (see Prob. 4-TF)

$$\mathcal{E}\{N(t)\} = \frac{\rho}{1-\rho} \tag{8.45}$$

A sketch of this result shown in Fig. 8.19 shows that the average length of the queue

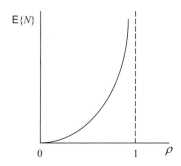

Figure 8.19 Average number of requests in an M/M/1/∞ system as a function of utilization factor ρ.

increases rapidly as the utilization factor ρ approaches one.

Consider some examples to illustrate the use of these formulas.

Example 8.5:   For a certain printer server, jobs arrive at a rate of three jobs per minute while the average printing time for a job is 15 seconds. What is the probability that there are nine or more requests in the queue? Does this probability change if we double the arrival rate but cut the average service time in half?

First, all information must be expressed in the same units. We have:

$$\lambda = 3 \text{ jobs per minute}$$
$$\mu = 60/15 = 1/0.25 = 4 \text{ jobs per minute}$$
$$\rho = \lambda/\mu = 0.75$$

The probability that there are nine or more requests in the queue is the probability that there are 10 or more requests in the system. From (8.44)

$$\Pr[[N \geq 10] = \rho^{10} = (0.75)^{10} = 0.056$$

This probability *does not change* if we double the arrival rate but cut the average service time in half.

□

Example 8.6:   What is the *average* number of requests in the system in the previous example? If the arrival rate remains at 3 requests per minute, what is the minimum average service time required to keep the average number of requests in the system less than or equal to 2?

The average number of requests in the system is given by (8.45):

$$\mathcal{E}\{N(t)\} = \frac{\rho}{1-\rho} = \frac{0.75}{1-0.75} = 3$$

To keep the average number of requests in the system less than or equal to 2 we require

$$\mathcal{E}\{N(t)\} = \frac{\rho}{1-\rho} \leq 2$$

or, solving for ρ,

$$\rho \leq 2/3$$

Then since

$$\rho = \lambda/\mu = 3/\mu$$

we find the average service time must satisfy

$$1/\mu \leq 2/9 \text{ (min)}$$

That is, the average service time must be less than or equal to $13\frac{1}{3}$ seconds.

□

8.4.2 Little's formula

Little's formula establishes a relationship between the expected value of the number of requests in the system and the average time that a request spends in the system. Little's formula can be stated as follows:

For systems that reach a steady state condition, the average number of requests in the system is given by

$$\mathcal{E}\{N(t)\} = \lambda\mathcal{E}\{T\} \tag{8.46}$$

where λ is the arrival rate and $\mathcal{E}\{T\}$ is the average time spent in the system.

Little's formula is very powerful since it does not depend on the details of the queueing process and nevertheless applies in a large number of cases. While a totally rigorous proof is well beyond the scope of this discussion, some of the main ideas can be presented here.

A typical set of requests is depicted in Fig. 8.20(a). The time of arrival of the ith request $(t_{A,i})$ is indicated by the beginning of the horizontal bar and the departure of the request is represented by the end of the bar. The length of the bar represents the total time t_i that the ith request spends in the system. Thus the departure time of the ith request is given by $t_{D,i} = t_{A,i} + t_i$. The bars are stacked up in the figure so that at any time 't' the upper envelope represents the total number of arrivals $A(t)$ up to that point and the lower envelope represents the total number of departures $D(t)$. The difference between the envelopes

$$N(t) = A(t) - D(t)$$

represents the number of requests in the system at time t. This is plotted in Fig. 8.20(b). Observe that neither $A(t)$ nor $D(t)$ is a stationary random process, since neither is ever decreasing. In most situations the difference $N(t)$ will be stationary

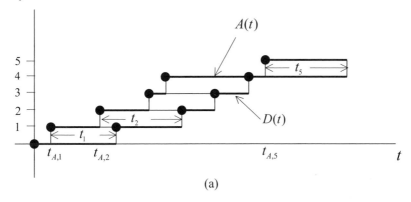

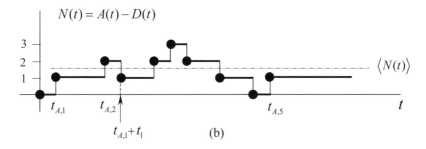

Figure 8.20 Processes in a queueing system. (a) Arrival and departure processes. (b) Requests in the system.

however. In fact, this is a *requirement* for Little's formula, since the value $\mathcal{E}\{N(t)\}$ given in (8.46) is independent of time.

To begin the proof, let us consider the time average of the process $\langle N(t)\rangle$ as shown in Fig. 8.20(b). By definition, this time average for any upper limit $t_{\max}$ is:

$$\langle N(t)\rangle = \frac{1}{t_{\max}} \int_0^{t_{\max}} N(t)dt = \frac{1}{t_{\max}} \int_0^{t_{\max}} (A(t) - D(t))dt = \frac{1}{t_{\max}} \sum_{i=1}^{A(t_{\max})} t_i \quad (8.47)$$

The last step is critical. It follows from observing that the integral of $A(t) - D(t)$ is the area between the upper and lower envelopes in Fig. 8.20(a), which is just equal to the sum of the times t_i that the requests are in the system. Now let us consider the *average* time spent by requests in the system. Since $A(t_{\max})$ is the total number of requests up to time $t_{\max}$, the average time is

$$\bar{t} = \frac{1}{A(t_{\max})} \sum_{i=1}^{A(t_{\max})} t_i$$

Finally, let us form an estimate for the arrival rate as

$$\hat{\lambda} = \frac{A(t_{\max})}{t_{\max}}$$

Combining these last two equations with (8.47) then yields

$$\langle N(t)\rangle = \hat{\lambda}\bar{t}$$

If $N(t)$ is ergodic (a step which we are not prepared to show) and we let t_{max} become sufficiently large, then $\langle N(t)\rangle$ and $\bar{t}$ converge to their expected values while $\hat{\lambda} \to \lambda$ and we arrive at (8.46). This is the essence of the proof.

Little's formula can also be applied to some of the other system variables. In particular, it is possible to show that

$$\mathcal{E}\{N_q(t)\} = \lambda\mathcal{E}\{W\} \quad (a)$$

$$\mathcal{E}\{N_s(t)\} = \lambda\mathcal{E}\{S\} \quad (b)$$

$$(8.48)$$

When (8.48) is combined with (8.37) and (8.38), the result is (8.46).

Some explicit expressions for these expectations can now be developed for the M/M/1 system. First, since the service process is Poisson with parameter μ, the expected service time is $\mathcal{E}\{S\} = 1/\mu$. Thus, from (8.48)(b),

$$\mathcal{E}\{N_s(t)\} = \lambda/\mu = \rho \tag{8.49}$$

To compute the mean total time in the system, we can combine Little's formula (8.46) with (8.45) to obtain

$$\mathcal{E}\{T\} = \frac{1}{\lambda}\mathcal{E}\{N\} = \frac{1}{\lambda}\frac{\rho}{1-\rho} = \frac{1}{\mu-\lambda} \tag{8.50}$$

where in the last step we substituted $\rho = \mu/\lambda$ and simplified.

Finally, using (8.38) together with (8.45) and (8.49) we can write

$$\mathcal{E}\{N_q(t)\} = \mathcal{E}\{N(t)\} - \mathcal{E}\{N_s(t)\} = \frac{\rho}{1-\rho} - \rho = \frac{\rho^2}{1-\rho} \tag{8.51}$$

Then applying Little's formula (8.48)(a) and simplifying yields

$$\mathcal{E}\{W\} = \frac{1}{\lambda}\frac{\rho^2}{1-\rho} = \frac{\lambda}{\mu(\mu-\lambda)} \tag{8.52}$$

These various results are summarized for convenience in Table 8.3. The average

$$\xrightarrow{\lambda}$$

$\mathcal{E}\{W\} = \dfrac{\lambda}{\mu(\mu-\lambda)}$	$\mathcal{E}\{N_q(t)\} = \dfrac{\rho^2}{1-\rho}$
$\mathcal{E}\{S\} \;=\; 1/\mu$	$\mathcal{E}\{N_s(t)\} = \rho$
$\mathcal{E}\{T\} \;=\; \dfrac{1}{\mu-\lambda}$	$\mathcal{E}\{N(t)\} \;=\; \dfrac{\rho}{1-\rho}$

Table 8.3 Expected values and relations for quantities involved in an M/M/1/∞ queueing system. Quantities in the right column are related to quantities in the left column according to Little's formula by the factor λ. Alternative forms can be obtained by using $\rho = \lambda/\mu$.

number of requests in the system, $\mathcal{E}\{N(t)\}$, and the mean time that a request spends in the system, $\mathcal{E}\{T\}$, are especially important quantities for evaluating the condition or performance of the system.

Let us consider an example involving the use of Little's formula and the results in Table 8.3.

Example 8.7: Consider the printer described in Example 8.5. From Table 8.3 we see that the average time spent in the system is given by

$$E\{T\} = \frac{1}{\mu - \lambda} = \frac{1}{4 - 3} = 1 \text{ (minute)}$$

This result can also be computed using Little's formula, as follows. In Example 8.6 it was determined that $E\{N(t)\} = 3$. Using this result with $\lambda = 3$ in Little's formula (8.46) yields

$$3 = 3 \cdot E\{T\}$$

Therefore $E\{T\} = 1$, as before.

Let us compute some of the other quantities in Table 8.3 for this example. The expected length of the queue is given by

$$E\{N_q(t)\} = \frac{\rho^2}{1 - \rho} = \frac{(0.75)^2}{1 - 0.75} = 2.25 \text{ (requests)}$$

The expected number of jobs in service is $E\{N_s(t)\} = \rho = 0.75$. Therefore the expected number of jobs in the system is $E\{N(t)\} = 2.25 + 0.75 = 3$, which is the same result found in Example 8.6. The average waiting time in the queue is

$$E\{W\} = \frac{\lambda}{\mu(\mu - \lambda)} = \frac{3}{4(4 - 3)} = 0.75 \text{ min}$$

Thus the total average time of 1 minute spent in the system is comprised of 0.75 minutes waiting in the queue and 0.25 ($1/\mu$) minutes receiving service.

□

The average time spent in the system is sometimes measured in units of mean service time. Using (8.50), this metric for the M/M/1/∞ system would be given by

$$\frac{E\{T\}}{1/\mu} = \frac{\mu}{\mu - \lambda} = \frac{1}{1 - \rho}$$

For the previous examples with $\rho = 0.75$, the average time that a request spends in the system is equal to $1/(1 - 0.75) = 4$ mean service times.

8.4.3 The single-server system with finite capacity

In some cases it is necessary to take into account that the service system may impose a limit on the number of requests that can be in the system; in other words, the system has *finite capacitiy*. For example, in computer and computer network systems, a finite amount of storage may be allocated to the queue. If this storage is very large so that the probability of exceeding the storage is small, then the M/M/1/∞ model is sufficent to describe the system. If the allocated storage is such that the probability of exceeding the storage is significant, then we need to represent the system by a model with finite capacity, the so-called M/M/1/C queueing model whose state transition diagram is shown in Fig. 8.21. In this model, the queue cannot exceed length $C - 1$; therefore the system can hold at most C requests. When the system is full, requests must be turned away.

Analysis of number of requests in the system

Developing the PMF for the number of requests in the system is identical to the procedure followed in Section 8.4.1 with the exception that the upper limit in the

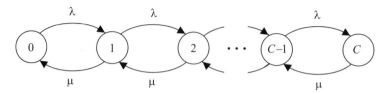

Figure 8.21 Birth and death process model for M/M/1 system with finite capacity C (M/M/1/C).

equation (8.42) that determines $\bar{p}_0$ is changed from ∞ to C. The effect of this change is that

$$\bar{p}_0 = \frac{1-\rho}{1-\rho^{C+1}} \tag{8.53}$$

and therefore

$$\boxed{f_N[n] = \frac{1-\rho}{1-\rho^{C+1}}\,\rho^n \qquad n = 0, 1, 2, \ldots, C} \tag{8.54}$$

Notice that for very large traffic intensity, $\rho \to 1$, the ρ^n term has little effect and the distribution approaches a discrete uniform PMF with $f_N[n] = 1/(C+1)$ for $0 \le n \le C$.

As with any queueing system, an important metric is the average number of requests in the system. An expression for this average number of requests is most easily obtained using the probability generating function (PGF) (see Prob. 8.28). The result is

$$\mathcal{E}\{N(t)\} = \frac{\rho}{1-\rho} - \frac{(C+1)\rho^{C+1}}{1-\rho^{C+1}} ; \qquad 0 < \rho < 1 \tag{8.55}$$

The first term is the result for an M/M/1/∞ system (let $C \to \infty$ and compare to Table 8.3), while the second term represents the reduction due to the finite-length queue. For $\rho = 1$ the distribution is uniform and we have $\mathcal{E}\{N(t)\} = C/2$.

A plot of $\mathcal{E}\{N\}$ versus C is given in Fig. 8.22 for some selected values of ρ. For ρ equal to 0.75 and 0.85 the average number of requests in the system is seen to level off at values of 3.0 and 5.67 respectively. These correspond to the values for a system with infinite capacity and can be computed from the formula in Table 8.3. For $\rho = 0.95$, which represents a very high traffic intensity, the average number of requests in the system increases almost linearly as the capacity C ranges from 1 to 30 as shown in the figure. The average number of requests eventually levels off at a value of 19 as the capacity of the system is increased, but this requires the system capacity C to be well over 100.

Mean time spent by requests in the system

In order to compute the mean time that a request spends in the system, we have to first look at the process of requests actually entering the system. Note that requests cannot enter the system and must "back off" whenever the system is full, i.e., whenever $N(t) = C$. Let us denote the probability of this event by P_b. The probability of back-off is thus found from (8.54) to be

$$P_b = \Pr[N(t) = C] = f_N[C] = \frac{1-\rho}{1-\rho^{C+1}}\,\rho^C \tag{8.56}$$

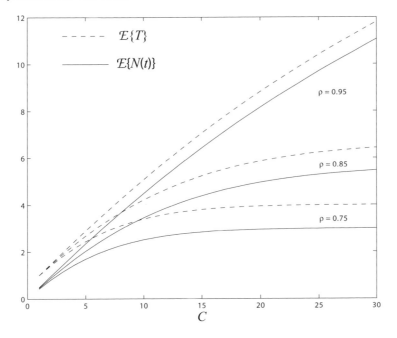

Figure 8.22 Average number of requests in a M/M/1/C system (solid lines) versus capacity C. Average time spent in the system (in units of mean service time) is also shown (dashed lines).

In any small increment of time $\Delta_t \to 0$ the probability that a request arrives and must back off is given by

$$(\lambda \Delta_t) \cdot P_b = (\lambda P_b) \Delta_t$$

Requests backing off therefore constitute a Poisson process with rate $\lambda_b = \lambda P_b$. Similarly, the probability that a request arrives and does *not* back off is given by

$$(\lambda \Delta_t) P_a = (\lambda P_a) \Delta_t$$

where $P_a = 1 - P_b$ and thus from (8.56)

$$P_a = 1 - \frac{1 - \rho}{1 - \rho^{C+1}} \rho^C = \frac{1 - \rho^C}{1 - \rho^{C+1}} \qquad (8.57)$$

Therefore requests actually entering the system are also a Poisson process, with rate $\lambda_a = \lambda P_a$. The two subprocesses constitute the original arrival process with rate $\lambda = \lambda_a + \lambda_b$.

The mean time that a request spends in the system, however, depends only on the requests actually entering the system. Thus we refer to λ_a as the *effective arrival rate* and λ_b as the *back-off rate*. The mean time that a request spends in the system is thus computed using Little's formula as

$$\mathcal{E}\{T\} = \frac{\mathcal{E}\{N\}}{\lambda_a} = \frac{\mathcal{E}\{N\}}{\lambda P_a} \qquad (8.58)$$

where $\mathcal{E}\{N\}$ is given by (8.55) and P_a is given by (8.57). The mean time (in units of mean service time) is shown as a function of system capacity C by the dashed lines in Fig. 8.22. For $C \to \infty$ the ratio between the average number of requests in the system and the mean time in the system is equal to the traffic intensity ρ.

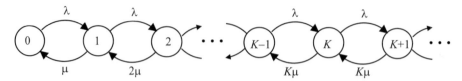

Figure 8.23 Transition diagram for an- $M/M/K/\infty$ multi-server system.

8.4.4 The multi-server system

In this final section of the chapter, we return to the general case of a queueing system with K servers depicted in Fig. 8.16. To begin the analysis of this system, suppose that k of the K servers are busy. The probability that one of these k servers completes its work within the time interval $(t, t+\Delta_t]$ is $\mu\Delta_t$. Since the servers act independently, the probability that any one of the servers complete is

$$\mu\Delta_t + \mu\Delta_t + \cdots + \mu\Delta_t = (k\mu)\Delta_t \tag{8.59}$$

The transition diagram for the system is depicted in Fig. 8.23. Notice that the backward transition probabilities for the first K states are not the same but are proportional to by $k\mu$ for $k = 1, 2, \ldots, K-1$. This follows from (8.59) above. For the remaining states, the backward transition probabilities are proportional to $K\mu$.

Following the discussion in Section 8.3.3, the limiting state probabilities for the $M/M/K$ system must satisfy the conditions

$$\begin{aligned} \bar{p}_0\lambda &= \bar{p}_1\mu \\ \bar{p}_1\lambda &= \bar{p}_2 2\mu \\ &\vdots \\ \bar{p}_{K-1}\lambda &= \bar{p}_K K\mu \\ &\vdots \end{aligned}$$

Solving this system of equations recursively leads to the expressions[5]

$$\bar{p}_n = \begin{cases} \dfrac{K^n \rho^n}{n!}\bar{p}_0 & 0 \le n < K \\[2ex] \dfrac{K^K \rho^n}{K!}\bar{p}_0 & K \le n < \infty \end{cases} \tag{8.60}$$

where it is conventional to *redefine* the parameter ρ for a K-server system as

$$\rho = \frac{\lambda}{K\mu} \tag{8.61}$$

By requiring that $\sum_{n=0}^{\infty} \bar{p}_n = 1$, the zero-state probability can be found to be

$$\bar{p}_0 = \left[\sum_{n=0}^{K-1} \frac{1}{n!}K^n \rho^n + \frac{K^K \rho^K}{K!(1-\rho)} \right]^{-1} \tag{8.62}$$

Thus all of the limiting-state probabilities can be computed from (8.60) and (8.62).

The probability that an arriving request finds all servers busy and therefore has to

---

[5] The details are left as an exercise (see Prob. 8.29).

wait in the queue is an important measure of performance. This will be denoted by Pr[queue] and is given by

$$\text{Pr[queue]} = \text{Pr}[N \geq K] = \sum_{n=K}^{\infty} \bar{p}_n \qquad (8.63)$$

Now observe from (8.60) that for $n \geq K$,

$$\bar{p}_n = \rho^{n-K} \bar{p}_K \qquad n \geq K \qquad (8.64)$$

where

$$\bar{p}_K = \frac{K^K \rho^n}{K!} \bar{p}_0 \qquad (8.65)$$

Thus the (8.63) becomes

$$\text{Pr[queue]} = \bar{p}_K \sum_{n=K}^{\infty} \rho^{n-K} = \bar{p}_K \frac{1}{1-\rho} \qquad (8.66)$$

Substituting (8.65) and (8.62) in (8.66) results in

$$\text{Pr[queue]} = C(K, \rho) = \frac{\dfrac{K^K \rho^K}{K!(1-\rho)}}{\left[\displaystyle\sum_{n=0}^{K-1} \frac{1}{n!} K^n \rho^n + \frac{K^K \rho^K}{K!(1-\rho)} \right]} \qquad (8.67)$$

This expression is known as the *Erlang C formula*.

A final measure of performance to be discussed is the mean number of requests in the queue. According to our notation, let N be the number of requests in the system. Then for $N \leq K$ the number N_q of requests in the queue is 0; while for $N > K$ we have $N_q = N - K$. Therefore the mean number of requests in the queue is given by

$$\mathcal{E}\{N_q\} = \sum_{n=K+1}^{\infty} (n-K)\bar{p}_n = \sum_{n=K+1}^{\infty} (n-K)\rho^{n-K}\bar{p}_K = \bar{p}_K \sum_{k=0}^{\infty} k\rho^k$$

where (8.64) was used to simplify the expression. Now observe that the term on the far right can be written as

$$\bar{p}_K \sum_{k=0}^{\infty} k\rho^k = \left(\frac{\bar{p}_K}{1-\rho} \right) \sum_{k=0}^{\infty} k(1-\rho)\rho^k$$

From (8.66) and (8.67) the term in parentheses can be recognized as $C(K, \rho)$. The rest of the expression is the formula for the mean of geometric random variable, which is given by $\rho/(1-\rho)$. With these observations, the equation for the mean number of requests in the queue becomes

$$\mathcal{E}\{N_q\} = \frac{\rho}{1-\rho} C(K, \rho) \qquad (8.68)$$

which is the desired final expression.

Using this expression and the mean service time $\mathcal{E}\{S\} = 1/\mu$ in conjuction with Little's formula, a number of other mean values can be computed for the $M/M/K$ system. These are listed in Table 8.4.

$$\xrightarrow{\lambda}$$

$$
\begin{array}{ll}
\mathcal{E}\{W\} = \dfrac{\mathcal{C}(K,\rho)}{K\mu + \lambda} & \mathcal{E}\{N_q(t)\} = \dfrac{\rho}{1-\rho}\mathcal{C}(K,\rho) \\[2.5ex]
\mathcal{E}\{S\} = 1/\mu & \mathcal{E}\{N_s(t)\} = K\rho \\[2.5ex]
\mathcal{E}\{T\} = \dfrac{\mathcal{C}(K,\rho)}{K\mu + \lambda} + \dfrac{1}{\mu} & \mathcal{E}\{N(t)\} = \dfrac{\rho}{1-\rho}\mathcal{C}(K,\rho) + K\rho
\end{array}
$$

Table 8.4 Expected values and relations for quantities involved in an $M/M/K/\infty$ queueing system. Quantities in the right column are related to quantities in the left column according to Little's formula by the factor λ.

The following example illustrates the application of the theory discussed in this section for the multi-server system.

Example 8.8: Technical support for a small company has three technicians on duty. Requests for support are placed in a queue and routed to the next available technician. Assume customers call in for support at a rate of 15 calls per hour and the average time for support is 10 minutes. What is the mean length of the queue and what is the mean time that customers can expect to be waiting in the queue?

The arrival rate is $\lambda = 15/60 = 0.25$ calls/minute. The mean service time is $1/\mu = 10$ minutes. The utilization factor is therefore $\rho = \lambda/K\mu = (0.25/3)10 = 5/6$.

The probability of a queue is given by the Erlang C formula (8.67):

$$
\mathcal{C}(K,\rho) = \frac{\dfrac{3^3\left(\frac{5}{6}\right)^3}{3!(1-\frac{5}{6})}}{\left[\displaystyle\sum_{n=0}^{3}\frac{1}{n!}3^n\left(\frac{5}{6}\right)^n + \frac{3^3\left(\frac{5}{6}\right)^3}{3!(1-\frac{5}{6})}\right]} \approx 0.7022
$$

The expected length of the queue is then given by (8.68):

$$
\mathcal{E}\{N_q\} = \frac{\frac{5}{6}}{1-\frac{5}{6}}\,0.7022 \approx 3.5 \text{ (customers)}
$$

The mean time spent waiting in the queue can be computed from the expression in Table 8.4.

$$
\mathcal{E}\{W\} = \frac{\mathcal{C}(K,\rho)}{K\mu + \lambda} = \frac{0.7022}{3/10 + 0.25} = 1.28 \text{ minutes}
$$

This is approximately 1 minute and 16 seconds.

□

8.5 Summary

Markov and Poisson models occur in many practical problems in electrical and computer engineering. Various aspects of the Poisson model have already been introduced in earlier chapters. This chapter brings all of these ideas together in the context of a single stochastic process known as the Poisson process. The process is defined by events whose probability of arrival is $\lambda\Delta_t$ and independence of events in non-overlapping time intervals (see Table 8.1). The number of events arriving in a fixed time interval of length t is described by the Poisson PMF with parameter $\alpha = \lambda t$ and the interarrival time T between successive events is described by the exponential PDF.

The Markov random process is a useful model for the data from many physical applications because it limits the conditioning of a random variable in the process to the random variable just preceding. When a Markov process takes on a (finite or infinite) discrete set of values the process is known as a Markov chain. Discrete-time Markov chains can be described algebraically by a state transition matrix or geometrically by a state diagram. Given a set of initial state probabilities, the probabilities of all the states can be computed as a function of time. For most cases of interest, these state probabililties approach limiting values called the "limiting state probabilities."

Continuous-time Markov processes are more challenging to analyze. The continuous-time Markov processes discussed in this chapter are based on the Poisson model. The "events" in this model are state transitions, which can be represented in a transition diagram. The time-dependent state probabilities satisfy a set of coupled first order differential equations. Fortunately, knowledge of this complete time dependence is not essential for many common engineering applications, and we can instead deal with the limiting-state probabilities. These probabilities can be computed from a system of *algebraic* equations called the global balance equations. These equations can be further simplified when the Markov model takes the form of a "birth and death" process. This is the case for most queueing systems.

A queueing system is characterized by random arrival of "requests" for service, and service times of random length. Requests that arrive and cannot be immediately placed in service are stacked up in a queue. We have introduced a few of these systems known as $M/M/K/C$ queueing systems. In this labeling, the two M's stand for Markov (Poisson) arrival and service models, K is the number of servers in the system, and C is the capacity of the system, determined by the number of servers and the maximum allowed length of the queue. $M/M/1/\infty$, $M/M/1/C$, and $M/M/K/\infty$ were studied explicitly. Little's formula was developed as a powerful tool that relates the average number of requests in the system to the mean time that a request spends in the system. Using the tools developed in this chapter, important engineering systems such as computers, telephone systems, and computer networks can be analyzed for their performance under various traffic conditions. Performance measures such as mean waiting time, mean length of queue, and others can be computed and plotted using the techniques developed here.

References

[1] Ross L. Finney, Maurice D. Weir, and Frank R. Giordano. *Thomas' Calculus*. Addison-Wesley, Reading, Massachusetts, tenth edition, 2001.

[2] Leonard Kleinrock. *Queueing Systems - Volume I: Theory*. John Wiley & Sons, New York, 1975.

[3] Leonard Kleinrock. *Queueing Systems - Volume II: Computer Applications*. John Wiley & Sons, New York, 1976.

[4] Howard M. Taylor and Samuel Karlin. *An Introduction to Stochastic Modeling*. Academic Press, New York, third edition, 1998.

[5] Sheldon M. Ross. *Introduction to Probability Models*. Academic Press, New York, eighth edition, 2002.

[6] Donald Gross and Carl M. Harris. *Fundamentals of Queueing Theory*. John Wiley & Sons, New York, third edition, 1998.

Problems

The Poisson model

8.1 Email messages arriving at Sgt. Snorkle's computer on a Monday morning can be modeled by a Poisson process with a mean interarrival time of 10 minutes.

(a) If he turns on his computer at exactly 8:00 and finds no new messages there, what is the probability that there will still be no new messages by 8:10?

(b) What is the probability that there will be *more than* one new message by 8:10? How about by 8:15?

8.2 The arrival of buses from the campus to student housing in the late afternoon is modeled by a Poisson process with an interarrival time of 30 minutes. Consider the situation where a bus has just arrived at 4:29 pm.

(a) A student arrives at 4:30 pm and misses the bus. What is the expected value of the time that the student will have to wait for the next bus to student housing?

(b) At 4:52 the next bus has not yet arrived and another student comes to the bus stop. What is the expected value of the time that these two students will have to wait for the next bus?

8.3 Two overhead projector bulbs are placed into operation at the same time. One has a mean time to failure (MTF) of 3600 hours while the other has an MTF of 5400 hours.

(a) What is the *probability* that the bulb with the shorter MTF fails before the bulb with the longer MTF?

(b) If the bulb with the shorter MTF costs the U.S. government $1.00 while the bulb with the longer MTF costs the government $1.45, which bulb is the "better deal?" That is, which choice would minimize the expected value of the cost of buying bulbs?

(c) What is the most that the U.S. government should pay for the bulb with the longer MTF given that the cost of the bulb with the shorter MTF is fixed at $1.00?

8.4 A web server receives requests according to a Poisson model with a rate of 10 hits per second.

(a) Suppose that the server goes down for 250 ms. What is the probability that the server misses no user accesses?

(b) In a 1 minute period, there are two outages: 0 to 10 ms and 12,780 to 12,930 ms. What is the probability that the server receives one hit during the first outage and no hits during the second?

(c) What is the probability that the server receives 6 hits during any 0.5-second period?

8.5 An Internet router receives traffic on two separate fiber cables. Arrivals on these feeds are independent and behave according to Poisson models with parameters α pps and β pps.

(a) Determine the PMF of the total number of packet arrivals at the router during an interval t seconds.

(b) What is the PDF of the interarrival period on the second feed (the one with β pps)?

8.6 Customers arriving at the 10th Street delicatessen are of two types. Type J are those customers ordering "Jaws" sandwiches and Type T are those ordering low-fat tuna. The shop owner knows that an hour before lunch the two types of customers can be modeled as independent Poisson processes with arrival rates $\lambda_J = 24$ customers/hour and $\lambda_T = 12$ customers/hour. That is, customers liking Jaws will never ask for low-fat tuna and vice-versa.

(a) If a customer arrives at any random time t, what is the probability that the customer will order Jaws (i.e., what is the probability that the customer is of Type J)?

(b) Given that a customer has just come in and ordered Jaws, what is the expected value of the waiting time to the next customer ordering low-fat tuna?

(c) What is the probability that the delicatessen sells exactly 5 Jaws and 3 low-fat tuna sandwiches in a 20 minute interval?

(d) What is the average interarrival time between customers? Two successive customers need not be of the same type.

(e) What is the average waiting time between two customers ordering different sandwiches (i.e., the expected value of the time between either a Type J and a Type T customer or a Type T and a Type J customer)?

(f) If a Jaws sandwich sells for $3.00 and a low-fat tuna sells for $3.50, what is the expected value of the delicatessen's gross earnings in one hour?

8.7 Up in the town of Gold Rush, California public transportation isn't what it used to be, so the arrival of buses headed for Los Angeles, San Francisco, and Durango, Colorado are independent Poisson processes with rates of 3, 2, and 5 buses per day, respectively.

(a) Given that you just missed the bus to Los Angeles, what is the probability that the next bus that comes along is a bus also headed for Los Angeles?

(b) If buses going to Los Angeles stop in San Francisco, what is the expected value of the time you must wait for the next bus that can take you to San Francisco?

Discrete-time Markov chains

8.8 A certain binary random process $X[k]$ that takes on values $+1$ and -1 is represented by a Markov chain with the state diagram shown below:

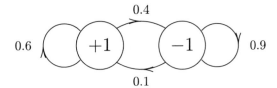

(a) What is the transition matrix for the process?

(b) Given that the process is in state '+1,' what is the probability that the process will be in state '−1' three transistions from now?

(c) Given that $X[k] = -1$, what is the probability that $X[k+3] = -1$?

(d) What are the limiting-state probabilities for the process?

(e) If the process is observed beginning at a random time 'k', what is the probability of a run of ten +1's? What is the probability of a run of ten −1's?

(f) Answer part (e) for a Bernoulli process that has the same probabilities as the limiting-state probabilities for this Markov process.

8.9 The discrete-time random process known as a *random walk* can be defined by the following equation:

$$Y[k] = \begin{cases} Y[k-1] + 1 & \text{with probability } p \\ Y[k-1] - 1 & \text{with probability } 1 - p \end{cases}$$

(a) Draw a state transition diagram for the random walk. Define state 'i' by the condition $Y[k] = i$. Be sure to label the transition probabilities.

(b) Do limiting-state probabilities exist for the random walk? If so, what are they?

8.10 In a discrete-time Markov chain, we have $p_{00} = 0.3$ and $p_{11} = 0.6$.

(a) Construct the state transition matrix.

(b) Draw the state transition diagram.

(c) Determine the limiting-state probabilities.

8.11 In a discrete-time Markov chain $p_{01} = 0.6$ and $p_{10} = 0.3$.

(a) Determine the limiting-state probabilities.

(b) Determine the probability of the occurrence of 8 consecutive 0's.

(c) Determine the probability of the occurrence of 8 consecutive 1's.

(d) Now consider a Bernoulli process with the same limiting-state probabilities as in (a). Repeat (b) and (c) for this Bernoulli process.

8.12 Refer to Problem 8.11.

(a) What is the probability of a sequence of 5 0's followed by 3 1's.

(b) Now consider a Bernoulli process with the same limiting-state probability as those used in (a). Repeat (a) for the Bernoulli process.

8.13 A state diagram for a discrete-time Markov chain is shown below.

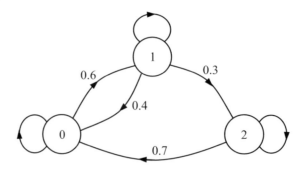

(a) Fill in the transition probabilities of the self-loops in the state diagram.

(b) What is the transition matrix?

(c) Which state sequence has the largest probability: $0, 0, 0$ or $1, 1, 1$ or $2, 2, 2$? What is the value of this largest probability?

Continuous-time Markov chains

8.14 A three-state continuous-time Markov chain is described by rates $r_{01} = 4$, $r_{02} = 5$, $r_{12} = 5$, $r_{10} = 3$, $r_{20} = 2$, and $r_{21} = 2$.

(a) Draw the transition rate diagram for the process.

(b) Find the limiting-state probabilities p_0, p_1, and p_2 for the process.

8.15 In a continuous-time Markov chain with four states, the following transition rates are given: $r_{01} = 2$, $r_{10} = 3$, $r_{02} = 1$, $r_{12} = 1$, $r_{21} = 2$, $r_{23} = 2$, $r_{31} = 2$, and $r_{32} = 3$.

(a) Draw the state transition diagram.

(b) Determine the limiting-state state probabilities.

8.16 In order to study the behavior of communication networks for transmission of speech, speaker activity is modeled as a Markov chain as shown below. The interpretation of the states is as follows: State 0 indicates silence; State 1 indicates unvoiced speech; and State 2 represents voiced speech. The transition rates are given to be: $r_{01} = 1$, $r_{10} = 1.5$, $r_{12} = 2$, and $r_{21} = 3$.

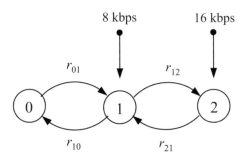

(a) Determine the limiting-state probabilities of this model.

(b) If unvoiced speech is transmitted at 8 kbps and voiced speech is transmitted at 16 kbps, find the average bit rate produced by the transmitter.

8.17 Yoohoo is an Internet traffic monitoring service, which counts the number of unique hits at a given web portal site. The arrival of unique hits is a Poisson process with a rate λ.

(a) Draw a state transition diagram for this process.

(b) In a 4-hour period, what is the probability that the number of hits is 12? Let $\lambda = 2$ hits/hour.

(c) Can you determine the limiting-state state probabilities for this Markov chain? Explain.

Basic queueing theory

8.18 In a certain M/M/1 queueing system the arrival rate is $\lambda = 4$ sec^{-1} and the service rate is $\mu = 5$ sec^{-1}.

(a) What is the utilization ρ?

(b) What is the probability $p_{10} = \Pr[N(t) > 10]$?

(c) What is the average total time in the system $E\{T\}$? What is the mean waiting time spent in the queue?

(d) To attain $p_{10} \le 10^{-2}$, what would have to happen to the arrival rate? What are ρ and $E\{T\}$ at the new arrival rate?

(e) To achieve $p_{10} \le 10^{-2}$, how should we change the service rate? What are ρ and $E\{T\}$ at the new service rate?

8.19 Email messages arrive at a mail server at the average rate of 30 messages per minute. The average time that it takes the server to process and reroute a message is 750 milliseconds.

(a) What is the expected value of the number of messages $N(t)$ waiting in the queue at any time?

(b) What is the average *time* a message spends from the time that it arrives until the time it is rerouted?

(c) How do the answers to parts (a) and (b) change if the arrival rate for messages is doubled, i.e., increased to 60 messages per minute?

8.20 Interrupts on a certain computer occur at an average rate of 10 per second. The average service time for an interrupt is 0.05 sec.

(a) What are the statistics $E\{N(t)\}$ and $E\{T\}$ for the queue?

(b) If you wanted to cut $E\{T\}$ in half, what average service time would you need? What would $E\{N(t)\}$ be in that case?

8.21 Compare the following two situations with respect to the average time a customer has to spend:

(a) Customers in a small bank with just two tellers have their choice of one of two lines (queues) they may wait in for service. Once they have chosen a line, they may not change to the other one. Both tellers are equally competent; the average service time for each teller is the same and equal to μ.

(b) Customers arrive and wait in a *single* line for two tellers. (Same tellers as before.) The customer at the head of the line goes up to the next available teller.

8.22 Consider an M/M/1 queueing system. The arrivals at the system are described by a Poisson model with $\lambda = 20$ packets per second (pps). The service rate is 25 pps.

(a) Is the system stable? What is the utilization factor ρ?

(b) The state probabilities are given by $p_k = \rho p_{k-1}$, $k \ge 2$, determine p_0.

(c) What is the average number of packets in the system?

(d) What is the average total delay in the system?

(e) What is the average queueing delay?

(f) Determine the number of packets in the queue.

8.23 In an M/M/1/3 system, the arrivals are Poisson with $\lambda = 150$ requests per second.

(a) Draw the state transition diagram.

(b) Determine the limiting-state state probabilities.

(c) If p_1 and p_2 are limited to $p_1 = 0.2$ and $p_2 = 0.1$, what average service rate μ is required?

8.24 Traffic arrives at a queueing system in a Poisson manner with parameter $\lambda = 24$ pps.

(a) Consider that the system is M/M/1 with an average service rate of 36 pps. What is the average delay in the system?

(b) Let the queueing system be actually two identical M/M/1 queues in parallel, and the traffic is divided equally among the two queues. The average service rate is 18 pps per queue. Determine the average total delay in the system.

(c) Consider an M/M/2 system for which the average service rate of each server is 18 pps. Find the average total delay in the system.

8.25 Fluorescent light bulbs in Spanagel Hall fail at the rate of 2 per week. (Assume that a week consists of 5 working days and that the probability of failure for light bulbs on weekends is negligible. Therefore weekends may be disregarded entirely.) It takes an average of 2 working days for Public Works to respond and replace a light bulb. Only one man works to replace the bulbs. If additional requests for service come in while he is out replacing a bulb those requests are placed in a queue and will be responded to on a "first come – first served" basis.

(a) What is the mean number of light bulbs that will be out of working condition in Spanagel at any given time?

(b) Assuming that an employee reports a bulb as soon as it fails, what is the average length of time an employee must put up with a burned out or blinking fluorescent bulb until it is replaced?

8.26 Requests on a network printer server are modeled by a Poisson process and arrive at a nominal rate of 3 requests/minute. The printer is very fast, and the average (mean) time required to print one of these jobs is 10 seconds. The printing time is an exponential random variable independent of the number of requests.

(a) What is the expected value of the waiting time until the *third* request?

(b) Assuming that the system can accommodate an infinitely long queue, what is the probability that when a job arrives there are no requests in the queue? (Be careful about answering this!)

(c) All real systems must have some finite capacity C. If the system (which includes the printer and its queue) is completely full, requests will be sent back to the user with the message "PRINTER BUSY: TRY AGAIN LATER." What minimum value of K is needed if the probability that a request arrives and is sent back is to be less than 0.001?

8.27 The color printer in the main ECE computing lab has a mean time to failure (MTF) of 11 days. The IT department (Code 05) has a mean time of response (MTR) of 2 days. Assume that both are described by a Poisson model; i.e., the failure time for the printer is an exponential random variable with parameter $\lambda = 1/11$ and the service time is also an exponential random variable with parameter $\mu = 1/2$. Assume that as soon as the IT technician arrives, the printer starts working again, so there is no extra waiting time while the technician services the machine. We can model the situation by a continuous-time Markov chain with two states

State 0:   printer is working
State 1:   printer is down

What are the limiting-state probabilities $\bar{p}_0$ and $\bar{p}_1$ that the printer is or is not working?

8.28 (a) Show that the PGF for the distribution (8.54) for a finite-capacity queueing system is given by

$$G_N(z) = \frac{1 - (\rho z)^{C+1}}{1 - \rho z}\, \bar{p}_0$$

where $\bar{p}_0$ is given by (8.53).

(b) Use this to show that the mean of the distribution is given by (8.55).

8.29 Show how the equations (8.60) follow from the conditions for the birth and death process preceding those equations. Also, by summing the limiting state probabilities and setting the sum equal to 1, derive the expression (8.62) for $\bar{p}_0$ for the K-server case.

8.30 In Example 8.8, what is the probability that at least one of the technicians is not busy?

8.31 One of the technicians in Example 8.8 goes on a coffee break, but does not come back for a long time. How does this change the mean waiting time that a customer calling in will experience?

8.32 Consider a queueing system with K servers and no queue. In other words, if requests arrive and all servers are busy, those requests are turned away. Show that the probability that an arriving request is turned away is given by the expression

$$\bar{p}_K = B(K, \rho) = \frac{(K\rho)^K / K!}{\sum_{n=0}^{K} (K\rho)^n / n!}$$

This expression is known as the *Erlang B formula*.
Hint: Begin by drawing the transition diagram for the system.

Computer Projects

Project 8.1

In this project you will generate random sequences of 0's and 1's and compute their relative frequencies. You will compare the cases where the 0's and 1's occur independently to the cases where dependency is introduced using a simple Markov chain model.

1. Simulate and plot a random sequence of 0's and 1's where each successive digit is chosen independently from a Bernoulli PMF with parameter $p = 0.5$. Call this Case A. Plot a sequence of $N = 50$ successive digits and then repeat the experiment; plot another set of 50 digits and demonstrate that it is different from the first.

2. Repeat Step 1 but now choose the parameter p of the Bernoulli distribution to be 0.8. Call this Case B. Again plot two realizations to be sure they are different. Cases A and B each represent a *Bernoulli process* but with different values for the parameter p.

3. Now assume that the digits are not independent; in particular, the sequence of 0's and 1's is generated by a Markov chain with state diagram shown below.

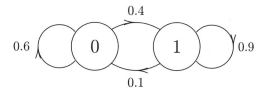

Plot a sequence of $N = 50$ digits for this Markov model of binary data assuming the initial value for the state probabilities is $\mathbf{p}[0] = \begin{bmatrix} 0.5 & 0.5 \end{bmatrix}^T$. Call this Case C. Repeat this step, but with different initial conditions; in particular, plot another realization of 50 binary digits assuming that $\mathbf{p}[0] = \begin{bmatrix} 0.2 & 0.8 \end{bmatrix}^T$. Call this Case D.

4. Now generate sequences of N points for each of Cases A, B, C and D and for the values of N specified below. Do *not* try to plot these sequences on paper. For each sequence of Cases A through D, estimate the probability of a 0 or 1 in the sequence by computing the relative frequency. That is, for a given sequence of length N the relative frequency of a 1 is the number of 1's appearing in the sequence divided by N. Summarize your results in a table for $N = 50$, 100, 1000, and 10,000.

5. Compare the relative frequencies of 0's and 1's for Cases A through D. Are there any differences you can observe in the characteristics of the sequences that have similar relative frequencies? Look at plots on your display monitor and zoom in on some of the longer sequences to better answer this question.

6. Compute the relative frequency of 0's and 1's for a Markov chain with initial condition $\mathbf{p}[0] = \begin{bmatrix} 0.8 & 0.2 \end{bmatrix}^T$ and compare it to Cases C and D above. What does this tell you about the properties of a Markov chain?

7. Solve for the limiting-state probabilities of the Markov chain and compare them to the relative frequency results.

MATLAB programming notes

See programming notes for Project 2.1 for generating the random binary sequence.

Project 8.2

In this project, you will simulate and study an M/M/1 queueing system.

1. Consider an M/M/1 packet queueing system with an arrival rate λ and a service rate μ. For $\lambda = 0.1$ packets per millisecond and $\mu = 0.125$ packets per millisecond, calculate the mean (expected value) number of packets in the system and in the queue, and the mean time spent by packets in the system and in the queue.

2. Simulate this M/M/1 queue by following these steps:

   (a) The arrival process $A(t)$ can be realized as a sequence of 0's and 1's where a 1 indicates the arrival of a packet. Assume that in a given 1 ms interval only one arrival can occur. The interarrival time (time between successive arrivals) is characterized by an exponential random variable with parameter λ.

   (b) Generate the departure process $D(t)$. Each arriving packet is served by the queue; the service time is characterized by an exponential random variable with parameter μ. At a given time, the system can only serve one packet on a first-come first-served basis. Additional packets must wait in the queue for service. If there are no packets waiting to be served, the server stays idle until the next packet arrives.

   (c) Measure the following quantities:

      i. Number of packets in the system $N(t)$ as the difference between the arrival and departure processes: $N(t) = A(t) - D(t)$. Use a time resolution of 1 ms.

      ii. The time spent by the ith packet in the system $t_i = t_{Ai} - t_{Di}$, where t_{Ai} is the arrival time of the ith packet and t_{Di} is the departure time of the same packet.

   (d) Plot the following for a simulation time period of 200 ms (200 points):

      i. The arrival process $A(t)$ and the departure process $D(t)$ vs. time on the same graph.

      ii. The number of packets in the system $N(t)$ vs. time.

      iii. A graph of the total time t_i spent by the ith packet vs. its arrival time t_{Ai}.

3. From your simulation and the measurements in Step 2(c), obtain the time average value $\overline{N} = \langle N(t) \rangle$ and the average time $\overline{t} = 1/K \sum_{i=1}^{K} t_i$ spent by K packets in the system. Compare these with the corresponding theoretical values $\mathcal{E}\{N(t)\}$ and $\mathcal{E}\{T\}$ found in Step 1. Repeat these calculations for simulation time periods of 200 ms, 1000 ms, 10,000 ms, 100,000 ms, and 1,000,000 ms.

4. Repeat Steps 2(a), 2(b), and 2(d) for a simulation time period of 10,000 ms, a fixed $\mu = 0.10$, and a range of values of λ from 0.005 to 0.095 in increments of 0.01. In each case, calculate $\overline{N}$ and $\overline{t}$ as in Step 3. Plot $\overline{N}$ and $\overline{t}$ as a function of $\rho = \lambda/\mu$. Do these experimental results obey Little's formula?

MATLAB programming notes

The function 'expon' from the software package can be used to generate exponential random variables.

A Basic Combinatorics

Elements of combinatorics (counting arrangements of objects) appear in several problems on probability. While advanced knowledge of this topic is not necessary for the study of probability, some rudimentary knowledge is essential. A summary of the minimum requirements for the main part of the text is presented in this appendix.

A.1 The Rule of Product

To begin this discussion, consider the problem of computing the number of 4-digit hexadecimal numbers that are possible, using the the the sixteen characters { 0,1, 2, 3, 4, 5, 6, 7, 8, 9, A, B, C, D, E, F }. A typical such hexadecimal number is

$$06FF$$

In forming these numbers, there are 16 choices for the first digit, sixteen choices for the second digit, and so on. Therefore the number of possible 4-digit hexadecimal numbers is

$$16 \cdot 16 \cdot 16 \cdot 16 = 16^4 = 65,536$$

The principle behind this computation is known as the Rule of Product and applies in many similar situations. The most general form of the rule is as follows.

> In an experiment where we form a sequence of k elements or k-tuple, and where there are N_i possible choices for the ith element, the number of unique k-tuples that can be formed is given by $\prod_{i=1}^{k} N_i$.

In many applications, such as in the foregoing example, the number of choices for all elements is the same (i.e., $N_i = N$). In this case the the number of unique k-tuples is simple N^k.

Let us consider a more complicated example.

Example A.1: A digital logic gate has four inputs. How many different Boolean functions can a four-input logic gate realize?

The question can be answered more generally. Let us denote the number of inputs by n. Then the input-output relation of the gate can be represented by a so-called "truth table" which is illustrated below for the case of $n = 4$.

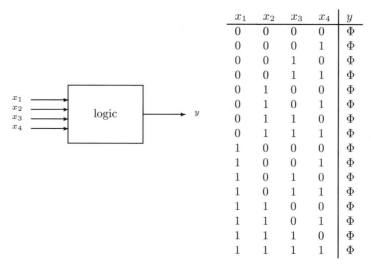

x_1	x_2	x_3	x_4	y
0	0	0	0	Φ
0	0	0	1	Φ
0	0	1	0	Φ
0	0	1	1	Φ
0	1	0	0	Φ
0	1	0	1	Φ
0	1	1	0	Φ
0	1	1	1	Φ
1	0	0	0	Φ
1	0	0	1	Φ
1	0	1	0	Φ
1	0	1	1	Φ
1	1	0	0	Φ
1	1	0	1	Φ
1	1	1	0	Φ
1	1	1	1	Φ

The 1's and 0's represent combinations of inputs and the Φ represents either a 0 or a 1 depending on the function that is being realized.

Notice that there are 16 rows in this table. That is because there are four inputs and two possible choices for each input (0 or 1). Thus, applying our principle, there are $2^4 = 16$ combinations of inputs or rows in the table. In general, for a logical function with n inputs, there will be 2^n rows in the truth table. This is our first application of the Rule of Product in this example.

Now consider the main question: how many possible functions are there? In defining a function, you can choose the output y for each row of the truth table to be either a 0 or 1. If $k = 2^n$ is the number of rows in the truth table, the number of possible functions is 2^k or 2^{2^n}. This is a second use of the Rule of Product. Therefore a logic gate with four inputs could implement $2^{2^4} = 2^{16} = 65,536$ possible different functions.

□

A.2 Permutations

A related combinatorial problem is the following. Eight people are to ride in a minivan with eight seats. How many different ways can the people be seated?

To answer this question, we first note that there are eight choices for the driver. Once the driver is chosen, there are seven choices for who will ride "shotgun," i.e., in the front passenger seat. After that, there are six choices for the passenger who will sit directly behind the driver, and so on. Thus, according to the Rule of Product, the number of ways of arranging the riders is

$$8 \cdot 7 \cdot 6 \cdot 5 \cdot 4 \cdot 3 \cdot 2 \cdot 1 = 8! = 40,320$$

(a surprisingly large number!)

A particular arrangement of the riders is called a *permutation*. If there are N riders (and N seats) then the number of permutations is given by[1]

$$P^N = N! \tag{A.1}$$

---

[1] The notation here is just P with a superscript N to indicate the order of the permutation. It does *not* mean that some number P is raised to the Nth power.

As a variation on this problem, suppose there are ten potential riders, but still only eight seats. Then the number of ways of arranging the people (with two left out) is

$$10 \cdot 9 \cdot 8 \cdot 7 \cdot 6 \cdot 5 \cdot 4 \cdot 3 = 1,814,400$$

Notice that this computation could be expressed as

$$\frac{10!}{2!} = 1,814,400$$

This kind of arrangements is referred to as a permutation of N *objects taken k at a time*. The number of such permutations is given by

$$P_k^N = \frac{N!}{(N-k)!} \qquad (\text{A.2})$$

It is interesting to compare the essential features of this problem to the one described in the previous section. In many probability studies it is traditional to develop the problems described here as problems of drawing colored balls from an urn (jug). The balls in the urn all have different colors so they can be distinguished. In the type of problem described in the previous section a ball is selected but then is replaced in the urn. Thus the total number of choices remains fixed at N for each drawing. In permutation problems, described here, the ball is not replaced in the urn, so the number of choices is reduced by one after each drawing. Because of this analogy, problems of these sorts are sometimes referred to by combinatorial mathematicians as "sampling with replacement" and "sampling without replacement" respectively.

The number of permutations of a moderate number of objects can become very large. Consider the following example taken from the arts.

Example A.2:   Abstract-expressionist painter Johnson Polkadot begins his daily work by selecting a can of paint from his supply of 50 different colors. This paint is applied randomly to the canvas and allowed to dry. He then selects another color and applies it over the first and continues in this way until he is finished. Assume he can apply only one color per hour. If he is now 35 years old and works for 12 hours a day every day of the year, can he live long enough to be forced to repeat a color sequence?

The number of sequences of 12 colors is given by

$$P_{12}^{50} = \frac{50!}{(50-12)!} = 50 \cdot 49 \cdots \cdot 39 \approx 1.01 \times 10^{19}$$

If Johnson is now 35 and lives to be 105, the number of paintings he will produce is $70 \cdot 365 = 25,550$ (assuming he rests one day a year on leap years). Since this number is nowhere close to the number of possible color sequences, he will never be forced to repeat the sequence of colors used in a painting.

☐

A.3 Combinations

The last topic to be described is that of *combinations*. This is an important combinatorial result probably used more than the others, but is also a bit more complicated to explain.

Let us consider the transmission of N binary data bits of information and ask, "In how many different ways can k errors occur in N bits?" To answer this question, let us pretend that errors are caused by a little demon who shoots colored arrows into the bits (see illustration below) which cause the errors. This demon is an excellent shot, so if it decides to attack a particular bit, it never misses. Assume the demon has k arrows in its quiver and that each arrow is a different color so we can distinguish them. (This mode of operation is known as NTSC for Never Twice the Same Color, which is also the analog color television standard in the USA.)

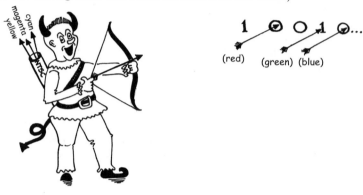

If the demon decides to cause k errors, the number of ways to place k arrows in N bits is given by (A.2). To be specific, pretend that there are $N = 5$ bits and the demon causes $k = 3$ errors; then this number would be

$$P_3^5 = \frac{5!}{(5-3)!} = 60$$

We can show that this number is *much* too large! To see this, assume that the arrows are colored red, green and blue. If we list all the sequences with $k = 3$ errors, one of the sequences would be (for example)

$$\Phi \quad R \quad \Phi \quad G \quad B$$

where Φ indicates that no error has occured and R, G, or B indicates that the bit has been destroyed by the corresponding colored arrow. There are other configurations with the colored arrows in the same set of positions however. These other configurations are:

$$
\begin{array}{ccccc}
\Phi & R & \Phi & B & G \\
\Phi & G & \Phi & R & B \\
\Phi & G & \Phi & B & R \\
\Phi & B & \Phi & R & G \\
\Phi & B & \Phi & G & R \\
\end{array}
$$

and these are included in the count given by (A.2).

If we are concerned with just the errors, all of these configurations are equivalent. In other words, we don't care about the color; an arrow is an error regardless of its color. So it is clear we are overcounting when we use (A.2). The factor by which we are overcounting is equal to the number of permutations of the colors; in this case where $k = 3$ the factor is $P^3 = 3! = 6$. (Check this out, above.) Thus, the number of errors when $N = 5$ and $k = 3$ is

$$P_3^5/3! = \frac{5!}{2!\,3!} = 10$$

The general result is given by

$$\boxed{C_k^N = \binom{N}{k} = \frac{N!}{k!(N-k)!}} \tag{A.3}$$

and represents the number of *combinations* of N objects taken k at a time. The expression $\binom{N}{k}$ is known as the *binomial coefficient* and is typically read as "N choose k." You can think of it as the number of ways that k errors can occur in N bits, although it applies in a multitude of other problems.

Example A.3:   A technician is testing a cable with six conductors for short circuits. If the technician checks continuity between every possible pair of wires, how many tests need to be made on the cable?

The number of tests needed is the number of combinations of 6 wires taken 2 at a time or the number of ways to choose 2 wires from a set of 6. This number is

$$C_2^6 = \binom{6}{2} = \frac{6!}{2!\,4!} = \frac{6 \cdot 5}{2} = 15$$

☐

B The Unit Impulse

The unit impulse is one of the most frequently used, but frequently misunderstood concepts in engineering. While well-founded in physical applications, this seemingly innocent idea led to consternation among mathematicians of the last century until given a proper foundation through the theory of Distributions and Generalized Functions (see e.g., [1, 2]). Although the unit impulse has its greatest importance in system theory, it also has significant applications in the study of random variables from an engineering perspective. This appendix provides an introduction to the unit impulse for students who may not yet have encountered this "function" in their other studies. For further discussion from an engineering perspective, you may want to read [3] and [4].

B.1 The Impulse

In the study of engineering and physics, you sometimes encounter situations where a finite amount of energy is delivered over a very brief interval of time. Consider modern electronic flash photography. The light produced by the flash tube is extremely bright but lasts only a very small fraction of a second. The light intensity arriving at the camera is integrated by the film or array of electronic sensors to produce the exposure. The light intensity can be modeled as shown in Fig. B.1. Although the flash output

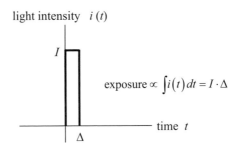

Figure B.1 Model for a pulse of light used in electronic flash photography.

may not be precisely a square pulse, this is not important, since the camera responds to the integral or total amount of light reaching the film plane.

As another example, consider the crash-testing of an automobile. The car is driven into a solid stationary wall at some velocity v. When the car hits the wall there is an immediate change of momentum, since the velocity goes to 0. Newton's Law requires that this change of momentum be produced by a force at the instant of time that the car hits the wall. This force, which exists for only a brief moment, must be very large so that the integral of the force over time produces the finite change of momentum dictated by the situation. The effect on the car is significant!

In these situations, as well as in others, there is a need to deal with a pulse-like function $g_\Delta(x)$ which is extremely narrow, but also sufficiently large at the origin so that the integral is a finite positive value, say unity. Such a function, while having clear application in these physical scenarios, turns out to very useful in other areas of electrical engineering, including the study of probability. Figure B.2 shows a few

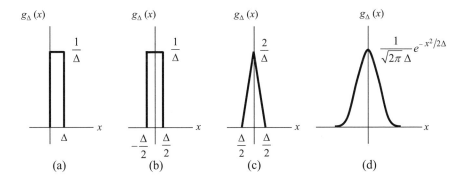

Figure B.2 Various approximations to a unit impulse. (a) Square pulse. (b) Centered square pulse. (c) Triangular pulse. (d) Gaussian pulse.

possible ways that this pulse-like function could be defined. In all cases the pulses have an area of 1, and we can write

$$\lim_{\Delta \to 0} \int_{-\infty}^{\infty} g_\Delta(x)dx = 1 \qquad (\text{B.1})$$

As Δ approaches 0, the details of the pulse shape ultimately do not matter. In engineering, we think of the *unit impulse* $\delta(x)$ as the idealized limiting form of the function $g_\Delta(x)$ as $\Delta \to 0$. We could write

$$\delta(x) = \lim_{\Delta \to 0} g_\Delta(x)$$

but this is at best misleading since this "function" assigns a nonzero value to only a single point ($x = 0$) where it assigns infinity.[1] A more correct statement is given by (B.1), and we can use any of the limiting forms above and the statement

$$\int_{-\infty}^{\infty} \delta(x)dx = 1 \qquad (\text{B.2})$$

to define the unit impulse.

Graphically, the impulse is represented as shown in Fig. B.3, i.e., as a bold thick

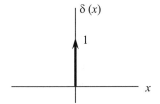

Figure B.3 Graphical representation of the unit impulse.

---

[1] The late professor Sam Mason was said to have quoted an M.I.T. engineering student as remarking, "You mean the impulse is so small everywhere that you can't see it except at the origin where it is so large that you can't see it either? In other words, you can't see it at all! At least I can't."

upward-pointing arrow at the origin. The number '1' which is optionally placed next to the arrow indicates that when integrated, the impulse has unit area.

B.2 Properties

Although the unit impulse is not a proper function according to the mathematical definition of a function, we often manipulate the impulse as we would manipulate any proper function. This makes the impulse extremely convenient to use in engineering applications. For example, a scaled impulse, i.e., an impulse with area A is simply represented by

$$A\delta(x)$$

In a graphical representation like Fig. B.3, we may write the area A next to the arrow instead of the value 1 to indicate a scaled impulse with area A.

Impulses may occur shifted along the x axis as well. For example, we could plot the expression

$$2.5\delta(x) + 1.5\delta(x - 2)$$

as shown in Fig. B.4. This is a set of two scaled impulses with the second impulse shifted

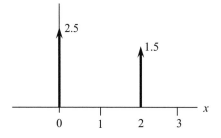

Figure B.4 Set of two impulses with different areas.

to the point $x = 2$. In the graphical representation it is common (but not universal) to represent the difference in areas by different heights of the arrows as illustrated in this figure. In the case of a negative scale factor such as $-2.5\delta(x)$, downward-pointing arrows are even used in the representation. While all of this may be convenient, you should remember that the actual height of the impulse is *infinite* and should not be mislead by the size of the arrows when they are drawn in this manner.

One of the most important properties of the impulse is known as the "sifting property." For any function $f(x)$ we may write

$$f(x) = \int_{-\infty}^{\infty} f(\xi)\delta(\xi - x)d\xi \tag{B.3}$$

which is valid at least for values of x where there is no discontinuity.[2] This equation, which may seem puzzling when encountered for the first time, is easily explained by the picture of Fig. B.5 where the integrands are plotted as a function of the dummy variable ξ. The product of the function $f(\xi)$ and the impulse $\delta(\xi - x)$ is 0 everywhere except at the point $\xi = x$. At this point the impulse is scaled by the value of the function, which is $f(x)$. Integration then produces the *area* of the scaled impulse, which is equal to $f(x)$. While the usefulness of (B.3) may not seem immediately apparent, the relation is extremely important in simplifying integrals involving impulses.

A final property we shall mention is that the unit impulse can be considered as

---

[2] Discontinuities can lead to inconsistency in interpretation of the integral. Although this is often ignored in casual use of the sifting property, it is well to be aware of the problem.

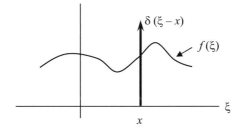

Figure B.5 Illustration of the "sifting" property of an impulse.

the formal derivative of the unit step function. The unit step function, conventionally denoted by $u(x)$, is defined by

$$u(x) = \begin{cases} 1 & x \geq 0 \\ 0 & x < 0 \end{cases} \tag{B.4}$$

and sketched in Fig. B.6. From Fig. B.6, it can be seen that $u(x)$ satisfies the relation

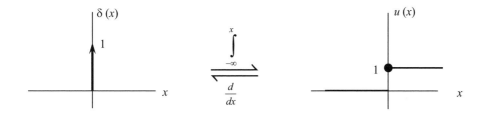

Figure B.6 Impulse as the derivative of the unit step function.

$$u(x) = \int_{-\infty}^{x} \delta(\xi) d\xi \tag{B.5}$$

In particular, for any negative value of the upper limit x, the integral in (B.5) is 0, while for any positive value of x the integral is 1. An ambiguity occurs for $x = 0$ which is generally resolved by defining the step to be continuous from the right. Since the integral of the unit impulse is found to be the unit step, the unit impulse can be regarded as the *derivative* of the unit step.

This relation, can be generalized to the cases where functions have simple discontinuities. Chapter 3 discusses how cumulative distribution functions which have such discontinuities result in probability density functions that have impulses at the corresponding locations.

References

[1] M. J. Lighthill. *In Introduction to Fourier Analysis and generalized Functions*. Cambridge University Press, New York, 1959.

[2] A. H. Zemanian. *Distribution Theory and Transform Analysis*. McGraw-Hill Book Company, New York, 1955.

[3] Athanasios Papoulis. *Signal Analysis*. McGraw-Hill, New York, 1977.

[4] William McC. Siebert. *Circuits, Signals, and Systems*. McGraw-Hill, New York, 1986.

Index

312

Table of Q function values

x	Q(x)	x	Q(x)	x	Q(x)	x	Q(x)
0.00	5.0000E-01	1.00	1.5866E-01	2.00	2.2750E-02	5.00	2.8665E-07
0.02	4.9202E-01	1.02	1.5386E-01	2.05	2.0182E-02	5.10	1.6983E-07
0.04	4.8405E-01	1.04	1.4917E-01	2.10	1.7864E-02	5.20	9.9644E-08
0.06	4.7608E-01	1.06	1.4457E-01	2.15	1.5778E-02	5.30	5.7901E-08
0.08	4.6812E-01	1.08	1.4007E-01	2.20	1.3903E-02	5.40	3.3320E-08
0.10	4.6017E-01	1.10	1.3567E-01	2.25	1.2224E-02	5.50	1.8990E-08
0.12	4.5224E-01	1.12	1.3136E-01	2.30	1.0724E-02	5.60	1.0718E-08
0.14	4.4433E-01	1.14	1.2714E-01	2.35	9.3867E-03	5.70	5.9904E-09
0.16	4.3644E-01	1.16	1.2302E-01	2.40	8.1975E-03	5.80	3.3157E-09
0.18	4.2858E-01	1.18	1.1900E-01	2.45	7.1428E-03	5.90	1.8175E-09
0.20	4.2074E-01	1.20	1.1507E-01	2.50	6.2097E-03	6.00	9.8659E-10
0.22	4.1294E-01	1.22	1.1123E-01	2.55	5.3861E-03	6.10	5.3034E-10
0.24	4.0517E-01	1.24	1.0749E-01	2.60	4.6612E-03	6.20	2.8232E-10
0.26	3.9743E-01	1.26	1.0383E-01	2.65	4.0246E-03	6.30	1.4882E-10
0.28	3.8974E-01	1.28	1.0027E-01	2.70	3.4670E-03	6.40	7.7688E-11
0.30	3.8209E-01	1.30	9.6800E-02	2.75	2.9798E-03	6.50	4.0160E-11
0.32	3.7448E-01	1.32	9.3418E-02	2.80	2.5551E-03	6.60	2.0558E-11
0.34	3.6693E-01	1.34	9.0123E-02	2.85	2.1860E-03	6.70	1.0421E-11
0.36	3.5942E-01	1.36	8.6915E-02	2.90	1.8658E-03	6.80	5.2310E-12
0.38	3.5197E-01	1.38	8.3793E-02	2.95	1.5889E-03	6.90	2.6001E-12
0.40	3.4458E-01	1.40	8.0757E-02	3.00	1.3499E-03	7.00	1.2798E-12
0.42	3.3724E-01	1.42	7.7804E-02	3.05	1.1442E-03	7.10	6.2378E-13
0.44	3.2997E-01	1.44	7.4934E-02	3.10	9.6760E-04	7.20	3.0106E-13
0.46	3.2276E-01	1.46	7.2145E-02	3.15	8.1635E-04	7.30	1.4388E-13
0.48	3.1561E-01	1.48	6.9437E-02	3.20	6.8714E-04	7.40	6.8092E-14
0.50	3.0854E-01	1.50	6.6807E-02	3.25	5.7703E-04	7.50	3.1909E-14
0.52	3.0153E-01	1.52	6.4255E-02	3.30	4.8342E-04	7.60	1.4807E-14
0.54	2.9460E-01	1.54	6.1780E-02	3.35	4.0406E-04	7.70	6.8033E-15
0.56	2.8774E-01	1.56	5.9380E-02	3.40	3.3693E-04	7.80	3.0954E-15
0.58	2.8096E-01	1.58	5.7053E-02	3.45	2.8029E-04	7.90	1.3945E-15
0.60	2.7425E-01	1.60	5.4799E-02	3.50	2.3263E-04	8.00	6.2210E-16
0.62	2.6763E-01	1.62	5.2616E-02	3.55	1.9262E-04	8.10	2.7480E-16
0.64	2.6109E-01	1.64	5.0503E-02	3.60	1.5911E-04	8.20	1.2019E-16
0.66	2.5463E-01	1.66	4.8457E-02	3.65	1.3112E-04	8.30	5.2056E-17
0.68	2.4825E-01	1.68	4.6479E-02	3.70	1.0780E-04	8.40	2.2324E-17
0.70	2.4196E-01	1.70	4.4565E-02	3.75	8.8417E-05	8.50	9.4795E-18
0.72	2.3576E-01	1.72	4.2716E-02	3.80	7.2348E-05	8.60	3.9858E-18
0.74	2.2965E-01	1.74	4.0930E-02	3.85	5.9059E-05	8.70	1.6594E-18
0.76	2.2363E-01	1.76	3.9204E-02	3.90	4.8096E-05	8.80	6.8408E-19
0.78	2.1770E-01	1.78	3.7538E-02	3.95	3.9076E-05	8.90	2.7923E-19
0.80	2.1186E-01	1.80	3.5930E-02	4.00	3.1671E-05	9.00	1.1286E-19
0.82	2.0611E-01	1.82	3.4380E-02	4.10	2.0658E-05	9.10	4.5166E-20
0.84	2.0045E-01	1.84	3.2884E-02	4.20	1.3346E-05	9.20	1.7897E-20
0.86	1.9489E-01	1.86	3.1443E-02	4.30	8.5399E-06	9.30	7.0223E-21
0.88	1.8943E-01	1.88	3.0054E-02	4.40	5.4125E-06	9.40	2.7282E-21
0.90	1.8406E-01	1.90	2.8717E-02	4.50	3.3977E-06	9.50	1.0495E-21
0.92	1.7879E-01	1.92	2.7429E-02	4.60	2.1125E-06	9.60	3.9972E-22
0.94	1.7361E-01	1.94	2.6190E-02	4.70	1.3008E-06	9.70	1.5075E-22
0.96	1.6853E-01	1.96	2.4998E-02	4.80	7.9333E-07	9.80	5.6293E-23
0.98	1.6354E-01	1.98	2.3852E-02	4.90	4.7918E-07	9.90	2.0814E-23